Aging and Cell Function

Aging and Cell Function

Edited by
John E. Johnson, Jr.

Department of Neurology
Johns Hopkins University School of Medicine
and
National Institute on Aging, NIH
Baltimore City Hospitals
Baltimore, Maryland

Plenum Press • New York and London

Library of Congress Cataloging in Publication Data

Main entry under title:

Aging and cell function.

Includes bibliographical references and index.
1. Cells—Aging. 2. Aging. I. Johnson, John E., 1945- . [DNLM: 1. Aging. 2. Cells. WT 104 A2663]
QH608.A36 1984 574.87′6 83-27167
ISBN 0-306-41420-1

A Division of Plenum Publishing Corporation
233 Spring Street, New York, N.Y. 10013

Printed in the United States of America

"But hereditary diseases and some other facts make me believe that the rule has a wider extension, and that, when there is no apparent reason why a peculiarity should appear at any particular age, yet that it does tend to appear in the offspring at the same period at which it first appeared in the parent."

—Charles Darwin, *The Origin of Species* 1859

This volume is dedicated to those
for whom the answers come too late.

Contributors

RICHARD G. CUTLER Gerontology Research Center, National Institute on Aging, Baltimore City Hospitals, Baltimore, Maryland 21224

ANGELOS C. ECONOMOS Genetics Laboratory, Catholic University of Louvain, Louvain-la-Neuve, Belgium

CHARLES R. FILBURN Laboratory of Molecular Aging and Cardiovascular Section, National Institute on Aging, National Institutes of Health, Gerontology Research Center, Baltimore City Hospitals, Baltimore, Maryland 21224

GERALD D. HESS Department of Natural Sciences, Messiah College, Grantham, Pennsylvania 17027

JOHN E. JOHNSON, JR. Department of Neurology, Johns Hopkins University School of Medicine, Baltimore, Maryland 21205, and National Institute on Aging, National Institutes of Health, Baltimore City Hospitals, Baltimore, Maryland 21224

EDWARD G. LAKATTA Laboratory of Molecular Aging and Cardiovascular Section, National Institute on Aging, National Institutes of Health, Gerontology Research Center, Baltimore City Hospitals, Baltimore, Maryland 21224

EDYTHE D. LONDON National Institute on Drug Abuse, Addiction Research Center, and National Institute on Aging, Gerontology Research Center, Baltimore City Hospitals, Baltimore, Maryland 21224

GEORGE S. ROTH National Institute on Aging, National Institutes of Health, Gerontology Research Center, Baltimore City Hospitals, Baltimore, Maryland 21224

Preface

The aging process, like most (all?) things in the universe, is a puzzle. It is such a fundamental occurrence, common to all living things, that it ironically may be the most complicated and perplexing puzzle to solve. There are, of course, puzzles sufficient in number to occupy all the scientists and philosophers who have ever lived a thousand times over. Yet what other phenomenon affects every one of us as much as growing old?

Curiosity about the ubiquity of aging as a phenomenon, boosted perhaps by an only natural fear of that same phenomenon as an undeniable manifestation of our own mortality, has led to the compilation of data on the aging process, which have accumulated as rapidly as the elderly who might benefit from those data.

Most of the scientific data on the biology of aging are at the biochemical and physiological levels, while comparatively little information has been available at the anatomical level. Because of this, a two-volume set called *Aging and Cell Structure* was conceived, the first volume having been published in 1981 and the second volume being published concurrently with the present one on cell function.

The emphasis on training of scientists and other individuals in advanced education has, during past decades, been one of increased specialization. Not merely do we specialize in physiology, anatomy, zoology, or literature, but in glomerular filtration, synaptic junctions, tree frogs of Brazil, or English poetry of the nineteenth century. It is a consequence of the mountain of information which would have to be learned if we halted our specialization at the more inclusive level. Perhaps it gives us security to know almost everything about one little corner of life, but our resultant ignorance of the remaining rather large percentage of the world is embarrassingly apparent. Educators are aware of this problem, and increasing numbers of interdisciplinary degrees are being offered by universities.

In keeping with the current emphasis on interdisciplinary training as a prophylaxis against the ills of overspecialization, *Aging and Cell Function* has been compiled. It is an attempt to bring together current data and viewpoints on a variety of salient experimental gerontological issues, focusing at the cellular functional level. Included in this volume, the reader will find chapters dealing with the evolution of the aging process, cell receptors, brain metabolism, heart cell function, and a systems analysis of cell aging.

The present volume on *Aging and Cell Function* will be of vital interest to cell biologists, pathologists, physicians, anatomists, biochemists, physiologists, and all other scientists who wish to have up-to-date interdisciplinary information which will provide an understanding of basic changes occurring in cells and tissues as they age.

John E. Johnson, Jr.

Baltimore

Contents

1

Evolutionary Biology of Aging and Longevity in Mammalian Species

RICHARD G. CUTLER

1. INTRODUCTION

The biomedical sciences have been remarkably successful in dealing with exogenously originating pathologies such as infectious diseases. This success, however, has uncovered a more formidable opponent of health and general well being, the intrinsically originating pathologies. These are the disabling or degenerative diseases, which are generally described as being caused by or related to the aging processes. The vast number of human diseases which have an intrinsic origin is just becoming appreciated, and the studies of these dysfunctions and the methods to be developed for their treatment are predicted to occupy the future efforts of much of the biomedical sciences (Ludwig, 1980).

This chapter reviews the effort of the author to help establish the scientific basis for the treatment of these intrinsic pathologies. This effort has centered on investigating the biological nature of human longevity. A general hypothesis on the nature of human aging and longevity has been developed which contains the conclusion that specific longevity determinant processes exist which act to preserve health and postpone the intrinsic pathologies of aging. This concept, if correct in principle, is expected to have an important impact on future biomedical care by directing a unique approach of scientific inquiry for the development of effective treatments against the general debilitations and diseases related to the aging process.

The testing of this hypothesis is just beginning, but the data that are in so far are very encouraging, suggesting that it is on the right track. The major

RICHARD G. CUTLER • Gerontology Research Center, National Institute on Aging, Baltimore City Hospitals, Baltimore, Maryland 21224.

objective of this chapter is to review the background leading to the hypothesis and to present some of the most recent experiments and experimental data used to test its validity.

Much of the present efforts taken to treat intrinsic pathologies still take the classical approach; that is, exogenous causes and/or specific causes unique to one specific type of dysfunction where conventional medical treatment would be applicable. Two good examples are the approaches taken to cure or reduce the debilitative effects of cancer and Alzheimer's disease (including senile dementia). Although there is clearly an interaction between exogenously and intrinsically originating factors in these two diseases, which underlies a principal difficulty in resolving cause from effect in age-related dysfunctions, the most unfortunate situation is that possible intrinsic causes of these diseases receive little, if any, attention. This is unfortunate because I believe there is a good chance that no effective treatment of cancer or senile dementia will ever result without an understanding of the biological basis of human longevity and aging.

Part of the problem of why aging research has not been taken seriously by the biomedical scientific community is the general belief of the hopelessness of dealing with dysfunctions having aging as their basic cause. Although it is true that aging is extraordinarily complex, with little hope of understanding all of its effects on the organism in the near future, the major message of this chapter is that the processes governing aging rate or longevity are separate and unique from the aging process. There is also good reason to believe these longevity determinant processes are considerably less complex than the aging process per se and are thus more readily understandable and subject to intervention.

2. AGING

I shall begin by briefly outlining some of the major aspects of human aging relevant to the development of the major theme of this paper. Figure 1 shows a percent survival curve for humans at different historical periods (Comfort, 1978; Kohn, 1978; Lamb, 1977; Strehler, 1978). This figure illustrates several important points. The first point is that the mean survival age (or 50% survival age) for most of human history has been somewhere between 18 and 35 years of age. Thus, few people lived much past the age where their general health and ability to function was decreased principally due to aging processes. For most of human history, there were few "aged" individuals to speak of, and the typical population essentially consisted of individuals in their most vigorous and healthy years of life. Under these conditions, lifespan was limited only by the

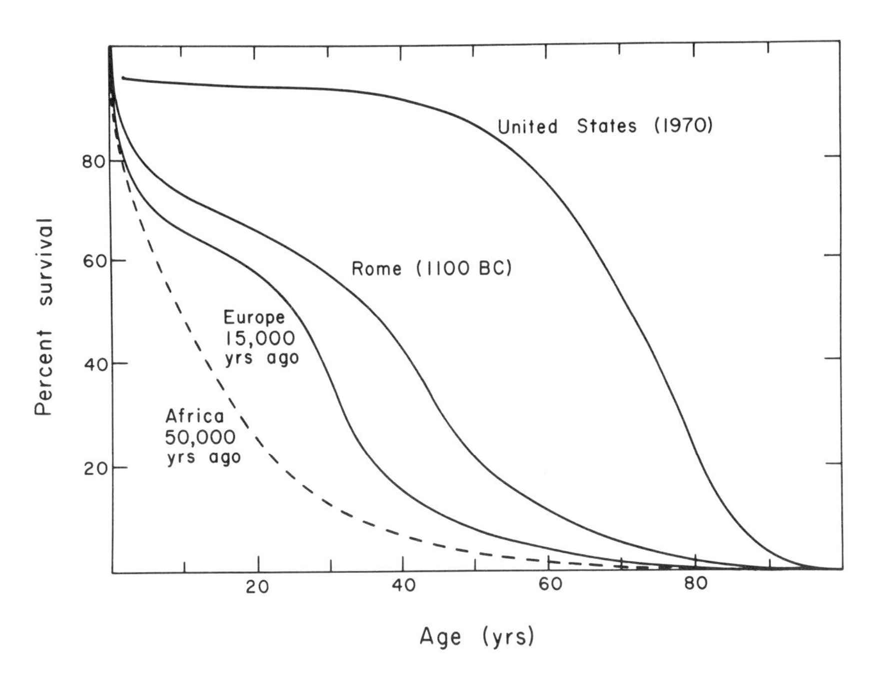

Figure 1. Percent survival curve for humans under different environmental hazard conditions. Note different 50% mean values but constant LSP values of about 100 years. Taken in part from Comfort (1956), Acsádi and Nemeskéri (1970), and Strehler (1978).

normal environmental hazards typical of that day and not by any loss of vigor or health due to the aging process (Acsádi and Nemeskéri, 1970).

The second point is that over the past 1000 years or so, and particularly over the past 400 years in the developed nations of the world, mean lifespan rapidly increased for the first time in the history of man (Comfort, 1978; Kohn, 1978; Strehler, 1978). This increase was due mainly to a significant reduction of environmental hazards of an exogenous nature. Recent medical advances against specific diseases like polio or diabetes have had little relative effect on extending human lifespan. This lower environmental hazard condition enabled humans to live more into physiological old age; that is, to live deeper into old age. Thus, the recent increase in mean lifespan of human population was not a result of slowing down the aging process, but rather a result of lowering the various exogenous hazards of life such that humans could remain alive longer and in a weaker, less healthy state of life.

The third point is that, in spite of the wide variation of mean lifespan evident under different degrees of hazardous environments, the maximum lifespan achieved for the human population was fairly constant, being about 100 years. This age, called maximum lifespan potential (MLP) or, as we shall use in this chapter, lifespan potential (LSP) is taken to reflect the innate biological aging rate or biological longevity potential of humans. Thus, the LSP of 100 years for humans is an innate biological characteristic of the human species, *Homo sapiens.* Throughout all human history (from about 15,000 to perhaps 50,000 years ago), human LSP has been the same. The mean lifespan of human populations is what has changed, and this has been determined by the various exogenous hazards of life. In support of the conclusion that there is no evidence of any person living much beyond 100 years, the oldest lifespan recorded for a human being is 113 years. Table I illustrates how definite LSP is by demonstrating the steepness of the death rate of individuals as they pass 98 years of age.

Another interesting point can be derived from the survival curve when these data are replotted as the normalized rate of death as a function of age as shown in Fig. 2. Doing this, we find that the point of minimum death rate (or maximum vigor and health) is near the age of sexual maturity and is independent of the various hazards of the environment (like LSP). After this age, for a number of different reasons, including physiological aging, the probability of death begins to increase in an exponential manner. There does not appear to be any plateau period to speak of where optimum vigor and general resistance to death from all causes remains constant. The probability of death is either decreasing up to the age of sexual maturity or increasing rapidly thereafter.

Table II presents some typical data documenting the point already made

Table I. Approach to Maximum Lifespan Potential for Man[a]

Age (yr)	Number of survivors
98	10,000,000
99	5,680,000
100	3,030,000
101	1,515,000
102	699,000
103	296,000
104	122,000
105	37,000
106	10,900
107	2,500
108	420
109	48
110	3

[a]Taken from Acsádi and Nameskéri (1970).

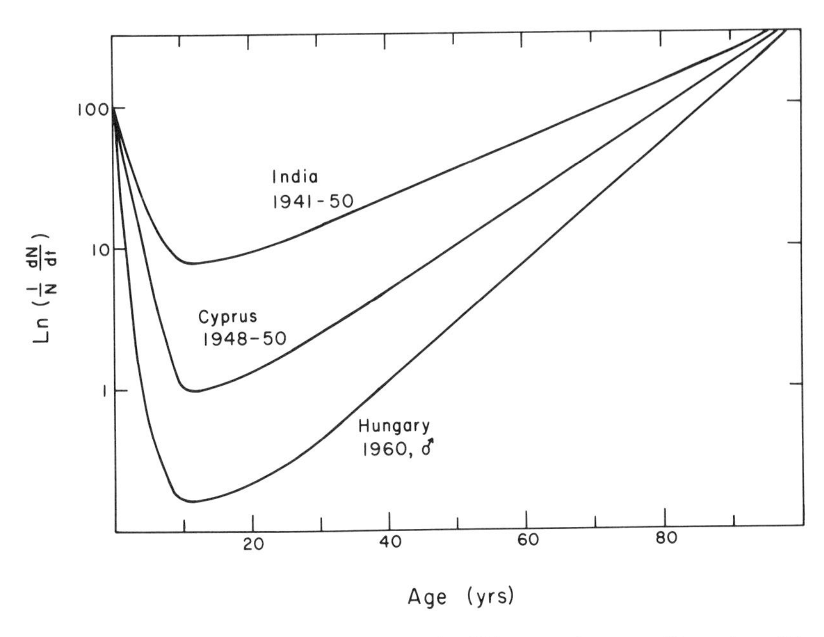

Figure 2. Rate of mortality curve for humans under different environmental hazard conditions. Note constant age where minimum probability of death occurs (about 15 years of age) and where the different slopes after that age become more similar. Taken in part from Acsádi and Nemeskéri (1970).

that the mean lifespan of humans for most of their history has been in the range of 30–35 years, whereas the LSP is predicted to remain constant.

The conclusion that the wild-type human population contained few "old" individuals also holds true for most other animal species living in their natural ecological niches. Some of the best data of this type come from studies of wild-type populations of mice, fish, and birds (Deevey, 1947; Blair, 1951; Lack, 1954; Caughley, 1966; Acsádi and Nemeskéri, 1970; Hirsch, 1979). In Table III, data are presented for a few bird species, where a good correlation is found between the species' natural environmental hazards and their innate ability to retard the aging rate, as measured by LSP in captivity. Such data suggest that, for each species, aging has been postponed in time to the point that most animals are killed by natural environmental causes and not by processes related to the aging process. Species that have developed the physical means to live a long lifespan in the wild (a low environmental hazard index) have developed, in direct proportion, the biological means to provide a lower innate aging rate. These data of survival in the wild compared to innate aging rates of species explain why species have different LSPs. Longevity has evolved to the point where a further increase would provide little significant survival advantage.

Table II. Average Mortality Rate of Present-Day Man in His Past and Present Environment[a]

Time period	Average chronoage at 50% survival (yr)	Maximum lifespan potential (yr)
Würm (about 70,000–30,000 yr ago)	29.4	69–77
Upper Paleolithic (about 30,000–12,000 yr ago)	32.4	95
Mesolithic (about 12,000–10,000 yr ago)	31.5	95
Neolithic Anatolian (about 10,000–8000 yr ago)	38.2	95
Classic Greece (1100 B.C.–1 A.D.)	35	95
Classic Rome (753 B.C.–476 A.D.)	32	95
England (1276 A.D.)	48	95
England (1376–1400)	38	95
United States (1900–1902)	61.5	95
United States (1950)	70.0	95
United States (1970)	72.5	95

[a]Data taken in part from Deevey (1960).

Table III. Correlation between Mortality Rate in the Wild and Maximum Lifespan Potential for Some Birds[a]

Common name	Average annual adult mortality (fraction that is killed)	Maximum lifespan potential (yr)
Blue tit	0.72	9
European robin	0.62	12
Lapwing	0.34	16
Common swift	0.18	21
Sooty shearwater	0.07	27
Herring gull	0.04	36
Royal albatross	0.03	45

[a]Data taken in part from Botkin and Miller (1974).

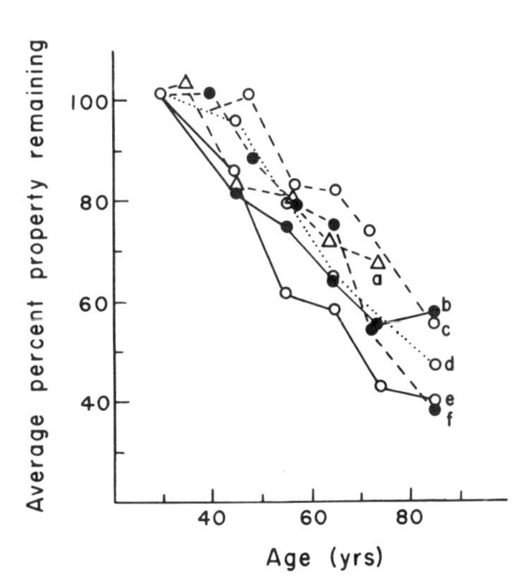

Figure 3. Decline of human physiological processes as a function of age. Maximum capacity to function is assigned 100% at age 30 years. (a) Cardiac index; (b) vital capacity; (c) standard glomerular filtration rate (inulin); (d) standard renal plasma flow (diodrast); (e) maximum breathing capacity; (f) standard renal plasma flow (PAH). Taken in part from Kohn (1978).

No information about the aging process per se can be gained from the analysis of survival curves (Maynard Smith, 1966). Thus, to determine how aging affects general health status, one of the best measurements has been to determine maximum functional capacity of physiological characteristics, such as cardiac index, vital capacity, or the standard glomerular filtration rate (Shock, 1952, 1961, 1970; Watkin, 1982). Typical data are shown in Fig. 3. Such cross-sectional data indicate that, in general, a linear decline in performance begins at the age of 20–30 years, and that most physiological functions (in terms of percent of maximum functional capacity) decline at a similar rate. Thus, the average slope of such a cluster of these declining curves would reflect the true aging rate of humans. This led to the important finding that the average slope of these curves is inversely related to a species' LSP. Thus, LSP appears to reflect, in an inverse manner, the physiological aging rate of a species.

From such studies of the physiological basis of aging, the tremendous complexity of the aging process becomes apparent; that is, aging appears to and probably very well does affect every tissue, organ, and physiological function of an organism. Nothing appears to get better with age, biologically speaking, after the age of 25–30, and in all probability, everything gets worse. There are, however, some psychologists reporting that many mental aspects do not change significantly and some mental functions even improve with age. I doubt very much if this really true, for it does not make sense evolutionarily or biologically.

Death is not usually the direct result of the age-related loss of physiological functions, but rather of the disease processes these physiological functions initiate (Kohn, 1982a). Figure 4 shows typical data for the age-dependent onset frequency of a number of different cancers (Kohn, 1978). Two interesting points can be drawn from such data. The first is that only a small fraction of people die of cancer under the age of 25–35 years, and that most die after the age of 25–35 years and mainly after the age of 60. The other point is the strong age-dependent increase of the onset frequency of cancer, increasing with the fifth power of age. Such a strong power function is typical of many of the other age-related degenerative types of diseases. This fact is demonstrated in Table IV, where the gain in lifespan is predicted if death due to specific types of diseases is completely eliminated. For example, the complete elimination of all the major cardiovascular–renal diseases at birth would result in a mean lifespan gain today for the people in the United States of about 11 years. This may not be as much of a gain in lifespan as might be expected, but the real surprise is that, if this cause of death is eliminated for those people at 65 years of age instead of at birth, a similar gain in lifespan (about 10 years) is also predicted. The reason for this is that most people who die of cardiovascular-renal disease are 65 years of age or older, so by eliminating this cause of death, this is the population of people that will benefit most. A similar result is found for the elimination of death from all forms of cancer. Here, the gain in mean

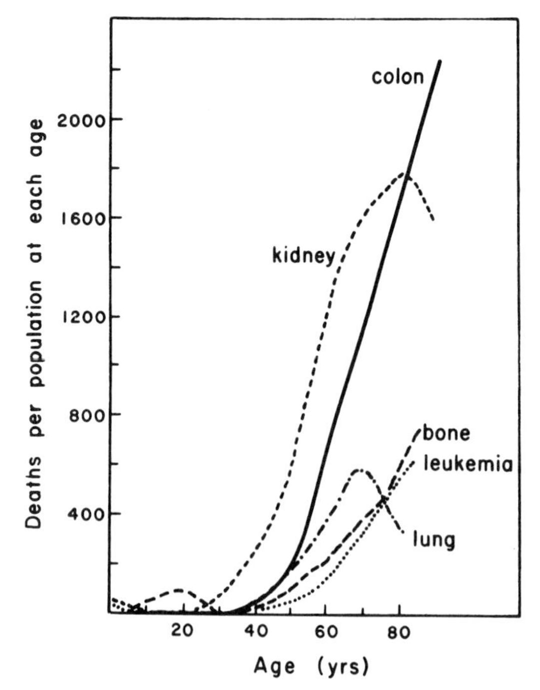

Figure 4. Age-specific death rates from several neoplasms: colon, lung, and leukemia per 10^6 people and for bone and kidney per 10^7 people. Taken in part from Kohn (1978).

Table IV. Gain in Expectancy of Life at Birth and at Age 65 due to Elimination of Various Causes of Death[a]

Cause of death	Gain in expectancy of life (yr) if cause were eliminated	
	At birth	At age 65
Major cardiovascular-renal diseases	10.9	10.0
Heart disease	5.9	4.9
Vascular diseases affecting central nervous system	1.3	1.2
Malignant neoplasms	2.3	1.2
Accidents other than by motor vehicles	0.6	0.1
Motor vehicle accidents	0.6	0.1
Influenza and pneumonia	0.5	0.2
Infectious diseases (excluding tuberculosis)	0.2	0.1
Diabetes mellitus	0.2	0.2
Tuberculosis	0.1	0.0

[a]From Life tables published by the National Center of Health Statistics, USPHS and U.S. Bureau of the Census, "Some Demographic Aspects of Aging in the United States," February 1973.

lifespan is only about 2 years. But if heart disease and cancer are diseases essentially of the aged, why does their elimination not result in a greater increase in lifespan? The answer is that, because of the ubiquitous nature of aging, the elimination of one cause of death simply uncovers another.

In evaluating the general benefit the population would have from such medical breakthroughs, the ubiquitous nature of the aging process must be kept in mind. The extra years of lifespan gained only allow one to live deeper into physiological old age, thus increasing one's probability of suffering from yet another age-related disease such as senile dementia.

These studies illustrate the fact not appreciated by many laymen, that the complete elimination of the major diseases (even if it were possible) will have a minimum impact on the extension of health, general well-being, and the productive lifespan of the average American. This is because the underlying cause of most disease and loss of general good health is due to aging, which affects every body function. Piecemeal repair or elimination of specific dysfunctions will have little impact. Only by a decrease in the aging rate of all physiological functions can a significant net gain in healthy good years of life be obtained.

To summarize the main points of this section, it is found that the recent increase in the mean lifespan of humans has been through a reduction of environmental hazards, and not through an increase of health maintenance or a reduction of aging rate. Human aging begins to become apparent at the age of 25–35 years, which is a characteristic of the human species, and few individuals lived much past this age until very recently in human history. Consequently, few people before this time died of age-related diseases such as cancer or cardiovascular diseases, or suffered from senile dementia.

These human data plus those of other animal species have offered an explanation of why humans begin to age after 25–35 years of life and also why species have different lifespans. Aging does not seem to be a normal characteristic of animals living in the wild. All mammalian species appear to have evolved the innate capacity to retard aging to the point where their death was largely due to extrinsic causes. For example, there would be no biological or evolutionary advantage for a mouse if it had the biological potential to live in an age-free health status to an age of say 10 years if the probability of the natural hazards of this species dictated an average lifespan of only one year. There is also likely to be a biological cost involved for greater longevity, such as greater levels of enzymes or a slower rate of development.

These conclusions have several important consequences in the designing of experiments to determine the biological nature of aging. One is that, because we now know that the large segment of the human population which is older than 35 years of age is not a natural situation, we do not have to invent a rationale for why aged individuals exist or what purpose they might serve. For example, a popular evolutionary reason given for why people 65 years or older exist in the population is because of their greater wisdom. Since people rarely lived to 65 years, however, this idea could not be true. Another reason given is that aging exists to get rid of the old, but of course if there were no aging there would be no old people to get rid of. Lastly, a popular reason for aging has been that it provides turnover in the population to make new jobs available or to bring people in with new fresh ideas or for evolutionarily selective purposes. However, normal environmental hazards appear sufficient for all these purposes. It has also been suggested that aging genes exist or that there is a genetic program of aging to age individuals for the good of the species. This type of genetic-programmed hypothesis of aging appears unlikely, for aging instead appears to be the lack of any evolutionary selective pressure to do anything about it. Aging may be related to differentiation and development, and there probably are genes that are importantly involved in aging, but these processes or genes were not selected specifically for the purpose of causing aging.

If aging is not genetically programmed we are left with only one alternative as to what the biological nature of aging might be. This is that aging must be the result of the side effects of normal biological processes that are necessary to ensure the survival of the individual in the period of maximum youth and vigor. Because of the positive correlation between aging rate of a species and the mortality rate due to the hazards of its ecological niche, it appears that aging rate can be decreased (evolutionarily speaking) to the point that it does not interfere significantly with the survival of a species. Thus, aging rate appears to be a characteristic that is subject to change like any other characteristic of a species if evolutionary selective pressure warrants it. The innate aging rate of a species does not then appear to be fixed due to some type of biological limitation to attain a greater lifespan. This means that the innate biological ability of humans to be able to maintain optimum health free of most effects of aging up to the age of 25–35 years exists simply because a further increase of this innate ability to retard the process of aging would have no significant benefit, not because it was biologically impossible for humans to have a longer lifespan.

There is one last point I would like to draw from the percent survival curves shown in Fig. 1. It is clear that all people do not live to the same age; the survival curve is not rectangular in shape. The shape of the human survival curve was usually close to an exponentially declining curve, implying equal probability of death independent of age; but over the last few hundred years of history, this curve has become increasingly rectangular in shape. This squaring of the survival curve is the result of the increase in mean lifespan, where maximum lifespan potential has remained constant. However, this rectangularization of the survival curve is also the result of the remarkable homogeneity of aging rates between different individuals in the population. When the survival curve was exponential, death among the population was completely random, but as environmental hazards were decreased and this random component of survival was being reduced, a nonrandom component shaping the survival curve became increasingly dominant. This component represents the probability of death due to aging. If aging rate between individuals varied greatly, the present survival curve today would show much less of a rectangularization effect. The fact that it is as square as we find it is good evidence that the aging rate between different individuals is largely genetically determined to about the same degree of precision as body weight or height.

But, then, what about people that have died in their forties and those that have been able to live to 100 years of age? Do these individuals have different innate aging rates? The answer to this question is, not likely. Instead, the difference in lifespan is probably related to different specific physiological weaknesses, such as heart disease or diabetes (Kohn, 1982a). People dying in their forties, for example, do so mainly from specific diseases, not aging. On the other hand, those individuals living to an age of 100 years have managed to avoid specific problems and instead have aged more uniformly, not slower. Thus, there is no evidence that long-lived individuals age at slower rates; instead, they simply do not have any particular physiological problem. Thus, on examining at death those few individuals that are able to live to 90 or 100 years of age, it is found that practically every organ and tissue of their body could not have functioned much longer. For these individuals, no heart or kidney transplant or other specific therapy would have increased their lifespan significantly. This is why elimination of specific diseases (as shown in Table IV) has little impact on lifespan extension for the population at large.

Thus, given the remarkable homogeneity of aging rate among the human population, and the fact that it is the aging process which underlies the cause of most diseases of mankind, then it becomes clear that the only means to prolong the healthy years of life appears to be a uniform reduction of the rate of aging throughout the body. Although this appears to be an impossibility (practically speaking) due to the great complexity of the aging process, there is at least no known evidence of a biological limitation that would theoretically prevent such a lifespan extension.

3. LONGEVITY

The shape of the survival curves, as shown for humans in Fig. 1, are similar for other mammalian species; they are exponential in shape for species living in their natural ecological niche and rectangular for the same species now living in protected or captivity conditions. In addition, each species has a characteristic LSP that is largely independent of environmental hazard conditions. From these comparative data, it becomes apparent that there is also a wide difference in aging rate or LSPs among the different mammalian species (Sacher, 1959; Lamb, 1977; Comfort, 1978; Cutler, 1976a,b, 1979; Washburn, 1981). The shortest-lived mammals are shrews and mice with LSPs of 1.5 to 3 years respectively. Some of the longest-lived mammals are the elephant with an LSP of about 80 and the great apes with LSPs of about 50 years. However, humans have by far the highest LSP, which is about 100 years.

The product of an animal's SMR and LSP is roughly a constant (Rubner, 1908; Sacher, 1959; Cutler, 1976a,b, 1979). Figure 5 illustrates this finding for a large number of different primate and nonprimate mammalian species. It is well known that, in general, larger species appear to have higher LSPs than smaller species. It is also known that these larger species have lower SMRs. This inverse correlation of body size and SMR is explained on the basis that less heat energy is required to maintain the body temperarure of animals as the surface area to total body weight decreases with increasing body size.

I have called the product of SMR and LSP a species' characteristic "lifespan energy potential" or LEP value. Thus, LEP is defined as

$$\text{LEP} = (\text{SMR})(\text{LSP})$$

where the units used are kcal/g, cal/g per d and years for LEP, SMR and LSP, respectively.

Figures 5 through 7 demonstrate that, for the mammalian species, there appear to be three different LEP categories. Most nonprimate mammalian species have LEP values of about 220 kcal/g on the average, and most primate species have LEP values of about 458 kcal/g on the average. The species with

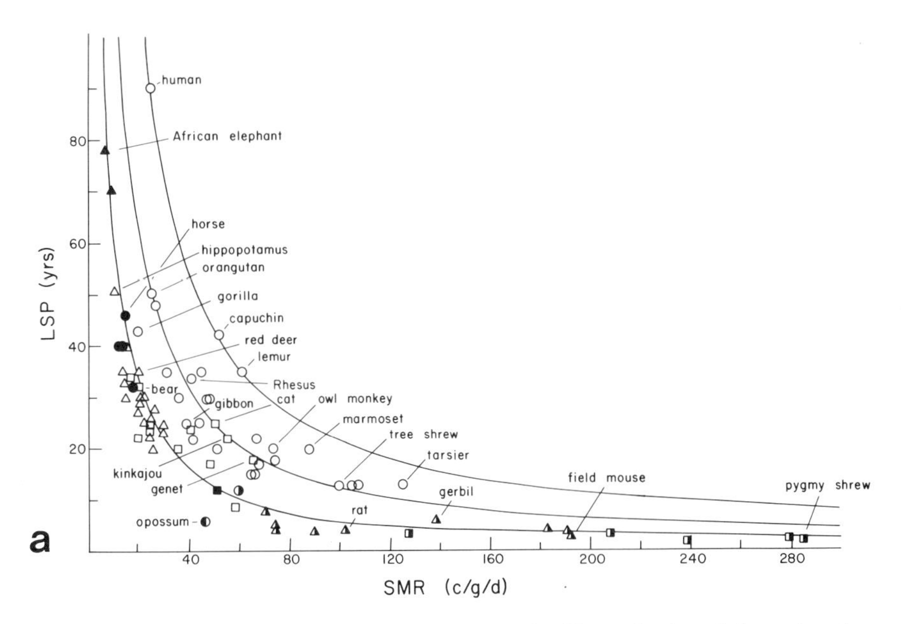

Figure 5. Lifespan energy potential of mammalian species. Figures 5a through 5c are based on specific metabolic rate (cal/g per day) and lifespan (years) where 90% mortality occurs. LSP taken from previously published data (Spector, 1956; Jones, 1962, 1968; Holliday *et al.*, 1967; Napier and Napier, 1967; Altman and Dittmer, 1968, 1972; Bauchot and Stephan, 1969; Jerison, 1973; Walker, 1975; Cutler, 1976a, 1979; Bowden and Jones, 1979; Tolmasoff *et al.*, 1980). Data base represents 77 different mammalian species chosen on the basis of reliable SMR and LSP values. Figure 5a represents the plot of LSP vs. SMR, indicating the three major classes of LEPs. Figure 5b represents these same data but for nonprimate species only, and Fig. 5c for primate species. LEP is 220 kcal/g for most nonprimate mammals, 458 kcal/g for most primate species, and 781 kcal/g for human, capuchin, and lemur. ○, Primates; △, artiodactyla; □, carnivora; ●, perissodactyla; ▲, proboscidea; ■, hydrocoidea; ◑, lagomorpha; ▲, rodentia; ◨, insectivora; ◐, marsupilia.

the highest LEP value is humans, with a value of about 815 kcal/g. Capuchins and lemurs are also outstanding with LEP values of about 700 kcal/g.

LEP reflects the average energy consumed on a per-gram-body weight basis over the lifespan of a species. SMR is usually measured as the basal rate of energy metabolism, but this value has been found to be proportional to the average metabolic rate of a species over a 24-hr period of normal activity. Thus, most nonprimate mammals, regardless of their LSP, consume about the same amount of energy per body weight over their lifespan.

This observation gave rise to one of the first hypothesis of aging, the wear-and-tear hypothesis (Rubner, 1908; Pearl, 1928; Cutler, 1972). According to this hypothesis, aging is a result of the metabolic "wear and tear" an organism

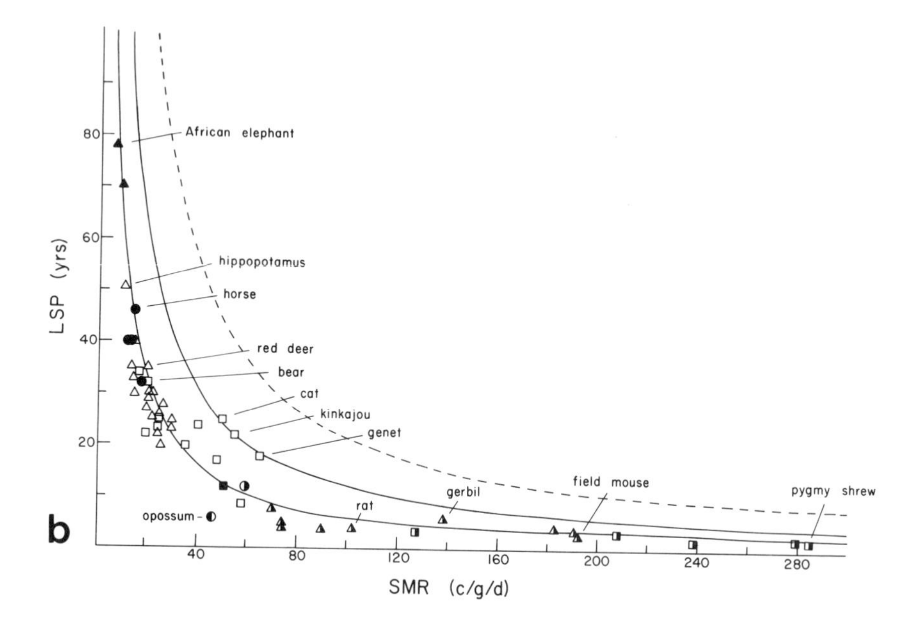

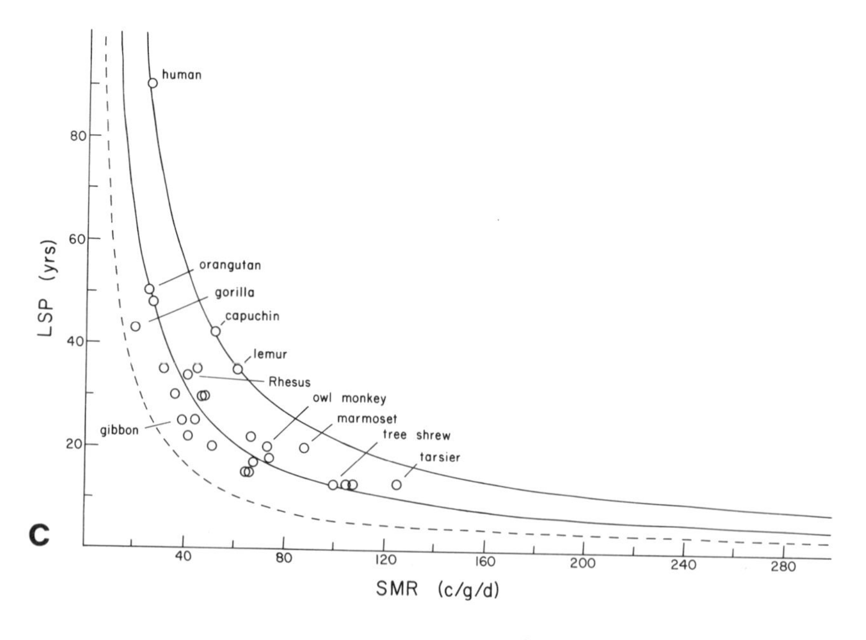

Figure 5. (*continued*).

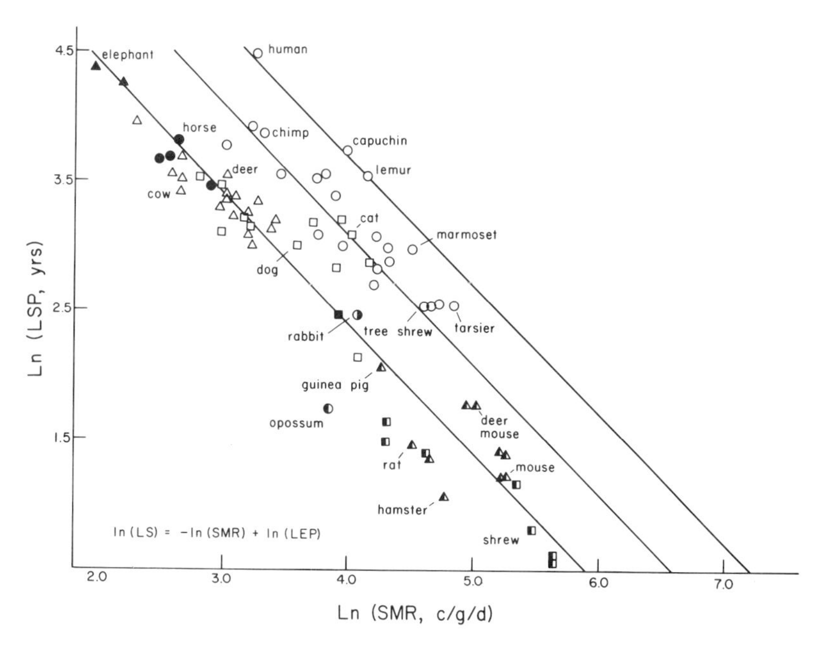

Figure 6. Correlation of LSP with SMR for the mammalian species. Data are similar as for Fig. 5. Correlation coefficient $r = -0.986$ and LEP V = 220 kcal/g for nonprimate mammals; $r = -0.925$ and LEP = 458 for all primates except human, capuchin, and lemur; $r = -0.999$ and LEP = 781 for human, capuchin, and lemur.

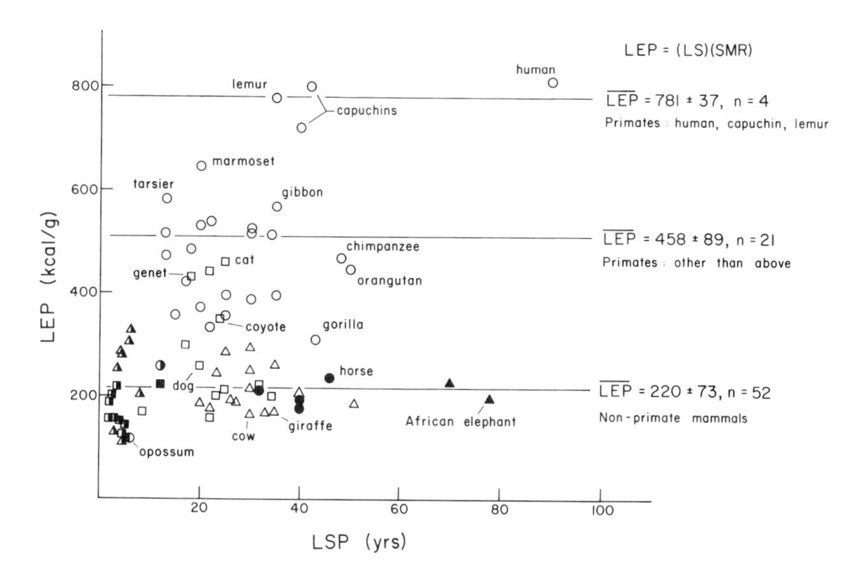

Figure 7. Three classes of LEP for the mammalian species. Data taken from Figs. 5 and 6.

undergoes during the utilization of energy. It follows then that species having lower rates of energy utilization (lower SMRs) would have in proportion longer LSPs. One difficulty of this hypothesis would be to identify which components of an organism actually do "wear out." However, the most important problem the hypothesis would have to explain is why some animals have different LEP values and what is the biological basis of the unusually high LEP value for humans. To do this, the wear and tear hypothesis suggests that the constituents of species with higher than normal LEP values are less subject to wear and tear; that is, humans would be predicted to be made up of longer-lasting parts.

A more straightforward interpretation of the constant LEP value found among most mammalian species is that it suggests that somehow the rate of energy metabolism is related to aging rate and that species with higher than normal LEP values like the primate species somehow have less sensitivity to whatever energy-related mechanism it is that causes aging (Cutler, 1972, 1976a,b, 1979, 1982a,b). Because most of the energy utilized by a mammalian species is involved in maintaining body temperature, it may be that the metabolic reactions leading to heat rather than kinetic energy generation are the most detrimental to the organism in terms of aging (Kleiber, 1975).

It is important to note that this new interpretation of LEP is consistent with the prediction presented in the previous section that aging is likely to be due to by-products of normal life processes necessary for the survival of the organism rather than by a genetically-programmed mechanism. Thus, support is gained for the concept that the cause(s) of aging are pleiotropic in nature, being the result of essential metabolic and developmental processes having long-term harmful effects.

But how can energy metabolism possibly cause aging? One answer coming from an evolutionary perspective is that essentially all metabolic processes result from a trade-off between beneficial versus harmful side-effects. Metabolic pathways in humans and other organisms are far from being the most efficient possible. There always appears room for improvement. An example is the utilization of oxygen, essential for high energy efficiency in the production of ATP. The toxic effect of oxygen is well known, and its metabolism produces not only beneficial products like ATP, but also a large number of active oxygen species such as the superoxide radical O_2^-, the hydroxy radical $\cdot$ OH, hydrogen peroxide H_2O_2, hydroperoxides ROOH, and aldehydes (Pryor, 1978; Chance *et al.*, 1979; Leibovitz and Seigel, 1980). Of importance in this context is that, if such side effects of metabolism are causing aging, then because most mammalian species are known to have remarkably similar metabolic processes, we must conclude that they must also have similar causes of aging. This then leads to the prediction that most mammalian species would be expected to age in a similar qualitative manner, and as far as we know this is the case. On the other hand, all mammalian species also appear to have a common set of repair, protection, or defense processes against the toxic effects of these active oxygen species. Examples of such processes are the DNA repair systems and antioxidants.

This reasoning led to the conclusion that species having similar LEP values would be expected to produce the same amount of toxic products over their lifespan (Cutler, 1972, 1976a,b). Thus, the net toxic effect of metabolism is the same, independent of LSP value. For those species having higher than normal LEP values, then, they are able to produce a proportionately larger toxic effect over their lifespan. This would be possible if LEP values were proportional to a species' repair and defense processes. Thus, at least part of the explanation of the unusually long human lifespan is that human tissues may have a higher level of protection against the long-term toxic effects of energy metabolism. This leads to a higher LEP value and consequently to a higher LSP.

From these types of analyses, LEP values appear to represent an important characteristic of a species' longevity potential. For this reason, LEP values probably should be taken into consideration in any study to explain the biological basis of aging or longevity. In Fig. 8, LSP and LEP values have been estimated for a number of primate species with reference to their evolutionary and phylogenetic relationships to each other.

Experimental evidence that LEP values are functionally coupled with aging rate or longevity is found for example when the average SMR over the lifespan of rats is increased by keeping them below ambient temperature. The LSP for the rats is found to decrease, and onset frequencies of diseases related to aging, such as cancer, were found to occur at an earlier age (Johnson *et al.*, 1963; Johnson and Pavelec, 1972). Other evidence is found with hibernating animals such as the hamster. If these animals are prevented from hibernating, they have a shorter lifespan. We can also explain the apparently high longevity of bats and the hummingbird (which live longer than their body size would suggest). For these species, body temperature and SMR undergo a dramatic decrease during their resting period, resulting in a 24-hr average LEP value little different from the mouse or elephant. Similarly, the long LSP of the Galapagos turtle (about 150 years) can be explained on the basis of its low average SMR, and not a high LEP value. Thus, on the basis of the LEP concept, these turtles do not appear to have evolved a longevity mechanism superior to that present in a mouse.

It is apparent then, that the differences in the longevity of many different species can be explained on the basis of their life-long SMR differences. These data also suggest a new interpretation of why some species undergo torpor, and hibernation. It not only conserves energy but also preserves or maintains LSP. Finally, considering the unusually high values of both LSP and LEP for humans, it becomes apparent that the human is probably one of, if not the, longest-lived species that ever lived on the earth.

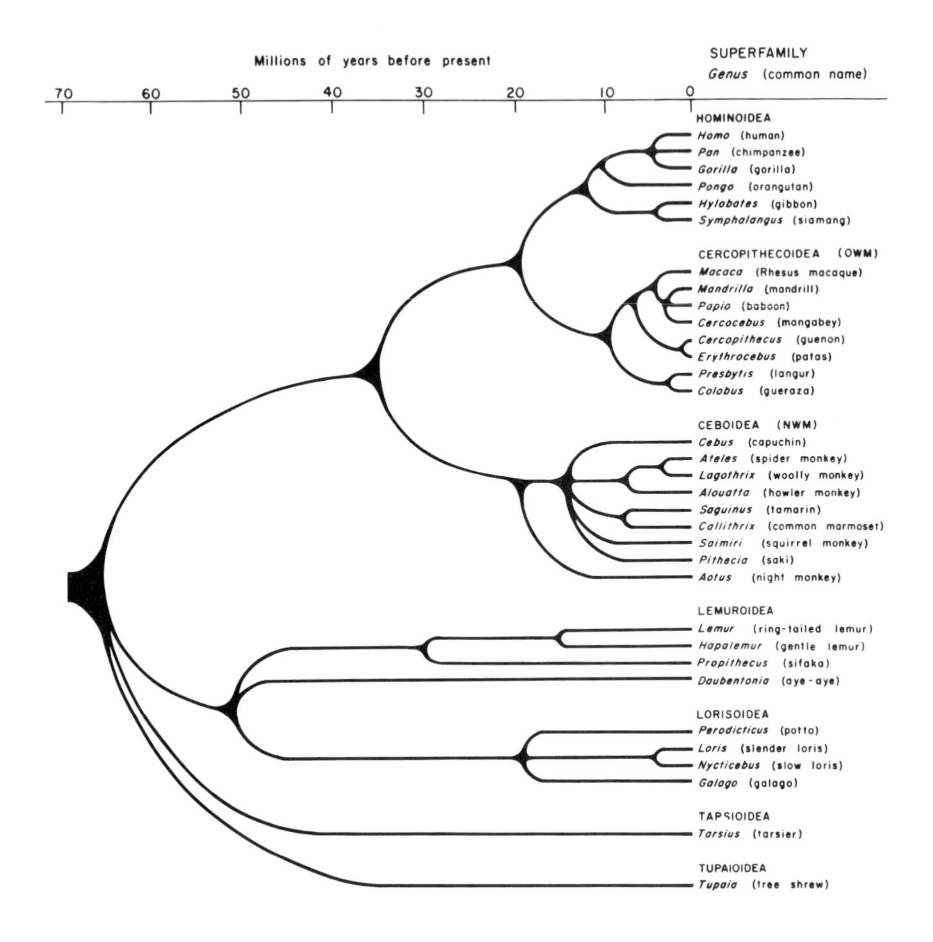

In addition to LEP, there is one other general physiological parameter that appears to have played a key role in determining the longevity of a mammalian species. This is the rate of development of the species from conception to the adult (Cutler, 1976a,b, 1980a,b). As with body size, it has long been observed that longer-lived species develop at slower rates. This correlation is illustrated for rodent species in Fig. 9 and for primate species in Fig. 10. In these two figures, overall rate of development is estimated by the age when sexual maturity is reached, but other markers of development can be used as well that result in a similar conclusion. A reasonably linear correlation is evident between LSP and age of sexual maturation. The slope of this correlation, however, is not the same between rodents and primates, indicating a basic difference in the relationships of developmental rate and longevity between these two groups.

LSP (yrs)	LEP (kc/g)
90	815
48	469
43	309
50	447
35	569
25	354
34	517
35	421
35	394
33	501
25	394
22	333
30	388
25	391
42	804
30	524
30	515
20	371
20	643
15	535
18	485
15	373
20	530
35	743
15	300
20	360
23	642
22	537
15	500
15	355
17	419
15	635
13	512

Figure 8. Phylogenetic relationship of LSP and LEP values for the primate species. Data represent estimates taken from literature values for LSP and SMR (Flower, 1931; Spector, 1956; Napier and Napier, 1967; Holliday *et al.*, 1967; Bauchot and Stephan, 1969; Altman and Dittmer, 1968, 1972; Jerison, 1973; Walker, 1975; Cutler, 1976a; Bowden and Jones, 1979).

But what is the meaning of this correlation between longevity and developmental rate? Is it related in a causal manner to longevity or to reproductivity, as explained by r and k strategies of population biology theory (Wilson, 1975). Without going into detail into all that may be implied, I would like to simply state here the argument that this correlation is at least in part related to a new set of causal factors of aging, apart from the energy metabolism concept of LEP (Cutler, 1976a,b, 1982a,b).

If aging is caused by by-products of otherwise essential living processes like oxygen metabolism, processes involved in differentiation and development might also have similar long-term harmful effects (Landfield, 1978, 1980; Landfield *et al.*, 1977, 1978, 1980, 1981). In seeking an answer to this question, many examples can be found for the toxic or aging effects of development. One of the most impressive is the effects of sexual maturity on the lifespan of the

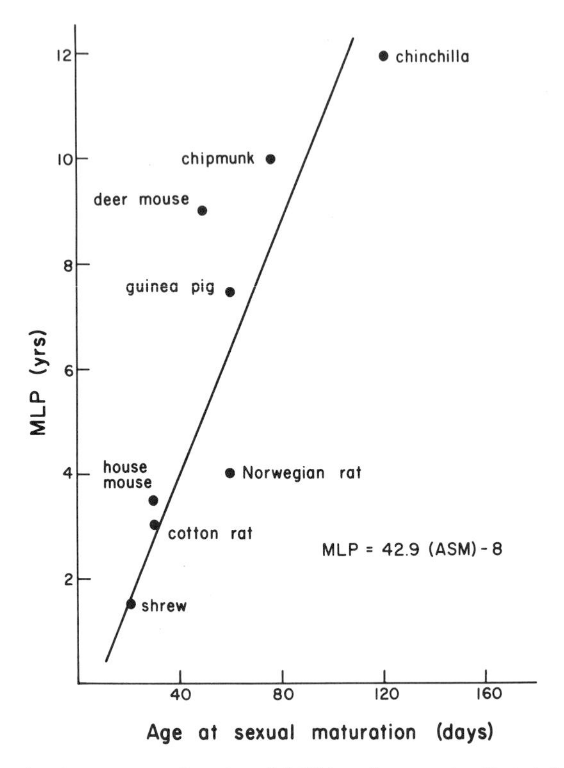

Figure 9. Rate of development as a function of LSP in rodent species. Data taken in part from Cutler (1976a,b; 1978).

Pacific coast salmon (Wexler, 1976) and octopus (Wodinsky, 1977). Here, the rapid aging of these animals can be accounted for by the appearance of hormonal factors (such as adrenal corticoids) that occur during the stages of sexual maturation. Prevention of sexual maturation by removal of the pituitary or castration of the animal prevents this rapid aging, and the animals live much longer. Accelerated aging can be induced in these animals by injection of these hormone factors.

Other evidence is that, when lifespans of *ad libitum*-fed rats are extended by placing them on a restricted-calorie diet at weaning age, lifespan is extended by prolonging the stages of development before sexual maturity (see Fig. 11). Thus, one of the mechanisms of lifespan extension of calorie-restricted animals may be in lengthening the period of time before sexual maturation is reached and thereby postponing the time when harmful hormone factors related to development are expressed.

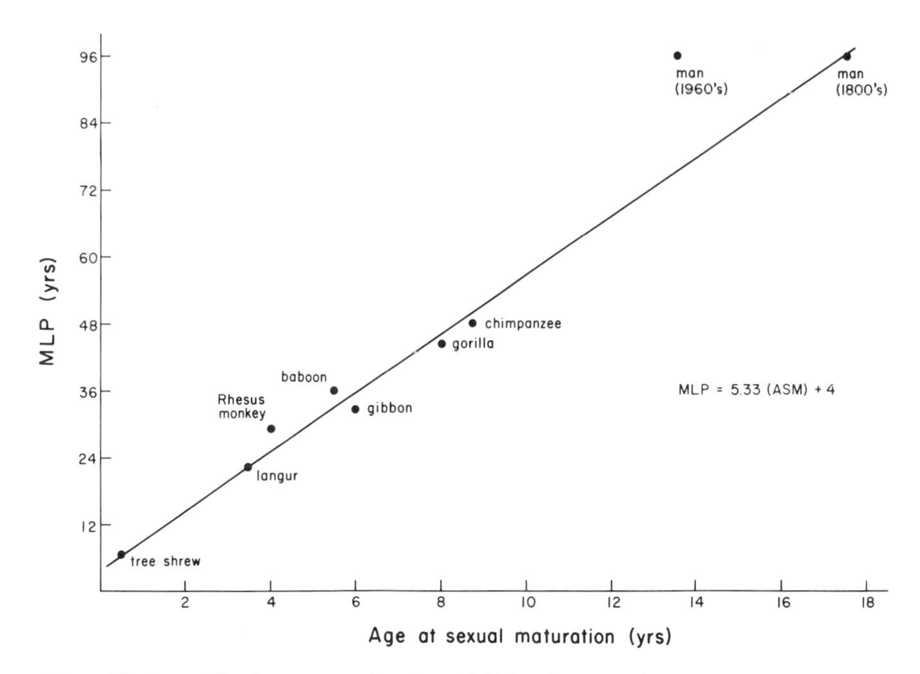

Figure 10. Rate of development as a function of LSP in primate species. Data taken in part from Cutler (1976a,b; 1978).

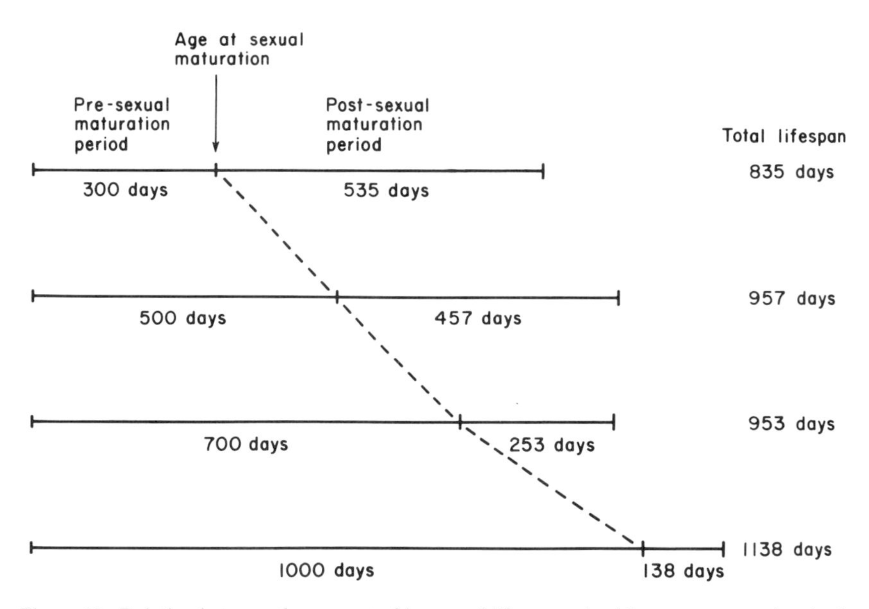

Figure 11. Relation between the amount of increased lifespan gained in rats on a restricted calorie diet in the presexual maturation period and the postsexual maturation periods of lifespan. Data taken in part from Cutler (1976a,b; 1978).

It should be pointed out, on using this example, that feeding animals *ad libitum* (feeding them all they want to eat all of the time) probably results in shortening the normal period of development and thus results in faster aging rates and shorter lifespans (Cutler, 1982b). Thus, the *ad libitum*-fed animals may be the test animals and the restricted animals, the normals or controls.

This and other evidence suggest that processes involved in differentiation and development may contribute a causative component to aging (Cutler, 1982a,b). Accordingly, the evolutionary strategy that appears to have occurred in reducing these developmental effects was not biochemical in nature, such as higher levels of protective and repair processes, as is the case for energy metabolism, but simply a postponement in time of their appearance. Clearly the decrease in the rate of growth and the increase in the age of sexual maturation has obvious disadvantage in relation to survival and reproduction. Thus, the very existence of this type of strategy to increase longevity emphasizes how important the advantage of increased longevity was to enhanced species' survival. Also, this longevity strategy may have been a last resort measure, the only one possible to deal with a vast array of widely different aging factors that are highly entrenched in the development and makeup of an organism where no readily available biochemical corrective possibilities exist. In this sense, then, the unusually slow developmental rate of the human is considered to be an essential mechanism accounting for his unusual longevity. Also, an important prediction of this concept is that any nontoxic method that decreases the overall rate of development would be expected to increase the species' LSP.

This interpretation of the longevity-enhancing effects of lengthened developmental period may give us some additional insight into whether a calorie-restricted diet might work in primate species as it clearly does in rodents. So far, there is no evidence that intermittent fasting or calorie-restricted diet can lengthen lifespan in primates or for humans (Cutler, 1982b). The only mammals tested so far are mice and rats. On examining Fig. 12, we find that not only do the rodents have a different correlation function between LSP and age of sexual maturation, as shown in Figs. 9 and 10, but they also have a different proportion of time of total lifespan potential spent before sexual maturity is reached. Thus, humans already have the highest fraction of lifspan before reaching sexual maturation of any mammalian species, and particularly so when compared to rodent species. It may very well be that rodent species are more sensitive to lifespan extension by caloric restriction because there is proportionally more room for extension of lifespan before sexual maturity occurs. In other words, it appears as if the strategy of lifespan extension by lengthening the developmental rate has already been utilized to a much greater extent in humans than in the mouse or rat. If so, then dietary restriction may be much less effective in humans as compared to rodents.

Further support for this argument is found on comparison of the onset

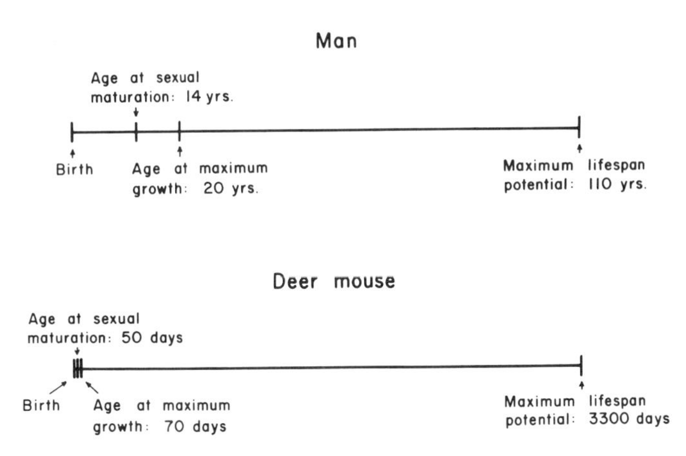

Figure 12. Relation of age at sexual maturation and age at maximum growth to LSP for human and the deer mouse. Taken in part from Cutler (1976a,b; 1978).

frequency of disease for human populations in developed nations, where feeding is essentially on an *ad libitum* basis. As shown in Fig. 4, the onset frequency of cancer does not increase dramatically until past the age of 35 years, which is what would be predicted from survival curve data of primitive human cultures where a calorie-restricted and intermittent-fasting diet was likely to be in effect. This match-up of primitite survival data and onset frequency of disease to what is found in modern cultures suggests that MLP was equal in the two cultures in spite of people today having an *ad libitum* diet. I also know of no data for primitive human cultures showing that LSP or disease frequency was different from today (Acsádi and Nemeskéri, 1970), and I doubt that this discrepancy can be explained by suggesting that malnutrition, was common for these primitive cultures.

However, dietary restriction also appears to extend lifespan if initiated after the age of sexual maturation, but not nearly as effectively (Cutler, 1981, 1982a,b). Thus the decrease in time of sexual maturation is not the entire explanation, but there are of course developmentally linked parameters other than sexual maturation that may be equally affected which continue on at a later age. Thus, although calorie restriction does extend longevity in rodents, it cannot be assumed to extend the present LSP in humans. Experiments designed to study the lifespan extension effects of calorie-restricted and intermittent-fasting diets in primate species are essential to answer this question.

To summarize the major points in this section, it is seen that, in spite of the vast complexity of the aging process, different species do have different LSPs and, accordingly, different aging rates. A comparison of some of the general physiological parameters that vary with species' LSP, such as specific metabolic rate (SMR), gives rise to the concept of lifespan energy potential (LEP) and the possibility that (1) aging is due in part to the by-products of energy metabolism and (2) a species' LEP value is related to its general level of defense and protection against the toxic effects of energy metabolism.

Other comparative data on the rate of development with LSP for different species suggest the possibility that (1) by-products of developmental processes may also cause aging and (2) these types of aging processes were dealt with in the longer-lived species by postponing the time of their appearance, a result achieved by slowing down overall developmental rate.

Thus, the initial hypothesis that aging is caused by normal by-products of processes necessary for life has been maintained and the potential causes of aging divided into two major categories: (1) the by-products of energy metabolism and (2) the by-products of developmental processes.

4. LONGEVITY DETERMINANT GENES

Studies of the biological nature of aging are of course important and required to understand aging. However, the difficulty of such studies making much gain is the extraordinary complexity of aging. Essentially all body functions are affected in a complex interacting web-like network, where a differentiation between cause and effect becomes extremely difficult. Also, once a particular aging process has been identified, of what value would this information be? If the long-term objective of studying aging was to eventually develop methods to reduce the impact aging has on causing disease and general incapacitation, then one is counting on the possibility that an understanding of a given aging process would immediately lead to possible therapeutic methods. Many researchers undertaking studies of aging have also followed the general approach used by most of the biomedical research community on non-age-related diseases, where it is hoped that some single factor would be found that changes with age, and a prevention of its change or loss would result in curing some specific aspect of aging.

However, I believe that such an understanding of specific aspects of aging is not likely to lead to practical applications. This conclusion is based partly on recognition of the ubiquitous nature of aging. Considering that all body processes are affected by aging in a detrimental manner, any significant lifespan extension would require the rejuvenation of most all functions of the body. The hopelessness of this possibility in the near future is obvious. Consequently, the approach taken by most in the health-related sciences has been to avoid age-related problems and to concentrate on specific diseases and dysfunctions. Some people appear to be so convinced of the impossibility of dealing with age-related diseases that there is now the trend to classify age-related dysfunctions (like senile dementia) as diseases not related to aging, thus placing them in the category of being potentially curable.

Thus, aging studies of both a descriptive and mechanistic orientation appear unlikely to yield any significant insight on how to deal with the effects of aging. It is the recognition of this problem that keep many of the top scientists out of the field of gerontology. The break from this type of thinking, began with the recognition that different species have true differences in longevities (LSPs) that cannot be explained away on the basis of diet or nutrition. It then became apparent that, if one could only understand the biological basis of longevity in these different species, it appears possible that such knowledge would also suggest practical means to affect longevity. But how is the study of longevity any different from a study of aging? The study of aging and longevity would be similar if the causes of aging determined longevity. Clearly, one can gain an insight of what determines how long a car will last by studying how it wears out. *But the causes of aging may not be involved in determining longevity.* Instead, it is proposed that aging processes are separate from those determining LSP. Thus, a study of the biological basis of longevity of mammalian species, as opposed to aging, represents a completely new approach to arrive more directly and efficiently at practical applications to extend lifespan and reduce the many detrimental effects of old age.

The study of the biological basis of longevity was first undertaken in a systematic manner by George Sacher (Sacher, 1959, 1960a,b, 1962, 1965, 1966, 1968, 1970). He rediscovered and refined an old observation that longevity of mammalian species can be predicted by an equation based only on the brain body weights of the young adult. This equation is written as follows:

$$\text{LSP} = (10.8)(\text{brain wt., g})^{0.633}\,(\text{body wt., g})^{-0.225} \qquad (1)$$

where LSP is in years.

This correlation suggested to Sacher that the most obvious feature longer-lived species have in comparison to shorter-lived species is a larger relative brain to body weight. Recognizing that the biochemistry of species with different LSPs were remarkably similar, Sacher also concluded that LSP differences could not be adequately explained by biochemical differences (Sacher, 1968). This led Sacher to the hypothesis that instead of biochemical differences, the brain is the "organ of longevity," for the mammalian species. The brain was postulated to govern the homeostatic functions of the organism, where the larger the brain was in relation to body size, the higher the degree of precision control. Aging was visioned as being the result of a gradual loss in an organism to maintain proper homeostatic regulation, and that species having higher LSPs have accordingly a qualitatively superior control system.

A different hypothesis from the one Sacher proposed can be constructed from the same data base (Cutler, 1972, 1973, 1975, 1976a,b, 1978, 1970). It is indeed true that a comparative analysis of the overall biological difference in design and biological makeup of mammalian species does not reveal any striking differences. This fact is remarkable, considering the great differences that exist in morphology as well as LSPs. However, the remarkable biochemical similarities among the mammalian species suggested the possibility that few biological differences may exist that underlie longevity differences. Also, there is no experimental evidence suggesting that longer-lived species have superior homeostatic control processes or that the brain provides some special functions which govern the longevity of species. It appears more reasonable to suppose that longer-lived species have larger brain size relative to total body weight because of some mutual beneficial factor. For example, larger relative brain size could have coevolved with increased longevity by mutually beneficial but separate noninteracting biological mechanisms. Indeed the coevolution of brain size with longevity can support an argument of why longevity evolved in the first place: to provide more time for the development of learned behavior, which is a distinctive characteristic of animals with higher brain to body weight ratios.

The idea that longevity may be governed by a few biological processes was pursued, and further support was found from studies in other laboratories on the biological basis of speciation (King and Wilson, 1976; Wilson *et al.*, 1974; Cherry *et al.*, 1979; Zuckerkandl, 1976b; Dover *et al.*, 1982; Flavell, 1982). These studies measured in a quantitative manner the remarkable similarities of genes and gene products between different species. In addition, a comparison of the rate of chromosomal changes versus DNA sequence changes was made. These data led to the conclusion that the appearance of new species is not related to the appearance of new gene products but instead is related more strongly to chromosomal alterations. Thus, in some way, chromosomal rearrangements rather than the appearance of new gene products seem to play the major role in the emergence of new species (Gillespie *et al.*, 1982).

On comparing enzyme levels in different species, it was found that, in spite of species having most gene products in common, species do differ greatly in the time when these genes are expressed and in the levels the products of these genes have in different tissues. This finding is supported by the genetic changes found in microorganisms on adapting to new nutritional environments. Bacteria adapt most frequently by altering the concentrations of existing gene products, not by the appearance of new gene products. This type of adaptation involves changes in regulatory genes, not structural genes.

These types of studies suggest that speciation is predominantly due to changes in the timing and degree of expression of a common set of structural genes; that is, speciation involves changes in the regulation of genes, presumably by DNA sequence changes (point mutational changes) of regulatory genes and/or by rearrangement of structural or regulatory genes within or in different chromosomes.

This mechanism of speciation can also apply to the genetic mechanisms of longevity differences, which of course is a species characteristic. Thus, longevity of the human and other primate species may be due not to qualitative differences, as Sacher had suggested, but to quantitative differences in levels of gene products held in common in all mammalian species. Longevity difference in mammalian species might then simply be a matter of more of the same gene products that all species have in common. This idea was of considerable interest, for it was consistent with the postulate that longevity differences between species may evolve few biochemical differences. The other exciting possibility this model of speciation suggests is that methods to extend longevity may not have to produce something better or longer lasting that is lacking in human biology but simply to enhance to higher levels those gene products that are already there. This possibility placed longevity intervention into the realm of practicality for the first time.

Still, the problem of understanding the biology of longevity could be extremely difficult, for thousands of genes may be involved in altering the timing and concentration of enzymes as a function of a species' LSP. A more accurate estimate was necessary to determine the upper limit of gene compexity involved in governing LSP. This estimate was accomplished by determining how fast longevity had evolved during the evolution of the hominid species leading to human (Cutler, 1975, 1976a,b). The basis of these studies was that the rate of appearance of a new characteristic along an evolutionary line of species is related (among other variables) to the number of genes involved in determining that characteristic.

Two different approaches were utilized in determining the rate of evolution of longevity in primate species. The first approach was based on the hypothesis that LSP of extinct primate species can be estimated by using the same equation as for estimating the LSP for living primate species [Equation (1)]. Evidence for this assumption is shown in Tables V and VI, where it is demonstrated that this equation predicts LSP equally well for living progressive species as for living fossil-like species.

Equally important to this study is the conclusion that the qualitative aspects of aging are remarkably similar in all primate species in spite of their large differences in LSPs. This implies that the differences in LSPs reflect true aging rate differences of the same aging processes between the primate species. If short- as well as long-lived species age in the same manner, and if they have the same general metabolic functions and physiology, then it is also reasonable to assume that extinct primate species also age qualitatively in a similar manner and by the same kinds of aging processes as the living species. It is therefore concluded that the evolution of longevity in primate species occurred by chang-

Table V. Estimation of Lifespan Potential for Living Fossil and Recent Mammals Living in a Mesozoic Niche[a]

Common name	Cranial capacity (cm^3)	Body wt. (g)	LSP (yr) Predicted	LSP (yr) Observed	Time of appearance ($\times 10^{-6}$ yr)
Short-tailed shrew	0.347	17.4	2.9	3.3	14–0
Pygmy shrew	0.11	5.3	1.8	1.5	14–0
European shrew	0.2	10.3	2.3	1.5	14–0
Long-tailed shrew	0.12	5.7	1.9	1.5	14–0
Tenrec	2.75	832	4.5	2	10–0
Madascar hedgehog	1.51	248	4.1	3	10–0
Streaked tenrec	0.83	110	3.3	3	5.5–0
Norwegian rat (wild)	1.59	200	4.4	3.4	5.5–0
Field mouse	0.45	22.6	3.2	3.5	5.5–0
Vole	0.66	23.7	4.1	2–3	3–0
Vole	0.74	27.9	4.2	2–3	3–0

[a]From Cutler (1980a).

ing the rate of aging of all physiological processes of the organism in an uniform manner.

The literature was found to have a substantial number of brain and body weight estimates of extinct primate species, particularly for the hominids, over the past 1 to 2 million years. To test the reliability of these data, they were first analyzed allometrically by a log (brain wt.) versus log (body wt.) plot. Such a plot is shown in Fig. 13. These data suggest that hominid evolution proceeds along the *A. africanus* to *H. habilis* sequences, where brain to body weight

Table VI. Prediction of Maximum Lifespan Potential on the Basis of Body and Brain Weights for Some Primates[a]

Common name	Cranial capacity (cm^3)	Body wt. (g)	Maximum lifespan potential (LSP, yr) Observed	Maximum lifespan potential (LSP, yr) Predicted
Tree shrew	4.3	275	7	7.7
Marmoset	9.8	413	15	12
Squirrel monkey	24.8	630	21	20
Rhesus monkey	106	8,719	29	27
Baboon	179	16,000	36	33
Gibbon	104	5,500	32	30
Orangutan	420	69,000	60	41
Gorilla	550	140,000	40	42
Chimpanzee	410	49,000	45	43
Human	1,446	65,000	95	92

[a]From Cutler (1980a).

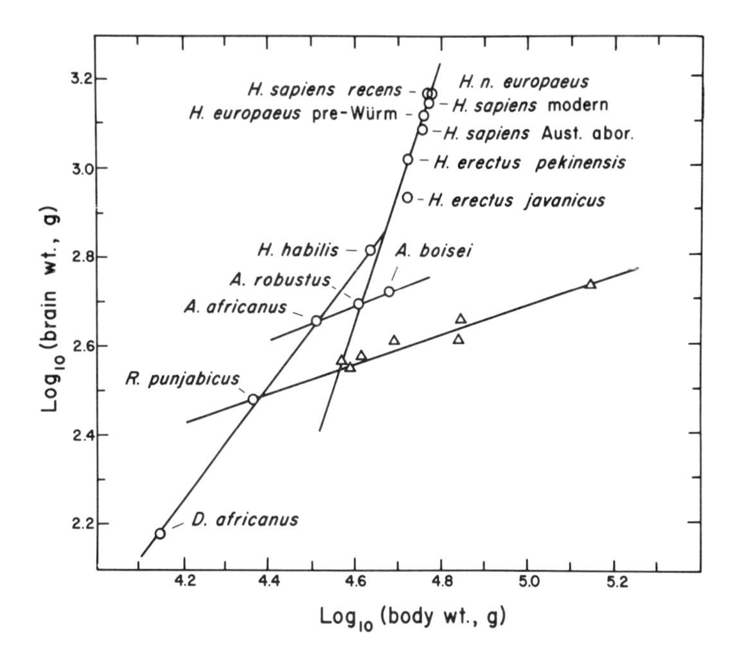

Figure 13. Allometric analysis of cranial capacities and body weights for living and fossil species of Hominoidea. Species expected to be on a hominid lineage (○); living species of Pongidae (△). Data taken from Cutler (1976a).

scaling was similar. However, near the *H. habilis* to *H. erectus* transition, a change in morphological scaling occurred, which was then maintained up to the appearance of *H. sapiens*. In addition to these conclusions, confidence was gained in the reliability of the brain and body weight data because of the constant scaling factor found along the *H. habilis–H. erectus* ancestral–descendant sequence leading to *H. sapiens*. This linear, nonrandom scaling factor did not appear likely to emerge from a randomized unreliable data base.

Using the brain and body weights of a wide range of extinct primate species, particularly hominid species, LSP was calculated. Taking these data with the evolutionary period when these species were present, a phylogenetic map of the evolution of primate longevity was constructed. This is shown in Fig. 14. The most general finding is that longevity always appeared to increase in primate species. This increase was found to be most rapid along the hominid ancestral–descendant sequence. This is shown in Fig. 15. A large number of estimates of brain and body weights for the homonid species were found in the literature for the past 1.5 million years of hominid evolution. These data were used in the construction of a plot of log (LSP) versus log (time) (Fig. 16) which was found to have a surprisingly high linear correlation coefficient. Using this equation for this line, the rate of evolution of longevity for the hominid species

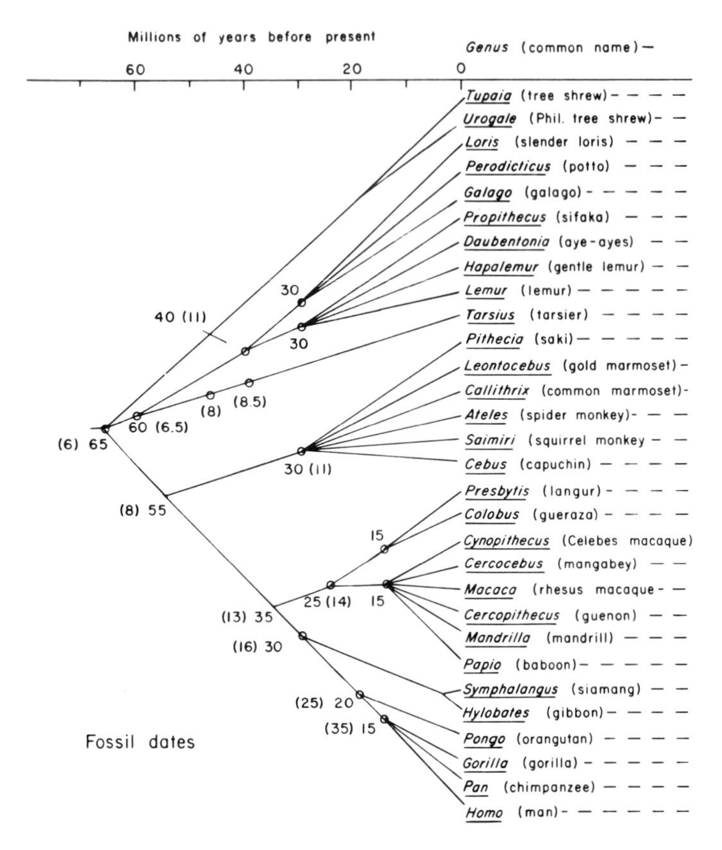

Figure 14. Evolution of longevity in a primate species. Numbers at the divergence points in the species based on geological (fossil) and molecular (albumin) methods. Numbers in parentheses

was calculated. These data were used to determine the rate of hominid evolution of LSP, as shown in Fig. 17. Here the maximum rate of LSP is found to occur about 100,000 years ago with the emergence of the *Homo erectus* species. The rate of increase of LSP 100,000 years ago is estimated to be about 14 years per 100,000 years.

Comparison of how fast amino acid substitutions in proteins or base sequence substitutions in DNA have appeared over a similar time period for primate evolution was used to place an upper limit to the number of new genes that could possibly have been involved in hominid evolution (Cutler, 1975, 1976a). The result was estimated to be about 0.6% of the total genes within the genome. If we assume 100,000 genes per genome, then no more than about 600 genes were altered at one codon each, resulting in an increase of 14 years of LSP.

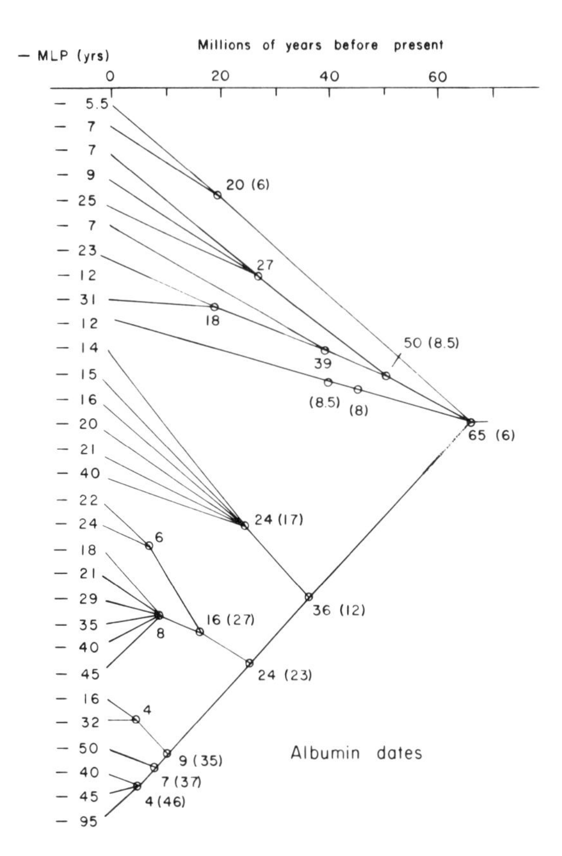

phylogenetic map without parentheses represent estimates of time of existance for the extinct are estimates of LSP for the extinct species. Taken from Cutler (1976a).

However, as pointed out, point mutational changes of DNA may not have played the most important role in speciation, but rather gene rearrangements within and/or between chromosomes (Cutler, 1980a, 1982a). If this is the case, then we do not yet know how to estimate how many such gene rearrangements could have accumulated over that 100,000 year period of hominid evolution. Such information is possible to obtain by studying the arrangement of genes in the chromosomes of primate species having different LSPs. Such studies are considered to be important not only to determine the possible role of gene rearrangement in species evolution but also to identify how many such rearrangements occurred and what genes were involved in the evolution of LSP.

It is also important to point out that, during the evolution of primate longevity, other changes in addition to increase of LSP were also occurring, namely, changes in overall morphology, together with larger brain to body

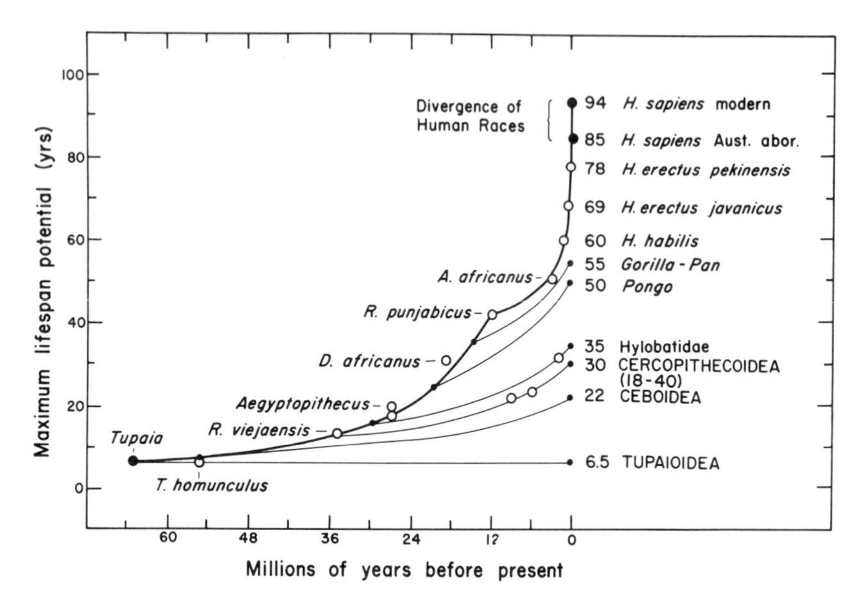

Figure 15. Evolution of longevity of the primate species along the hominid ancestral–descendant sequence leading to humans. Extinct species (○); living species (●). Numbers beside species are LSP estimates. Taken from Cutler (1976a).

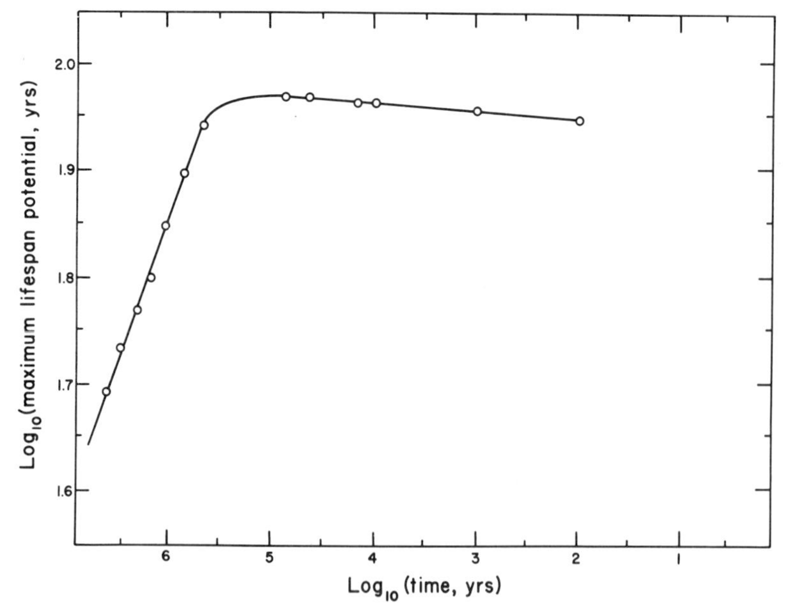

Figure 16. Evolution of longevity of the hominid species over the past 10 million years. Taken from Cutler (1976a).

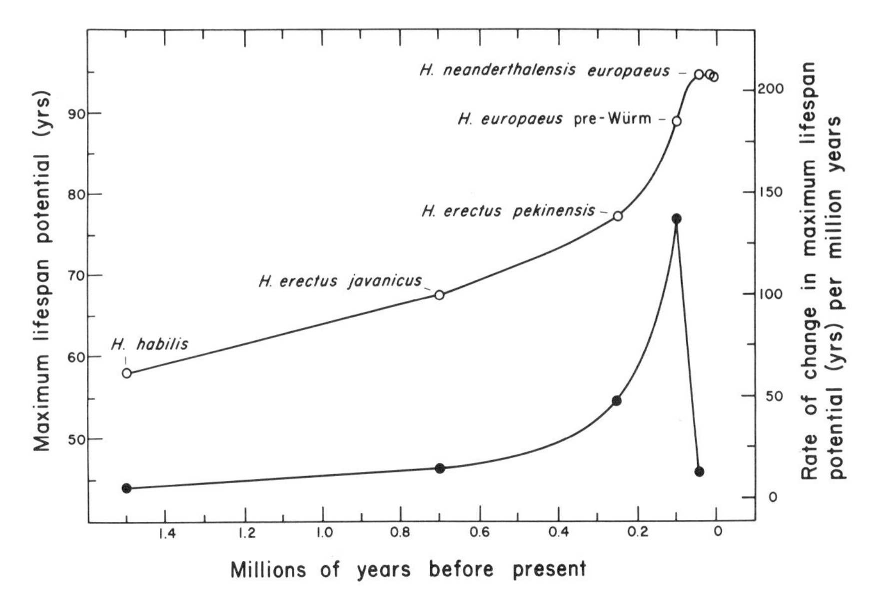

Figure 17. Evolutionary rate of increase of LSP for the hominid species. LSP (○) and rate of change of LSP (●). Taken from Cutler (1976a).

weight ratios, slower rates of development, and of course a higher level of intelligence. How many of these changes played a causative role in determining longevity is not known, but larger body size and slower rates of development are likely to be involved in this manner, as already pointed out. It is therefore of interest to learn if changes in morphology, body size, and rates of development are governed by regulatory genes, like the proposed genetic processes governing LSP. Only a few genetic changes do indeed seem necessary to alter body size (both in relation to body size and brain size) and developmental rate. These changes, like the predicted changes governing the biochemical basis of longevity, appear to be governed by regulatory changes in genes.

Much of the morphological changes occurring in human evolution can be explained by the principles of neoteny (Gould, 1977). The important feature of neoteny is that an increase in brain to body weight ratios, and decrease in developmental rate, are brought about by relatively simple changes occurring in regulatory genes governing a common scaling factor. Thus, additional evidence is gained that human evolution has occurred by relatively few changes in regulatory genes. So to understand the genetic basis of longevity we may also need to understand the genetic basis of neoteny. Perhaps these genes and their basis of operation are similar in their mode of action to the postulated longevity determinant genes. (See Section 9 for additional comments.)

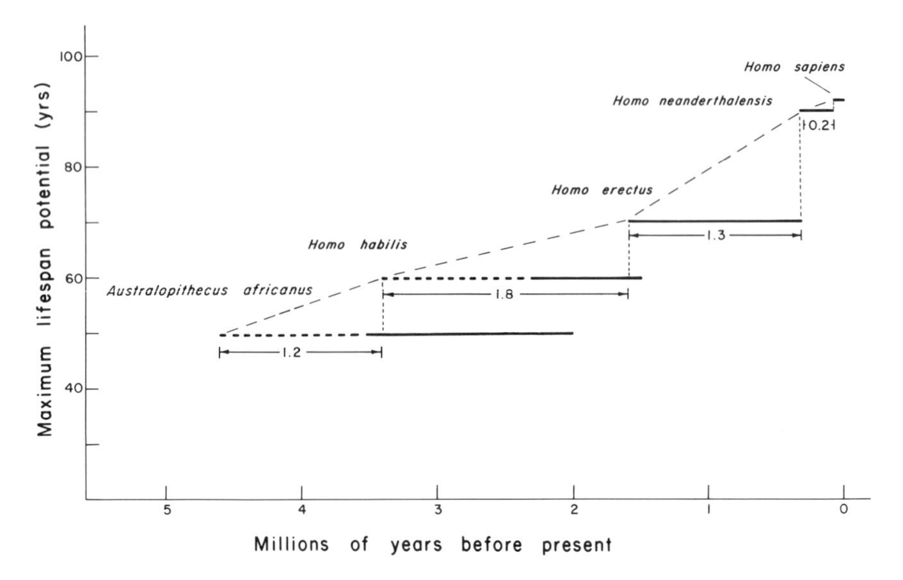

Figure 18. Macroevolution model of the evolution of hominid longevity. Dotted-plus-solid horizontal lines represent estimated maximum length of time a hominid species was in existence. The solid portions of the lines are more substantiated by the fossil data. Evolutionary appearance of new hominid species lies between the period of time of the two extremes, represented by the dashed-sloped and dashed-vertical lines connecting the different species. Numbers below horizontal lines represent the maximum length of time before appearance of new hominid species. Length of time species were in existence was taken from Stanley (1979).

The model of hominid evolution was initially constructed as a gradualistic model of evolution as compared to the macroevolutionary model. However, because of the recent interest in macroevolution, it is necessary to comment on how the new macroevolutionary model may affect the estimated rate of hominid evolution (Stanley, 1979, 1981). Estimates of the length of time the various hominid species were in existence were taken from Stanley's book (1979), and on using the LSP estimates for these hominids, a new plot was constructed of the evolution of LSP, as shown in Fig. 18. This plot represents the macroevolutionary model for the evolution of hominid longevity. The exciting result of these data is that the evolution of longevity may have occurred even faster (by short quantum jumps) than that estimated using the gradualistic model. Thus, even fewer or less complex genetic changes may be involved. However, a possible weakness of Stanley's data is that the occurrence of long time periods of stabilization of hominid species, which he assumes, is not supported by continued changes in the relative brain to body weight ratios I have found for these species. These ratios, as shown in Figs. 13 and 16, clearly indicate a gradual neotenylike change, not a stabilization. Some of this discrepancy with Stanley's data could be with the naming of species such as *H. habilis* to cover a broad range of brain to body weight ratios. Thus it would appear as if the species existed unchanged, whereas brain and body weight ratios would indicate a clear progressive change. However, one does expect gradualistic evolution to be made up of a series of small quantum jumps of increase of LSP that would, when smoothed out, look like the curve shown in Fig. 18.

Also, because it appears as if the recent evolution of hominids involved changes in the same genes for each quantum jump of increased LSP over the last million years or so, and not by an accumulation of different types of genetic changes over this time period, then the genetic complexity governing the recent speciation of the hominid may in fact be extraordinarily simple, involving changes in only a few key or master regulatorylike genes.

It could be argued that the validity of predicting LSP for extinct species by using estimates of their brain and body weights cannot be proven, so the conclusions drawn from these studies should not be taken seriously. To answer this criticsm, another approach was taken to estimate the evolutionary rate of longevity for the primate species (Cutler, 1976a). This was done by analyzing the differences in LSP of living primate species in relation to their evolutionary relationships to one another. Typical data using this approach are shown in Table VII. The results found using this method agreed with the predictions using brain and body weights to estimate LSP; that is, longevity is seen to increase during primate evolution and to be most rapid for the hominid species. Although the detail is not as great, the same general result is found using a completely different approach.

Table VII. Estimates of the Evolutionary Rate of Increase in Maximum Lifespan Potential within the Superfamilies of the Primates[a]

Order	Suborder	Superfamily	Mean MLP within a superfamily	MLP differences within a superfamily	Common divergence time ($\times 10^{-6}$ yr): Fossil dates	Common divergence time ($\times 10^{-6}$ yr): Albumin dates	Change in MLP per 10^6 yrs: Fossil	Change in MLP per 10^6 yrs: Albumin	Rate of increase in MLP in millidarwins[b]: Fossil	Rate of increase in MLP in millidarwins[b]: Albumin
Primates	Prosimii	Tupaioidea	6	1.5	65	65	0.02	0.02	1.2	1.2
		Lorisoidea	14	18	30	27	0.6	0.66	16	17
		Lemuroidea	19	24	30	39	0.8	0.62	17	13
	Anthropoidea	Ceboidea	22	25	30	24	0.83	1.04	14	17
		Cerocopithecoidea	30	27	25	16	1.1	1.7	14	21
		Hominoidea	52	63	30	9	2.1	7.0	15	50

[a]Table from Cutler (1976a).
[b]Darwins for this table are calculated as the change in MLP per 10^6 yr per mean MLP of the superfamily per *e* darwin units.

These two methods were also used to study the evolution of longevity of other mammalian species (Cutler, 1979). For ungulates and carnivores, an increase of LSP was also found to be the general rule. Another finding of interest is that LSP did not increase nearly as dramatically, if at all, in those species whose lines later became evolutionarily extinct. Similar results were found for the hominid species along the *A. africanus–A. robustus–A. boisei* line, which also became extinct (Cutler, 1976a). These results suggest that for mammals in general, but particularly for the primate species, the evolution of LSP has played an important role in their evolutionary success. In fact the evolution of longer LSP may have been equal in importance as the evolution of higher level of intelligence in determining evolutionary success of the primate species, a thought that anthropologists appear not to have considered adequately. A remarkably good positive correlation of LSP with performance in a number of learned behavioral parameters is found in different primate species. Tests such as determining the rank order of ability of primate species to perform learning and perception of oddity problems, observational learning, and successive discrimination reversal learning all correspond remarkably well with their LSP (Ingram and Cutler, in preparation).

In general, it appears that species with high LSPs show a higher ratio of learned behavior relative to instinctive behavior (Cutler, 1976a,b, 1979). This suggests that the driving force behind the evolution of the primate species, and particularly the hominids, are those processes developed to take advantage of this new trait of brain function. This new trait enables an individual to learn from the environment and to become more adaptable to changing environmental situations involving climate and the appearance of other species. To take advantage of this new trait of learned behavior, however, a given amount of healthy lifespan is required to learn and to pass along this adaptive information. This time was provided by increasing the LSP as this new trait continued to develop.

In conclusion, comparative and evolutionary studies, ranging from the morphological, chromosomal, and genetic, to the molecular biological level of different species, suggest that the genetic mechanism of speciation is mostly determined by changes in levels of existing gene products, not new gene products. These data also suggest that speciation is largely a matter of changes occurring in regulatory genes, altering the timing and tissue levels of gene products that are found in almost all species.

These concepts are consistent with a similar genetic process underlying the biological basis of the evolution of longevity in mammalian species, particularly for the primate species. The remarkable similarities of metabolic, developmental, and aging processes suggest common underlying mechanisms of aging and common processes that operate to govern aging rate. Many of the changes occurring during hominid evolution, such as those processes involving neoteny, are consistent with the genetic basis of speciation involving changes in regulatory genes.

Finally, the extraordinarily high evolutionary rate of increased longevity leading to the appearance of the hominid species suggests that few genetic alterations are likely to have been involved in this process. This conclusion is supported by the remarkable similarities of the proteins and nucleic acids of different primate species as well as the qualitative similarities of the biological processes of different mammalian species having widely different LSPs.

All of these studies point to the existence of a special set of genes, separate from most of the genes in a cell, that govern the LSP or the general health maintenance of all body functions. Instead of longevity being determined by the entire biological makeup of an organism or by all of its genes, it appears that only a fraction of the biological makeup of the organism is involved in determining how long all of its biological makeup will last against the effects of the aging process.

The general hypothesis that emerges is as follows. Mammalian species have common aging processes because they have common physiology and a common set of antiaging processes acting against the primary aging processes. This gives rise to the same spectrum of harmful by-products of development and metabolism and thus the same types of aging processes in all mammalian species. The common set of antiaging genes found in all organisms is controlled in its degree of effectiveness against aging by regulatory genes. Few genetic alterations of these regulatory genes, be it by mutation or gene rearrangement, appear necessary to uniformly decrease the aging rate of all the physiological functions of an organism. The hypothetical set of antiaging genes is termed longevity determinant genes (LDGs), and the processes they control or govern, longevity determinant processes (LDPs). The major conclusions related to this hypothesis can be summarized as follows.

1. Different mammalian species have different innate capacities to maintain good health in the most general sense and to postpone the global degenerative effects known as the aging process for a given length of time.
2. This innate capacity, known as longevity, is positively related to a species' lifespan potential and is a species characteristic.
3. Speciation during primate evolution is most closely related to changes occurring in the timing and degree of the same structural expression of genes found in all species, rather than in the appearance of new structural genes.
4. Longevity evolved at an extremely rapid rate during recent hominid evolution, resulting in the uniform extension of the length of time gen-

eral health is maintained. This result is most consistent with few genetic alterations being capable of decreasing the entire organismic complex of aging processes in a uniform and coordinate manner.

5. It is proposed that these few genetic alterations involved principally regulatory genes which resulted in a change in the timing and level of expression of a common set of longevity determinant structural genes found in all primate species.

5. DYSDIFFERENTIATION

With the prediction that specific longevity determinant processes exist, the task now is to identify what they are and how they determine longevity. To help accomplish these objectives, a working hypothesis of the biological nature of aging is required.

Aging also does not appear to be due to a loss of cells. Nor has any cellular change been found that suggests an impairment of function due to some type of accumulation of damage (Rosen, 1978a,b, 1981; Kohn, 1982b). These results reflect some of the general frustration of gerontology research, in that one starts for example with an old mouse of 24 months, and on analyzing deeper into the cellular and biochemical properties of the animal, the obvious macroscopic manifestations of age begin to diminish to the point that most cellular and biochemical properties appear unchanged.

There are of course many changes that do occur with aging, but these do not appear to be the result of an intrinsic inability of cells to synthesize proteins or make ATP, RNA, or hormone membrane receptors, or of an error catastrophe, an accumulation of altered protein, or some type of toxin. Instead, most of the evidence suggests that those changes that are found are the result of changes in gene expression not related to cell impairment of an intrinsic nature (Cutler, 1982a,b). This suggests the possibility that aging may be the result of cells simply drifting away from their proper state of differentiation. That is to say, aging may be like an ontogeny in reverse, and almost as well ordered.

If we think of the processes of differentiation and development that are thought to take place during the creation of a human, beginning with a single fertilized egg (the necessary cell-to-cell communication and progressing through the many states of cell differentiation), then aging may be the reverse of these processes, beginning with the higher-order states of differentiation and working backwards. In this model, aging is thought to begin with higher-order processes such as homeostasis control, the ability to adapt and to respond to stimulation. The more complicated or complex the function, the earlier this process begins to undergo the process of decline in function or aging. Thus, to a large extent, the aging of the organism can occur without any substantial changes detected at the less involved states of tissue differentiation or at the cellular level (Rosen, 1978b). However, changes are indeed occurring at the cellular level that underlie all aging processes, but the higher-order dysdifferentiation of organisms and tissue homeostasis involves only extremely small cellular changes, and what we observe at the macroscopic level of the organism is the amplification of these small changes.

These small cellular changes that first result in the loss of higher orders of homeostasis and then proceed slowly to lower orders of function are the result of a natural drifting away of cells to a less effective state of differentiation. The basis for this dysdifferentiation occurring is that the proper differentiated state of cells is naturally unstable and requires positive efforts (energy-requiring processes) for its maintenance.

If aging is a time-dependent function of dysdifferentiation, it follows that longevity is determined by those processes that act to stabilize the differentiated state of the cell or to slow down the rate of dysdifferentiation. Thus, cells from longer-lived species would be predicted to have an intrinsically higher stability against those processes that cause their drifting away from their proper state of differentiation, as compared to cells of shorter-lived species. *The dysdifferentiation hypothesis of aging narrows down the problem of aging to understanding the slow, time-dependent changes of dysdifferentiation and the problem of longevity to understanding the processes acting to stabilize the differentiated state of cells.*

What then are the processes which act to dysdifferentiate cells? Two good possibilities have already been suggested in the previous section. These are the by-products of normal energy metabolism and development. Energy metabolism produces various active oxygen species, such as peroxides and aldehydes. These toxic products can act as epigeneticlike inducers of dysdifferentiation at concentrations far below the point where cell damage would occur (Fahmy and Fahmy, 1980a,b; Braun, 1981; Freeman and Crapo, 1982). A good analogy of such dysdifferentiation processes would be the combined effects of mutagens and tumor promotors causing the initiation and promotion events leading the cancer (Trosko and Chang, 1981). In addition to by-products of energy metabolism, products of differentiation and development can themselves act to dysdifferentiate cells, such as the various hormones that are expressed during development and at sexual maturation (Barrett *et al.*, 1981).

All cells in an organism are not expected to be equally sensitive to dysdifferentiation. Instead, certain cell types would be expected to be more prone to dysdifferentiation than others, leading to an orderedlike cascading effect involving all cell populations. Thus, there are likely to be hot spots of cell dysdifferentiation in critical homeostatic control centers of the body that are the most important in initiating the early stages of aging. Such areas are likely to be in the regulatory tissues of the brain such as the pituitary, hypothalamus, and the cells making up the nucleus basalis of Meynert.

Two classes of experimental evidence form the major support for the dysdifferentiation hypothesis of aging. These are (a) the general lack of any evidence that aging is a result of an accumulation of damage leading to impairment of general cell function, and (b) the ability to explain much of what is known about aging in terms of a dysdifferentiative mechanism. Examples of this latter class of evidence follow (Cutler, 1982a,b).

1. Age-dependent changes in gene expression. Increased expressions of genes are found in tissues where their expression is not expected. An example is the appearance of alpha and beta hemoglobinlike RNA in the liver and brain tissues of mice with increasing age (Ono and Cutler, 1978). Similarly, mouse leukemia virus (MuLV) and mouse mammary tumor virus (MMTV) RNAs have been found to increase with age in brain and liver tissues of mice (Ono and Cutler, 1978; Florine *et al.,* 1980; Getz and Florine, 1981). Changes in the types and amount of RNA in tissues have also been found with increasing age (Richardson, 1981; Cutler, 1982a). Most enzyme levels in tissues vary with age. Some increase while others decrease, and about one-third remain unchanged (Finch, 1972).

2. Age-dependent changes in chromatin. Changes in the biochemical state of chromatin have been found (Cutler, 1982a). These changes would be expected to affect gene expression. Examples of these chromatin changes are alterations in DNA methylation, increase of DNA-protein binding complexes, and increase in amplified circular copies of the ALU sequence in human DNA (Shmookler Reis *et al.,* 1983). This latter observation is of particular importance, considering the recent information suggesting the highly dynamic nature of DNA in terms of migratory DNA elements or transposons.

3. The differentiated state of cells can be changed by low levels of mutagenic agents. There is much information suggesting that mutagenic and carcinogenic agents, which are capable of mutating and killing cells at high dosage, can act to alter the differentiated state of cells at extremely low dosages, where their mutagenic effect appears negligible (Fahmy and Fahmy, 1980a,b; Barrett *et al.,* 1981). Because many of these mutagenic and carcinogenic agents act via free radicals, these data support the concept that toxic by-products of energy metabolism such as the hydroperoxides (ROOH) and the hydroxyl free radical (·OH) may act most importantly to dysdifferentiate cells.

4. The age-dependent expression of endogenous viruses. The basis of slow virus diseases could be a result of a slow age-dependent relaxation in the repression of endogenous viral genes (Huebner and Todaro, 1969; Gajdusek, 1972; Tovey, 1980; Todaro, 1980; Weinberg, 1980). Experimental data indicating

the age-dependent derepression of the endogenous mouse viruses MuLV and MMTV is consistent with this suggestion (Ono and Cutler, 1978; Florine *et al.*, 1980). A number of mice strains show an early age-dependent derepression of such viruses, and these strains may represent a good model to study the mechanisms of these accelerated aginglike changes (Getz and Florine, 1981).

5. Age-dependent quantitative change in cellular hormone receptor levels. There is much evidence that hormone receptor levels change with age (Roth, 1979). There appears to be no particular pattern, in that some receptors increase, others decrease, and still others are not affected by age. These changes in receptor levels may be the result of dysdifferentiation rather than cell impairment. Proper cellular response to hormones is necessary for maintaining a general good health status, and if this is lost via changes in hormone receptor levels, this effect can give rise to further dysdifferentiation effects.

In addition to hormone receptors being located on the cell membrane and within the cytoplasm, they also exist as part of the chromatin structure. Age-dependent changes in the chromatin thyroid hormone receptor levels were measured, and preliminary data suggest that a decrease with age may be occurring (Cutler, 1981b). It is also important to emphasize that most of the age-related changes found for hormone receptors are quantitative in nature; that is, a change in number of receptors per cell occurs most frequently with age, not a qualitative alteration in the structure of the receptors, as would be evidenced by a change in binding affinities. This is consistent with the concept that no errors in protein synthesis or altered proteins underlie the change, but rather, a change in the expression of genes involved in determining the differentiated state of the cells.

6. Age-dependent increase in metaplasia. With increasing age, an increase of abnormal cells is found within a given tissue (Hartman, 1979, 1982; Farber and Cameron, 1980; Ponder, 1980) For example, intestinal cells appear in the lining of the stomach with increasing age. These cells do not have the protection they require against the acid environment of the stomach as normal stomach cells do, and as a consequence appear to have a higher transformation frequency to the cancerous state. Such age-dependent changes of cells going from one state of dysdifferentiation that is benign to another state much more harmful is found to be a common theme in the progression of age-dependent dysdifferentiation states.

7. Age-dependent appearance of foreignlike proteins. The appearance of neurofibrillary tangles and senile plaques may be a result of the cell's changing differentiated state (Terry, 1980; Wisniewski and Soifer, 1979). The cause of such changes in the cells of the cortex may be related to changes occurring in other cells of the brain such as in the band of Broca, the nucleus basalis of Meynert, or the medial septal nuclei (Coyle *et al.*, 1983). The cells in these regions of the brain are known to transport the neurotransmitter acetylcholine

to the frontal, parietal, and occipital cortex of the brain, which shows an accumulation of neurofibrillary tangles with age. In persons with Alzheimer's disease and senile dementia, the output of acetylcholine is greatly diminished, mainly because of a loss of these acetylcholine-producing cells. One key question has been why these cells are lost as opposed to the other cells of the brain. One possible answer is that perhaps most of the cells making up the brain are highly interdependent for their proper state of differentiation on the proper differentiated state of other cells making up the brain. Thus, a small drift of one type of cell from its optimum output say of a given neurotransmitter product or growth factor could conceivably cause something like a dysdifferentiative catastrophe or cascade through positive feedback mechanisms.

Thus, small changes beginning in the nucleus basalis of Meynert could affect the proper function of cells in the cortex, which in turn might further alter the cells back in the nucleus basalis of Meynert. Now, it is well known that the metabolism (degradation) of catecholamines and other neurotransmitters is associated with the formation of free radicals and other toxic products such as hydrogen peroxide. Thus, it could be that certain cell populations or the brain are more susceptible to dysdifferentiation because of the active oxygen products they produce. These cells would be protected normally enough for humans to live the necessary 25 to 36 years, but after that period the toxic by-products of their unique metabolism would lead to their early drifting away from their proper state of differentiation. This could trigger the dysdifferentiation cascade of the rest of the brain.

Finally, it is pointed out that cell death is also a well-known result of differentiation where, in the normal developmental process of an organism, many cells undergo what is called genetic-programmed death. Thus, dysdifferentiation could result equally in cell death or change in function.

Alzheimer's disease may represent an inheritedlike disease where critical brain cells have abnormally low levels of protection against the normal toxic by-products of neurotransmitter metabolism. Thus, this disease develops earlier than normal, but it is the same disease that eventually occurs in normal individuals. In these models, it is not likely that senile dementia or Alzheimer's disease will be any more responsive to specific therapy than any other nonneurological age-related dysfunction. The aging of the brain is likely to involve the slow dysfunction of most of its cells, although some cell types may be pacesetters. As a consequence, the prevention of the loss of say the cells of the nucleus basalis of Meynert by an artificial addition of antioxidants to these cells or the substitution of these cells by the addition of their neurotransmitter products is not likely to result in a long-term practical solution to these types of age-related diseases.

8. The age-dependent frequency of autoimmune disease. Some types of autoimmune disease appear to have a dysdifferentiative basis (Cutler, 1981a,

1982a). For example, if proteins are synthesized in the adult that were not synthesized during the early periods of life when the immune system was developing, then these proteins would appear to be foreign and would stimulate an immune response against the cells that produced them. This effect would represent another case where a rather small benign change of dysdifferentiation, the derepression of a gene, could lead to the death of the cell even though the presence of this foreign gene product per se is not initially harmful to the cell (Nandy, 1976).

9. Age-dependent changes in hair growth. A decrease in the growth rate of hair, loss of hair in some areas, and an increase in abnormal types of hair such as in the ear and nose occur with increasing age (Rook, 1965; Adachi, 1973). Many of these changes are known to be governed by extracellular hormones. For example, testosterone is known to be associated with balding, which could be an example of a developmentally linked pleiotropic aging process, the undesirable side effect of an otherwise beneficial hormone. Balding does not lead to the death of an individual, but if testosterone can inhibit hair growth what else might it be doing to the organism? Postponement of the appearance of such hormones in time was probably necessary to achieve a physiologically uniform increase of longevity. One effect of castration, as seen in the eunuch, is a lack of balding and a longer lifespan by about 6–10 years (Hamilton and Mestler, 1969).

10. Age-dependent increase of cancer. Many properties of cancer cells are consistent with a dysdifferentiated-state condition (Sugimura *et al.,* 1972; Taketa *et al.,* 1976; Fahmy and Fahmy, 1980a,b; Braun, 1981; Huberman *et al.,* 1979; Cairns, 1979). This is shown by the appearance of new cell functions, such as the synthesis of foreign proteins, extopic hormone-producing tumors (Samaan, 1979), the loss in sensitivity to normal controlling elements of cell division, altered patterns of isoenzymes (Criss, 1971; Schapira, 1973; Knox, 1976), and changes in properties of the cell membrane (Cone, 1971, 1974, 1980; Cone and Cone, 1978). It is also well known that the probability of cancer appearing as a function of time increases with an individual's age. What is not as well known is that the probability of cancer appearing as a function of time is also a function of LSP; that is, longer-lived species are able to live longer before the probability of cancer appearance is equal to that of shorter-lived species (McClure, 1973, 1975; Cairns, 1979) These data lead to the prediction that the same processes governing the aging rate of a mammalian species are also likely to govern the time-dependent probability of normal cells transforming to cancer cells. The causive link between aging and cancer being proposed is that: (a) both are caused by the same processes, (b) both are the result of dysdifferentiation processes, and (c) the probability of both occurring with time is governed by LDPs.

The cancer cell is postulated to be a special type of aged cell, and as such, is part of and not separate from the general aging process that is occurring in all cells with time.What makes cancer cells special is that their particular type of dysdifferentiation becomes amplified by their nonrestricted proliferation properties. Other aged cells remain normal in this respect when they drift from the proper state of differentation. For example, a small amount of hemoglobin synthesis in neurons does not kill a person, but the genetic mechanism leading to this synthesis is likely to be similar and have the same origin as that which leads to the transformation of cells. Thus, the same processes governing aging rate, the proposed differentiation stability processes, act also to reduce the probability of cell transformation to cancer. If this is the case, most types of cancer will not likely be preventable to a significant degree without also altering the processes governing aging rate. (Totter, 1980).

11. Age-dependent changes in cell morphology. Gross changes in cell morphology (Tomlinson, 1977) like the degeneration of dendrites in neurons (Scheibel *et al.,* 1975; Feldman, 1976) or changes in cell membrane composition could be reflective of changes occurring in the cell's differentiated state. The loss of dendrites is usually thought of as being the result of some damaging process. However, it could just as well be the result of a repression of the genes necessary to produce dendrite structures.

12. Age-dependent alteration of membranes. The qualitative characteristics of membrane composition are known to play an important role in determining the differentiated state of cells. Membrane changes are known to be associated with the occurrence of cancer (Cone, 1971, 1974, 1980; Cone and Cone, 1978) and with aging (Zs.-Nagy, 1979; Hegner, 1980). Some of these changes involve specific proteins, such as the sulfated proteoglycans (Pacifici *et al.,* 1981a) and other lipid fluidity changes (Pacifici *et al.,* 1981b, Heron *et al.,* 1980). This latter observation may be quite important, for the age-dependent decrease in fluidity (increase in viscosity) has been shown to modulate the binding of serotonin. Thus, changes in this type of membrane property could change a cell's response to specific hormones by altering receptor binding site availability. This might be what is occurring when a loss of hormone receptor is found. Also, such changes would be consistent with dysdifferentiation occurring in these cells. Changes in membrane fluidity could be a result of a change in the relative amount of its constituents. However, there is evidence that peroxidative damage of membranes may also increase as a function of age in rat adipocytes, which may reflect a loss of protection and/or a decrease in repair (turnover) of the membrane (Hughes *et al.,* 1980).

13. Age-dependent decrease in the rate of cell proliferation. Control of cell division is known to be regulated to a large degree by extracellular factors (Holley, 1980). The age-dependent decrease in cell proliferation rate known to occur in most dividing cells of mammals (Cameron and Thrasher, 1971) is not likely to be intrinsic in nature but due to altered extracellular signals to main-

tain this state of differentiation. An accumulation of DNA errors, abnormal proteins, or damaged mitochondria, for example, seems less likely as an explanation.

14. Age-dependent decrease in the rate of RNA and protein synthesis. It appears clear that most evidence for mice and rats indicates an age-dependent decrease in the rate of RNA and protein synthesis (Richardson, 1981). Such changes are likely to be adaptive in nature and not the result of intrinsic cell impairments. A large number of different enzymes vary in their activity with increasing age, with about equal number increasing and decreasing in activity (Finch, 1972). These changes are also likely to reflect either adaptive or dysdifferentiational processes.

In summary, by-products of metabolism and development are predicted to cause aging through the mechanism of dysdifferentiation. Basic to this prediction is that aging is considered to be the random disorganization of the same orderly processes that gave rise to the creation of the organism. Starting with a single cell, an organism is created by differentiation and developmental processes, one event followed by another in a highly organized manner. The mechanisms determining when and how a cell differentiates during development are not known but are thought to involve position effects, ionic strength gradients, electrical effects, cell-to-cell contact, hormones, growth factors, and concurrent intracellular changes in the processing of information contained within the genome of each cell. Whether different differentiated cells have qualitatively and/or quantitatively different sets of DNA in terms of amount, composition, and/or organization is not known. Although all cells in an organism are currently thought to have the same DNA content, chromosome structure, and chromosome number, as far as it is now possible to determine, the possibility is now seriously being considered that DNA may change during development, at least in linear organization patterns.

But what happens after the adult reaches maturity? Is the differentiated state of the cell locked in and irreversible or is a positive energy-requiring effort required to maintain a given differentiated state, some states being more energy-requiring than others? Does the differentiation and developmental program really come to an end, followed by a steady-state period (a genetic program with a beginning and an end) or are developmental processes still operative in an open-ended-type genetic program (a genetic program with a beginning but no end)? Such an open-ended program would be more consistent in view that individuals seldom live much beyond the age of sexual maturity, so little selective pressure would exist to evolve an end to the program. This would be true particularly if compensational processes could operate effectively for sufficient time periods (Dilman, 1979).

With these considerations in mind, it was postulated that the maintenance of proper differentiated states does require a positive energy effort, and that the genetic program of development is open-ended in nature (Cutler, 1978, 1979a,b,1980; Tolmasoff *et al.,* 1980). This reduces the problem of understanding the biological mechanisms of aging to that of understanding the dysdifferentiative process and its effects at the cell, organ, and organism levels and the biological mechanisms of longevity to that of understanding the processes governing the time-dependent stability of differentiated cells.

New questions then arise as to what mechanism(s) maintain the stability of the differentiated state, if they are related to the process determining the state of differentiation, and how might these processes be studied best. Since much of biology is now concerned with differentiation and development and the genetics and biochemistry of gene-controlling processes, an understanding of dysdifferentiation processes and the processes governing the stabilization of differentiation may not be too distant in the future. Unfortunately, the question of the stability of the differentiated state of cells, once the state of differentiation has been achieved, has not received much attention except in the field of cancer research.

Loss of proper differentiation could be caused by a mutational and/or an epigenetic mechanism. Although normal changes in differentiation during development were thought to be of an epigenetic nature (Dover and Doolittle, 1980; Straus, 1981), this view is now being challenged by new insights implicating the highly dynamic nature of DNA and chromatin (Harris, 1979). Gene amplification or migratory DNA elements such as the transposon elements may play an important role in differentiation and developmental processes, and thus also in dysdifferentiation (Fahmy and Fahmy, 1980a,b; Braun, 1981; Pall, 1981). Clearly we no longer have to propose rare mutations or errors to accumulate during DNA, RNA, or protein synthesis to create a nonfunctional cell to explain aging. Indeed, the problem now appears how to explain the great stability of the differentiated cells in light of the significant dynamic and fluid nature of a cell's genetic apparatus.

Also, in view that a substantial amount of DNA in mammalian cells (the intermediate repetitive class) may be of the selfish type (Doolittle and Sapienza, 1980; Orgel and Crick, 1980; Ohno, 1980), its migration could conceivably interfere with proper gene regulation. Thus, longer-lived species might have selected against the type of DNA parasite in terms of reducing its rate of migration or amount of this DNA per genome. This is a prediction that can be tested with present techniques. It is also possible that both epigenetic and mutational events play an important role in aging, similar to the mechanisms proposed for cancer induction (Trosko and Chang, 1981). For example, the aging process concerned with energy metabolism may be more mutational in nature (like the initiating agents) and the aging process concerned with devel-

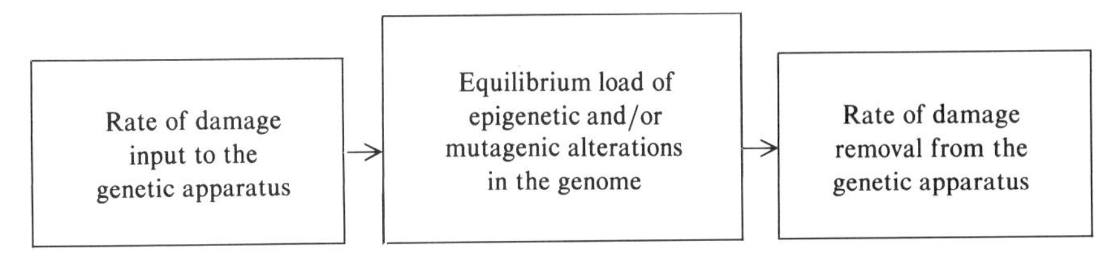

Figure 19. Dysdifferentiation model of cellular aging.

opment may be more promotor in nature (like the epigenetic effects of tumor promotor agents).

Finally it should be pointed out that an equilibrium may exist within a cell between the rate of damage produced by pleiotropic aging processes and the rate this damage is removed by degradation and repair processes (see Fig. 19).

The equilibrium mutational load carried by a cell would then be determined by the difference between the rate of damage input and the rate of damage output. On this basis, it would be predicted that cells from longer- lived species would not necessarily accumulate less damage with time but would instead have a lower equilibrium mutational load than shorter-lived species due to a relatively shorter dwell time of damage in the genome. If the probability of a cell drifting away from its proper state of differentiation is proportional to its steady-state mutational load, then we have a mechanism of how the level of a cell's mutational load could determine the aging rate of an organism.

In conclusion, the following proposals have been made concerning what causes aging, what is the aging process, and what is the genetic basis of longevity.

1. The cause of aging: Aging is the result of pleiotropic effects of development, differentiation, and metabolism. These processes act to accelerate the destabilization of the differentiated state of cell, cell organization in tissues, and the general homeostasis state of the organism.
2. What is aging: Aging is dysdifferentiation, the time-dependent drifting away of cells, cell organization and homeostatic control from their most optimum state of function.
3. What is the genetic basis of aging: Longevity (LSP) is determined by the processes governing the time-dependent stability of differentiation. These stabilization processes are encoded by longevity determinant genes (LDGs), and their levels of expression are governed by a relatively few regulatory genes. The same longevity determinant genes are found in all mammalian species independent of their LSP.

6. THE WORKING HYPOTHESIS

This section integrates the information and ideas that have been developed with the data that has been developed in the previous sections on the biological nature of human aging and longevity. All of these studies have been based on an evolutionary and comparative viewpoint, as presented in the writings of Haldane (1947), Medawar (1955, 1957), Williams (1957), Hamilton (1966), Maynard Smith (1959,1962, 1966), de Beer (1958), Schultz (1936, 1949, 1950, 1960, 1966, 1968), Wilson and co-workers (Wilson, *et al.,* 1974; King and Wilson, 1974; 1976; Bush *et al.,* 1977; Wilson *et al.,* 1977; Cherry *et al.,* 1979), Gould (1971, 1975, 1976a,b, 1977), Sacher (1959, 1960a,b, 1962, 1966, 1968, 1970), Monod (1971), and Wilson (1975).

Usually, hypotheses of aging do not account for longevity differences between species (Strehler, 1978; Lamb, 1977; Kohn, 1978; Comfort, 1978). For example, proponents of the error hypothesis (Medvedev, 1966; Orgel, 1970, 1973), the free radical hypothesis (Harman 1956, 1962, 1969; Leibovitz and Siegel, 1980), the somatic mutation hypothesis (Curtis, 1966, 1971), or the immunological hypothesis (Walford, 1974) have not proposed mechanisms of how their hypotheses could account for species differences in aging rates. They are accordingly hypotheses only of aging, not of longevity. As far as I am aware of, Sacher's homeostasis hypothesis (1968) and the hypothesis described in this chapter are unique in this respect by offering an explanation to account for both aging and longevity of different species.

The major postulates describing the general aspects of the hypothesis as it concerns all forms of life are as follows:

1. Life is any process capable of preserving a set of information which ensures its existence to be able to preserve this same set of information (Cutler, 1972, 1976a,b,c, 1978, 1980). This set of information is encoded within a physical structure which has a finite lifespan. A living system is the physical entity where the life process is carried out. The characteristics of living systems are the physical means they use to preserve this set of information in its physical state. The most primitive and primary means for this preservation has been the duplication of the set of information before it is lost. Other means are the use of defense and repair processes. Physical elements which make up the organism provide and maintain these functions. Any time-dependent intrinsic alteration in the information-containing physical structure and/or its supporting elements leading toward an increased probability of a loss of its essential properties of self-preservation is considered in a broad sense to be an aging process. On the other hand, all intrinsic processes which have evolved to preserve this information and the information-containing structure are considered in a broad sense to be antiaging processes. From the above proposal, a definition of life

can be derived, of how it exists, and why it exists (Cutler, 1972, 1976a,b,c, 1978, 1980).

DNA is an example of a physical structure where such information is stored, and the duplication of DNA is an example of one of the most primitive and still most essential information preservational processes which has evolved. It should be noted that life is a process involving entities having a physical makeup, not an entity itself. The function of life (why it exists and what keeps it going) and its purpose is simply the maintenance and propagation of information that has this property. Thus, information is contained in but is not part of any physical entity. A fundamental characteristic of life is its ability to propagate information indefinitely within physical entities having themselves a limited lifespan. In this sense, the living process is potentially immortal.

These arguments serve as a definition of life and describe its innate characteristics. They were first described in 1972 and in later publications (Cutler, 1972, 1976a,b,c, 1978, 1980). From this definition, an endogenous virus genome, even though it may never live as a separate free entity, would be considered a living system in its own right. In fact, any system whose existence depends on the preservation of information which acts to preserve itself is considered a living system. This definition would also include the so-called jumping genes or transposon elements that may make up a considerable portion of nonessential DNA in mammalian species (Doolittle and Sapienza, 1980; Orgel and Crick, 1980), episomal particles, and all other intracellular information-containing entities, regardless if their evolutionary development is governed by molecular events within the cell or by macroevents involving the whole organism.

2. From the origin of life and throughout the following evolution of all living systems, the interaction of aging and antiaging forces has prevailed. The continued preservation of life is not assured by any means, but the fact that life still exists on Earth today about 4 billion years after its origin indicates that antiaging processes have so far been remarkably successful, but this may be more from accident than by design. The entire biological makeup and behavior of all forms of life existing today are taken to reflect the complex history of the interactions and trade-offs molded by the actions and counteractions of these two forces. *Consequently, it is considered impossible to reach a fundamental understanding of an organism's behavior, morphology, or its genetic and biological makeup such as we find them today without also seeking an understanding and identification of the interactions of these aging and antiaging forces and how they shaped the organism.*

3. The longevity of an individual of a given life form (species) is only one of many different types of strategies that have evolved to ensure a continued preservation of its characteristic set of information (Orgel *et al.*, 1980). A great number of other strategies exist, such as an increase in the number of individ-

uals. Clearly, however, reproduction should not form part of the definition of life, as is frequently done. Although reproduction appears to be an essential protective means of life, it is not an absolute necessity. In addition, the success of a species is not measured by the number of individual members existing but by the length of time its particular set of information has been able to survive and to evolve new more successful sets of information.

4. Since the origin of life, longevity determinant processes have evolved. These processes operate to ensure the continuity of life against any threat to destroy it. However, as organisms evolved, life preservation processes became increasingly complex, and so for simplicity in dealing with multicellular organisms, longevity determinant processes have been defined here as those dealing only with endogenously originating threats. Thus, longevity determinants here are antiaging processes, and the endogenously originating threats, the aging process. These endogenous threats or aging processes are by-products of metabolism and development. Aging processes were not evolutionarily selected for, but arose as a result of trade-offs in the selection for beneficial factors.

5. The cause of aging is of an endogenous origin and is pleiotropic in nature; that is, the result of by-products of the developmental and metabolic processes which have evolved to maintain the continuity of the living process. Thus, a primary cause of aging is postulated to exist. Two major classes for the origin of aging process are (a) energy metabolism (producing a wide spectrum of toxic active oxygen species) and (b) differentiation and developmental processes (processes controlled by hormones and growth factors) which have nonspecific detrimental effects on nontarget cells.

From this general hypothesis of aging and longevity, a special hypothesis has been derived which relates only to mammalian species. Some of the major points of this special hypothesis follow.

1. Few individuals of a given species living in their natural ecological niche are able to live much beyond a chronological age where their physiological aging processes seriously decrease their health status or physiological functions (Cutler, 1978, 1980). Support for this argument is obtained from animal populations today, where a good correlation is seen between biological longevity potential of a species (as measured in captivity) and its longevity in the wild (Cutler, 1976a,b, 1978; Hirsch, 1979; Washburn, 1981). The individuals in these wild-type populations were always vigorous and and healthy and were killed only by natural external factors, rarely by endogenous dysfunctions (Hirsch, 1979). The endogenous ability of a species to maintain good health and vigor rarely appears to have been in vast excess or in serious deficiency in relation to its innate ability to survive its external environmental hazards. These natural hazards were sufficient to provide enough turnover in the populations for natural selection processes to operate.

For most species living in the wild, death occurred well before signs of senescence began to show. It was not just the old, feeble, or sick that were killed by predators because there were not enough of these individuals to feed them. Healthy individuals at all ages have a similar probability of death. Thus it is considered unlikely that aging genes or a genetic program of aging existed to age individuals for the good of the species (Denckla, 1975).

2. For biological longevity potential to have increased, the organism must be physically able to live longer. For this to occur, the effective environmental hazards have to decrease. This can be achieved by the evolution of appropriate morphological and behavioral characteristics, resulting in a higher probability of survival and thus a lower effective environmental hazard. Thus, innate ability to survive longer against external originating hazards coevolved with the innate ability to survive longer against endogenously originating environmental hazards.

3. The driving force behind the evolution of increased longevity in the mammalian species, and particularly for primates and the hominids, is the extraordinary success of a new type of function or trait, learned-adaptive behavior (Cutler, 1976b; Lovejoy, 1981). Because of an apparent biological limit of how fast a brain can mature, gather and process information and the benefits of long exposure of the organism to the environment, for learned behavior to function best (obtainment of information and its processing), increased longevity was essential for the realization of the potentials of this new trait. The evolutionary success of mammalian species is viewed as representing largely the evolutionary success of learned-adaptive behavior over instinctive behavior. Thus, the evolution of longevity was essential for the success of all mammalian species.

Learned-adaptive behavior evolved to its most advanced state during the evolution of primates. Humans represent the species taking the most advantage of this new trait, having the highest degree of learned-adpative behavior, minimum degree of instinctive behavior, and the longest potential lifespan of all mammalian species. Thus, fundamental characteristics determining the evolutionary success of humans involved not only the brain, hand, thumb, ability to walk upright, and all of the cultural or biosocial aspects, which are emphasized by anthropologists, but also the characteristic of unusually long innate longevity potential.

4. The unusually long lifespan of humans is not considered to have just happened as a by-product of other processes that were more essential (Lindstedt and Calder, 1981). Instead, the evolutionary success of the human requires an unusually long period of good health and vigor to take advantage of the newly evolved potential of the brain in terms of intelligence or learned-adaptive behavior. Thus, longevity was selected for directly.

5. A number of different biological strategies were utilized during the evo-

lution of longevity. Examples are an increase in body size and a decrease in developmental rate, both of which are of particular interest because they appear to be governed by changes occurring in only a few regulatory genes. The increase in body size is a general feature of mammalian evolution. Body size increase resulted in both a decrease in the effective external environmental hazards for the organism and an increase in biological longevity potential due to the decrease in metabolic rate. Thus the general increase in body size during mammalian evolution is interpreted as a longevity strategy.

Many other examples of a change in a single factor, resulting in both a decrease in effective external environmental hazards and an increase in biological longevity potential, are likely to exist. On the other hand, the decrease in developmental rate has many obvious disadvantages, suggesting that a more important benefit was provided. This benefit was the postponement of the aging effects of the various factors associated with development, a benefit that was probably essential for longevity to have evolved in the mammalian species.

6. The biological and genetic complexity related to mammalian species' differences in longevity was estimated by determining the rate longevity evolved (Cutler, 1972, 1976a, 1979). Two methods were used. One was the use of the equation developed by Sacher (1960), which has the remarkable ability to predict the LSP of both progressive and living fossil species by using only the adult body and brain weights of the species. Using published results of body and brain weights for extinct species, LSPs were calculated, and an evolutionary tree of longevity for mammals was constructed. The results clearly indicate that (1) longevity increased throughout the evolutionary period of all mammalian species and (2) longevity increased at an ever-increasing rate along the hominid ancestral–descendant sequence leading to humans.

The second method to estimate the direction and rate of evolution of longevity compared the LSPs of living primate species. Through a knowledge of the phylogenetic relationships of these species, a rough estimate of the direction and magnitude of species longevity during primate evolution was determined. The results found were similar to the first method; that is, longevity was found to increase and at a very high rate during the recent evolutionary history of the hominids.

7. A decrease in the rate of development played a causative role in the evolution of longevity. Through arguments reviewed by Gould (1971, 1975, 1976a,b, 1977), the evolution of larger brain size and many other aspects of primate morphology are governed by constant allometric scaling factors. Such scaling factors appear to place a limit or constriction on what patterns of development are most likely to evolve. It is also found that changes in morphology and developmental rate appear to be governed by a few hormonally related processes and the qualitative nature of the changes taking place determined by scaling factors (Schultz, 1936, 1949, 1950, 1960). In this regard, the processes

of neoteny (Schultz, 1936; Gould, 1971, 1976), underlying the increase in relative brain to body size, decrease in the rate of development, and the increase in longevity all appear to be linked together by common genetic mechanisms involving regulatory genes. Thus, these processes are likely to have evolved coordinately at high rates. The ability to predict LSP of all mammalian species by brain and body weight, using only three constants, also suggests very rigid constrictions on how longevity relates to neoteny processes (Cutler, 1972, 1973).

8. The remarkable similarities of the primate species in terms of morphology, physiology, general biology, biochemistry, and genetics, suggest that the evolutionary appearance of new species was accomplished mainly through changes in the timing and degree of expression of structural genes and not by changes in structural genes. These changes in expression of the structural genes were a result of changes occurring in their regulatory gene counterparts. Such regulatory genes may determine not only the developmental patterns and morphology of all mammals but also their longevity. These genes are called longevity determinant genes (LDGs). The role postulated for these LDGs is to provide the necessary duration of good health and vigor to ensure the evolutionary success of the species.

The same set of LDGs are to be found in all mammalian species regardless of their LSP. Only the timing and degree of their expression makes the difference of whether a species would live 10 or 100 years. LDGs are structural genes. The genes governing the timing and degree of expression of the LDGs would be their corresponding regulatory genes. It is these regulatory genes that were calculated to be few in number, not the structural genes. Thus, aging rate, developmental rate, morphology, and intelligence are all controlled in a coordinated manner by regulatory genes, and the changes occurring in their properties during hominid evolution appear to have only a few highly probable genetic alterations.

9. The primary causes of aging are pleiotropic in nature; they are by-products of normally beneficial metabolic and developmental processes (Cutler, 1972, 1976a,b,c, 1978). Two basic types of pleiotropic aging processes are postulated: (1) continuously acting biosenescent processes (CABPs) and (2) developmentally linked biosenescent processes (DLBPs). The evolution of longevity in the mammalian species is predicted to have occurred by the reduction of these two major classes of biosenescent processes. Without a coordinated reduction of both, a significant increase in lifespan may have been impossible.

Examples of the DLBPs are the various hormones and growth factors associated with growth, cessation of growth, and sexual maturation (the aging of the Pacific coast salmon, eels, and octopus being good examples). Examples of the CABPs are by-products of oxygen metabolism, such as the superoxide free radical O_2^-, the hydroxyl radical $\cdot OH$, and hydrogen peroxide H_2O_2.

10. To counteract these two classes of biosenescent processes, two classes of corresponding longevity determinant processes (LDPs) evolved. Many of these LDPs were probably in existence in early prokaryotic and eukaryotic cells. For example, by the time the primate species appeared, pleiotropic aging effects of the DLBPs were so strongly entrenched by a background of a billion years of evolution that few alternatives were probably available for reducing their aging effects. Therefore, the primary means for the effective reduction of the DLBPs was largely a postponement in time of their appearance during development. The decrease in developmental rate with increased longevity observed in the primates and other mammalian species is support for this concept and is likely to involve only a few regulatory genes (Cutler 1976, 1976a,c). In turn, as a result of the constrictions of what changes are most likely to occur, decreases in developmental rate are coupled with the morphological changes observed in neoteny processes; that is, the general morphology of present-day humans, such as the relatively large brain and unusually slow rate of development, could be a result of changes of the same regulatory genes. All of these changes could therefore be genetically and functionally linked, serving in a synergistic manner the same objective and all occurring by changes in the timing and regulation of a common set of genes.

The reduction of the CABPs during the evolution of mammalian longevity is predicted to have occurred by the increased expression of genes (more enzymes or products being produced) which evolved over billions of years to counteract the more ancient pleiotropic side effects of energy metabolism (as compared to the pleiotropic processes associated with development). For example, the superoxide free radicals $O_2^{\overline{\cdot}}$, a pleiotropic by-product of oxygen metabolism, would be predicted to be counteracted more effectively in longer-lived species by higher levels of the enzyme superoxide dismutase (SOD) being synthesized (Cutler, 1972, 1976c; Tolmasoff *et al.,* 1980). However, SOD has probably been in existence since organisms began to utilize oxygen about 2 billion years ago. Another example is the higher levels of DNA repair found in longer-lived mammalian species to remove more completely and in less time the damage to DNA that does occur. In general, longer-lived mammalian species would be expected to have higher levels of antioxidants and higher levels of DNA repair than shorter-lived species, but all life would require antioxidants, DNA repair processes and protective processes against toxic products of metabolism to survive (Cutler, 1973).

7. LONGEVITY DETERMINANT PROCESSES

It has been proposed that: (1) the aging process in mammals has a primary cause and this is the pleiotropiclike by-products of essential developmental and metabolic processes (CABPs and DLBPs); (2) this primary cause has a primary effect, to accelerate the drifting-away of the cells from their most optimum and proper state of differentiation—this process is called dysdifferentiation; and (3) there are specific processes governing aging rate, and these act to decrease the rate of dysdifferentiation. These are called the longevity determinant processes (LDPs).

Very little is known about the processes which act to destabilize the differentiated state of cells. In terms of the DLBPs, hormones involved in differentiation and development do have aging effects. The adrenocortical hormones accelerate the senescence of the Pacific coast salmon, trout (Hane *et al.,* 1966), and the octopus (Wodinsky, 1977). There is also evidence that these effects might involve a change in the methylation of DNA (Berdyshev and Protsenko, 1972). Steroid metabolism is also associated with toxic effects, as demonstrated by Hornsby (Hornsby *et al.,* 1979; Hornsby, 1980; Hornsby and Gill, 1981a,b). This is a particularly interesting example, in that antioxidants appear to protect the cells in terms of stabilizing the cell's differentiated state. These data also imply that steroid metabolism produces free radicals. Relevant to this observation is that the pituitary, the hypothalamus, and the adrenocorticoid tissues have unusually high levels of antioxidants such as vitamin C, vitamin E, and glutathione.

Postponement of aging by hypophysectomy, as reported by Everitt (Everitt and Burgess, 1976; Everitt *et al.,* 1980; Everitt, 1980) and Denckla (Bilder and Denckla, 1977; Denckla, 1978; Parker *et al.,* 1978; Scott *et al.,* 1979; Miller *et al.,* 1980) support the toxic effects of hormones. The compensation-threshold concept of Dilman (1979) is also supportive of the open-ended nature of the genetic program of development and the long-term toxic effects of hormones. Recent studies by Landfield and co-workers (Landfield, 1978, 1980; Landfield *et al.,* 1977, 1978, 1980a,b) are based on the concept that hormones such as the glucocorticoids may play a role as a cause of aging. Thus, the possibility of a long-term accumulatory effect of hormones causing dysdifferentiation is gathering support.

Other data show a good correlation of the rate of development and aging rate. This correlation is an important part of the argument that postponement of the appearance of DLBPs is an important longevity determinant process (Cutler, 1976b; 1980). Evidence indicating that nontoxic methods which reduce developmental rate also increase longevity lends support to this proposal (Cutler, 1980; Goodrick, 1980). Thus, a straightforward prediction is that any nontoxic process that reduces rate of development and growth would also be expected to lengthen lifespan. The increased lifespan achieved by dietary restriction could operate in part by this mechanism.

Other LDPs that evolved to counteract the DLBPs could be the tissues associated with the control of development and sexual maturation. In this case, it would be of value to determine if neuroendocrine tissues such as pituitary, hypothalamus, and adrenal of longer-lived species have higher levels of antioxidants. Ascorbic acid levels are known to be unusually high in brain tissue.

The androgen, dehydroepiandrosterone (DHEA), has been implicated recently as a possible longevity enhancement agent and to be able to inhibit the increase of certain types of cancer, as well as to protect cells from carcinogenic agents (Schwartz and Perantoni, 1975; Schwartz, 1979). In primates and other mammalian species, the plasma and tissue levels of DHEA are very low shortly after birth and up to sexual maturity (Cutler *et al.*, 1978; Townsley and Courtois, 1981). During sexual maturation, DHEA levels dramatically increase, and shortly afterwards show a slow but progressive age-dependent decline (De Peretti and Forest, 1976). One mechanism of action of DHEA may be its inhibitory effects on glucose-6-phosphate dehydrogenase. By inhibiting this enzyme, DHEA could act to reduce detoxification processes, DNA replication, cell division, and a number of other functions (Oertel *et al.*, 1970; Pashko *et al.*, 1981). Another possibility is that DHEA increases during sexual maturation because it is required as a protective agent against the pleiotropic effects of the hormones and other growth factors that are also increasing in levels at this time. Support for this possibility is that DHEA does appear to have antioxidant properties, as do a number of other steroids (Demopoulos *et al.*, 1980).

Explanations of how CABPs may act to dysdifferentiate cells are more straightforward. Here, it is postulated that by-products of energy metabolism common to all mammalian species act to dysdifferentiate cells. There are many toxic by-products of energy metabolism that can be listed, and in turn a number of defense and repair processes acting against these toxic agents are also known. Some of these defense and DNA repair processes that may be involved as LDPs are as follows (Cutler, 1982a):

1. Detoxification processes: Some of the enzymes known to be involved in detoxification (Jakoby, 1980) are (a) mixed function oxidases (cytochrome P-450 reductase, microsomal flavin-containing monoxygenases), (b) oxidative-reduction enzymes (alcohol dehydrogenases, aldehyde reductase, aldehyde-oxidizing enzymes, ketone reductase, xanthine oxidase, aldehyde oxidase, superoxide dismutase, glutathione peroxidase, monoamine oxidase, quinone reductase), (c) conjugation and hydrolic processes (glucuronidases, *N*- and *O*-methylases, glutathione, glutathione S-transferases), (d) peptide bond and mercapturic acid formation, cystein conjugate b-lyase, thiol S-methyltransferase, acetylase, arylhydroxamic acid actyltransferase, sulfotransferase, glyoxalase, epoxide hydrolase, carboxylesterases, and amidases. Also, levels of gastric mucoproteins that may play protective roles against gastric carcinogens could be investigated (Hartman, 1981).

2. DNA repair processes: many different DNA repair processes are likely to involve common enzymes: (a) enzymes involved in X-ray repair, (b) enzymes involved in UV light repair, (c) enzymes involved in repair of damage due to different types of chemical mutagens, (d) enzymes involved in repair of endogenous damage of a spontaneous nature, as in depurination.

3. Natural antioxidants, free radical scavengers, and heavy metal chelators (Forman and Fisher, 1981): (a) superoxide dismutase (CuZn and Mn types and relative amounts of their isoenzymes), (b) glutathione peroxidase (Se and Se-independent types), (c) catalase, (d) glutathione, (e) cysteine, methionine, histidine, tyrosine, (f) ascorbic acid, (g) α-tocopherol, (h) cytochrome peroxidase, (i) ascorbate free radical reductase, (j) glutathione reductase, (k) dehydroascorbate reductase, (l) ceruloplasmin, (m) unsaturated fatty acids, (n) polyamines, (o) urate and related derivatives, (p) carotenoids, retinoids, carotene, and xanthrophylls, (q) metallothionine, (r) ubiquinone (coenzyme Q_{10}), (s) glucose, and (t) cholesterol.

A particularly good organ to study in regard to relative levels of natural antioxidants and free radical scavengers is the eye (Reddy *et al.,* 1980; Zigler and Goosey, 1981). It is obvious that the function of the eye, including the lens, lasts for a much longer period of time in longer-lived species. It is also well known that proper maintenance of the eye requires comparatively high levels of antioxidants such as ascorbate, α-tocopherol, and glutathione. In addition, superoxide dismutase and glutathione peroxidase are likely to play important protective roles. Oxidative damage to a number of structural proteins is known to occur, such as to aldolase and pyruvate kinase (Banroques *et al.,* 1980). Also, there appears to be a relation between incidence of diabetes and formation of cataracts. Thus, a correlation of the amount of antioxidants and free radical scavengers in the various tissues of the eye with LSP would be expected.

Other factors that may act importantly to stabilize the differentiated state of cells are the following (Cutler, 1982a):

1. Cellular renewal processes: (a) nonselective protein and lipid degradation (nonselective implies equal rate of removal of abnormal protein versus normal protein), (b) selective protein degradation (abnormal proteins selected for degradation), (c) organelle renewal, such as mitochondria, peroxisomes, and lysosomes, (d) whole cell turnover rates (mitosis), (e) protein repair processes. The level of methionine sulfoxide reductase may be involved in the removal of methionine sulfoxide in proteins (Reiss and Gershon, 1979; Brot *et al.,* 1981).

2. Redundancy process: (a) molecular level, extra chromosomes in diploid and tetraploid cells, providing multiple copies of critical structural genes, (b) cellular levels, reserve factors in organelles and cell number, (c) organ level, overlap of function between different tissues, separate organs may compensate for same function.

3. Specific metabolic rate: Efficiency of energy utilization: (a) relative efficiency of ATP production in terms of O_2^- and ·OH radicals produced, (b) increased body size to decrease SMR, (c) biochemical pathway of heat energy production to reduce free radical production in comparison to the production of kinetic energy, (d) sleeping to reduce energy requirements and body temperature, and consequently to lower free radical production over a lifespan (Cutler, 1972, 1976c).

4. Gene control systems: Intracellular mechanisms may act to stabilize the differentiated state of cells, such as (a) critical gene control elements may be redundant, (b) genetic reprogramming during each cell division cycle may be more complete in longer-lived species (Cutler, 1972), (c) eukaryotic cells have no operons and the structural genes in a given pathway are frequently dispersed. Lack of operons could act as a protective process against one mutation inactivating whole clusters of genes. Genes and their controlling element may be arranged in the chromosomes in a special manner in longer-lived species to further decrease the probability of such mutations affecting the differentiated state.

5. Relative enzyme levels: Shifts in the relative levels of enzymes in metabolic pathways may have been involved to increase the efficiency of pathways and to reduce toxic by-products, such as the change in the ratio of the isoenzymes of lactate dehydrogenase. Preliminary results have indicated a good correlation of the a/b types to longevity in primates, particularly in different tissues of the brain (Cutler, 1976c). Other similar isoenzyme correlations with longevity may exist. Levels of enzymes in the mixed function oxidase system may also be altered. Many alterations in the levels between different species are known, but the reasons are not understood. Antioxidants appear to alter these ratios towards greater efficiencies of detoxification. Levels of enzymes in the mitochondria may be altered to reduce possible leakage of O_2^- and ·OH radicals. For example, the structure of mitochondria in longer-lived species may be different to reduce leakage of free radicals.

6. Structural components of a cell: (a) Lower ratio of saturated to unsaturated fatty acids or other constituents of the membrane may occur, providing more resistance to lipid peroxidation reactions. (b) Mechanisms might exist to reduce the age-dependent increase in membrane viscosity (decrease in membrane fluidity). This age-dependent change may be related to species' aging rate and, if so, what are the mechanisms governing the rate of this change? (c) Changes may occur in the amino acid content of proteins, making them less susceptible to oxidation reactions (Reddy *et al.*, 1980; Zigler and Goosey, 1981; Banroques *et al.*, 1980). A decrease in the methionine and other amino acids might decrease the sensitivity of proteins to oxidation or racemization types of alterations.

The search for longevity determinants has just begun. However, from what data are available, the results are very encouraging. Many of the data used to test the hypothesis have been taken from the existing literature, where I have searched for papers that give the tissue levels of potential longevity determinants in different mammlian species. The basic approach taken then in screening for potential longevity determinants has been to determine if a correlation exists in their levels as a function of lifespan of different mammalian species. Once such potential longevity determinants are identified, then cause and effect type of experiments should be undertaken for their further testing. Final proof of the existence of LSPs will depend on showing that alterations of their tissue levels in an organism will result in changing their innate aging rate.

I shall review here some of the most recent findings of lifespan correlation in our search for LDPs, particularly for the antioxidants.

1. Genetic stability: One of the first experimental results indicating that the genetic apparatus is becoming destabilized with age is from the early work of Curtis (see Table VIII). He found that, with increasing age of mice, guinea pig, and dog, there is an increased frequency of chromosomal aberrations in the liver cells when these cells are stimulated to divide. Most importantly, however, he found that the rate of increase of chromosomal aberrations was correlated with the species' aging rate. These data are shown in Table VIII and Figure 20.

2. DNA repair: DNA excision repair levels have been found to correlate with species' lifespan for both skin fibroblast (Hart and Setlow, 1974; Francis *et al.,* 1981) and *in vivo* lens epithelial cells (Treton and Courtois, 1982). Although much more work of this type needs to be done in terms of looking at other types of DNA repair processes and assays of repair enzyme levels *in vivo* (Cutler, 1982), the results in general are consistent with the hypothesis made in 1972 (Cutler, 1972) that such a correlation is expected.

3. Activation of mutagenic agents: A number of mutagenic agents appear to be activated to lesser extents in longer-lived species (Schwartz and Peran-

Table VIII. Rate of Chromosome Aberration Frequency vs. Lifespan Potential and Lifespan Energy Potential[a]

Species	LSP (yr)	LEP (kcal/g)	RCA $\times$ 10^{-3}
Mouse	3.5	182	74.0
Guinea pig	8.0	204	20.0
Dog, beagle	20.0	255	1.88

[a]LSP, lifespan potential; LEP, lifespan energy potential; RCA, rate of chromosome aberration frequency. Nonlifespan data taken from Crowley and Curtis (1963), Curtis (1966), and Curtis and Miller (1971).

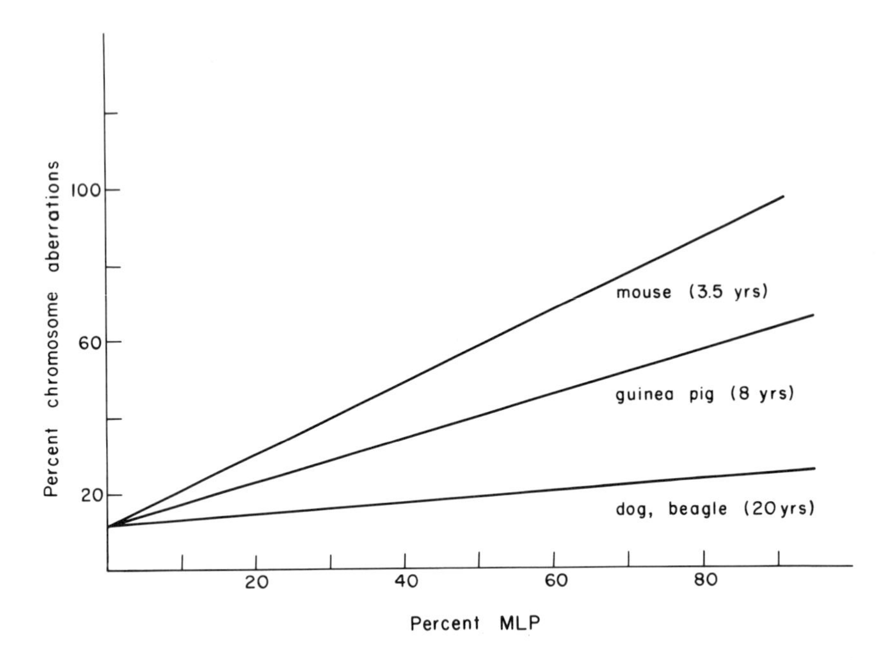

Figure 20. Percent chromosome aberration frequency as a function of percent of maximum lifespan potential. Data taken from Table VIII.

toni, 1975; Schwartz, 1979). This finding has now been extended by showing that longer-lived species have less of the cytochrome P-445 and P-448 enzymes (Pashko and Schwartz, 1982). These enzymes are also known to produce free radicals in their metabolism of various substances, so although they are involved in detoxification processes, their decrease in levels with increasing LSPs may be a means to decrease this source of active oxygen species.

4. Tissue concentration of free radicals: Longer-lived species should have lower tissue levels of free radicals than shorter-lived species. This prediction of course assumes that, although free radicals are known to be essential in many metabolic pathways, most are detrimental.

Although looking at whole tissue levels of free radicals is a difficult task, there is one report where such a study was done as a function of lifespan of different species (Maréchel *et al.*, 1973). They reported that longer-lived species (including human) do indeed have lower levels of free radicals in their tissues. Such experiments need to be repeated and extended to include other tissues and species.

5. Tissue levels of antioxidants: Instead of measuring directly the tissue levels of free radicals and other oxidizing agents, an indirect approach to obtain

the same information is to measure tissue levels of antioxidants. The rational for this indirect approach is as follows:

a. The measurement of normal *in vivo* tissue concentrations of the highly reactive free radicals which are likely be involved in producing cellular damage is extremely difficult and subject to many pitfalls.
b. The measurement of the normal *in vivo* tissue concentrations of antioxidants which are known to protect cells from these highly reactive free radicals is relatively easy and straightforward.
c. The concentrations of antioxidants in tissues are not likely to be in vast excess of what is required to protect the organism—but instead to reflect the minimum concentration sufficient to maintain proper biological function over a certain time period. Because of biological costs, unnecessarily high or low concentrations of any gene product may result in an evolutionary disadvantage.
d. Thus, tissue concentrations of antioxidants in different species are likely to reflect, in an inverse manner, the normal *in vivo* concentration of the reactive free radicals the organism needs to be protected against.

There is, however, an important problem that must be dealt with in using this comparative approach to evaluate possible longevity determinants. This is where an enzyme activity or any other factor that is being measured might be found to correlate with LSP but have nothing to do with or is very indirectly related as a causative factor in determining longevity (Kunkel *et al.,* 1956; Stahl, 1962; Emmett and Hochachka, 1981). For example, body size, SMR, or rate of development correlates with LSP and, accordingly the activities of some enzymes are expected to vary with the intensity of these parameters. Indeed, it is possible that most of the enzymes that do correlate with species' longevity are not involved as LDPs (Lindstedt and Calder, 1981).

To evaluate how special the case is when a correlation is found with LSP or LEP, a number of enzymes and other factors that were not thought to be potential longevity determinants were evaluated using values found in the literature for different mammalian species (Albritton, 1952; Altman and Dittmer, 1961, 1962, 1972, 1974; Dixon and Webb, 1964; Mattenheimer, 1971; Bernirschke *et al.,* 1978; Mitruka and Rawnsley, 1981). No significant correlation in the concentration of the following substances was found with lifespan potential (LSP) or lifespan energy potential (LEP):

1. Tissue enzymes: random assortment of 57 analyzed.
2. Vitamins (whole blood): Thiamine, riboflavin, pyridine, cyanocobalamine, nicotinic acid, pantothenic acid, retinol (vitamin A)**.
3. Antioxidants (tissue and blood): Ascorbate, glutathione*, glutathione peroxidase*, ceruloplasmin, glutathione S-transferase*.
4. Blood chemistries: Albumin, alpha and beta globulins, cholesterol**,

Table IX. Control Enzyme Activity Correlations[a]

Enzyme	r-Correlation coefficient					
	Liver		Kidney		Brain	
	LSP	LEP	LSP	LEP	LSP	LEP
3-hydroxybutyrate dehydrogenase	−0.494 $n = 7$	−0.381 $n = 7$	—	—	—	—
D-aminoacid oxidase	−0.0309 $n = 7$	−0.235 $n = 7$	−0.0425 $n = 7$	−0.209 $n = 7$	—	—
Diamine oxidase	0.541 $n = 7$	0.516 $n = 7$	−0.176 $n = 6$	−0.513 $n = 6$	—	—
Cytochrome oxidase	−0.504 $n = 5$	−0.528 $n = 5$	−0.575 $n = 5$	−0.584 $n = 5$	−0.589 $n = 5$	−0.608 $n = 5$
Alkaline phosphatase	−0.0770 $n = 7$	−0.205 $n = 7$	−0.0330 $n = 7$	−0.194 $n = 7$	—	
Cytochrome c	−0.575 $n = 5$	−0.352 $n = 5$	−0.653 $n = 5$	−0.355 $n = 5$	−0.720 $n = 5$	−0.441 $n = 5$

[a]Enzyme data from Dixon and Webb (1962). Activity is relative base of 100 assigned to most active tissue of group. LSP is lifespan potential, LEP is lifespan energy potential, n is number of species.

Table X. Human vs. Chimpanzee: Comparative Blood Biochemistry[a]

Name	Human		Chimpanzee	
Sodium mEq/l	140	(138–140)	140	(135–143)
Potassium mEq/l	4.15	(3.9–4.4)	3.2	(2.8–3.4)
Chloride mEq/l	101	(99–104)	103	(100–108)
Glucose mg %	87.5	(70–105)	84	(84–85)
BUN mg %	13.5	(9–18)	8	(6–10)
Creatinine mg %	0.8	(0.6–1.0)	0.9	(0.8–1.1)
Calcium mg %	10	(9.5–10.5)	9.3	(9–9.6)
Phosphorus mg %	4.75	(2.5–7.0)	3.7	(2.8–4.5)
Magnesium mg %	2.05	(1.8–2.3)	2.0	(1.6–2.2)
Bilirubin mg %	0.4	(0–0.8)	0.5	(0.2–0.8)
Amylase (Somogyi units)	95	(40–150)	42	(15–65)
Lipase mg %	0.5	(0–1.0)	0.1	—
LDH mU/ml	300	(100–500)	734	(621–837)
Alkaline phosphatase mU/ml	5.0	(2–8)	4.1	(3.7–4.4)
Uric acid mg %	4.5	(3–6)	4.1	(3.8–4.7)
Albumin g %	4.3	(3.8–4.8)	2.7	(2.6–2.8)
Cholesterol mg %	190	(160–220)	201	(136–249)
Total lipids mg %	500	(200–800)	615	(495–735)
Total protein mg %	6.75	(6.3–7.7)	7.0	(6.5–7.4)

[a]Data taken from Sparberg (1981). Values are mean with normal range in parentheses. Chimpanzee data based on five determinations for each of five different individuals of age range 8½ to 10 yr.

glucose**, blood urea nitrogen**, minerals (Na, K, Ca, Mg), dehydroepiandrosterone (DHEA).

Those substances marked with an * were found to decrease with increasing LSP, and those with ** to increase with LSP for the short-lived species but not for species with lifespans over 40–50 years.

Typical data determining the possible correlation of a random assortment of enzyme activities with LSP and LEP are listed in Table IX. A comparative analysis of blood chemistries between human and chimpanzee is shown in Table X. These data indicate that most enzyme levels and levels of other biological factors do not appear to correlate significantly with LSP or LEP. Thus, those that do correlate can be classified as being good potential LDPs.

Some of the antioxidants that were found to correlate significantly with LSP or LEP are the following:

1. Superoxide dismutase: Levels of the sum of CuZn-SOD and Mn-SOD activities were found to increase linearly with LEP values for different primate species (Tolmasoff *et al.*, 1980). Two types of correlations were found. The first is illustrated in Fig. 21, showing that SOD per SMR is a linear function with

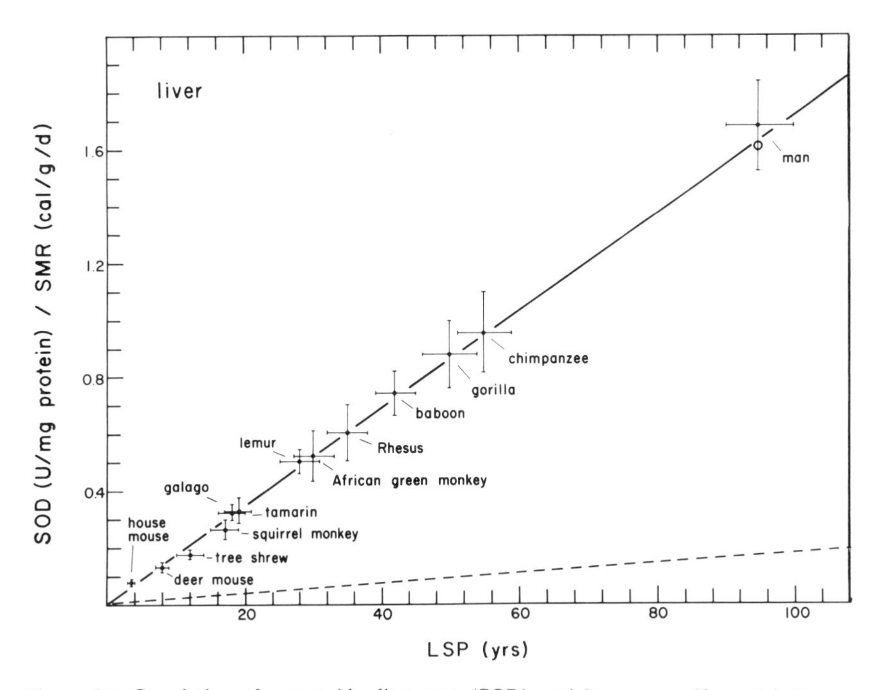

Figure 21. Correlation of superoxide dismutase (SOD) activity per specific metabolic rate (SMR) against lifespan potential (LSP) in liver of primate and rodent species. Data taken from Tolmasoff *et al.* (1980).

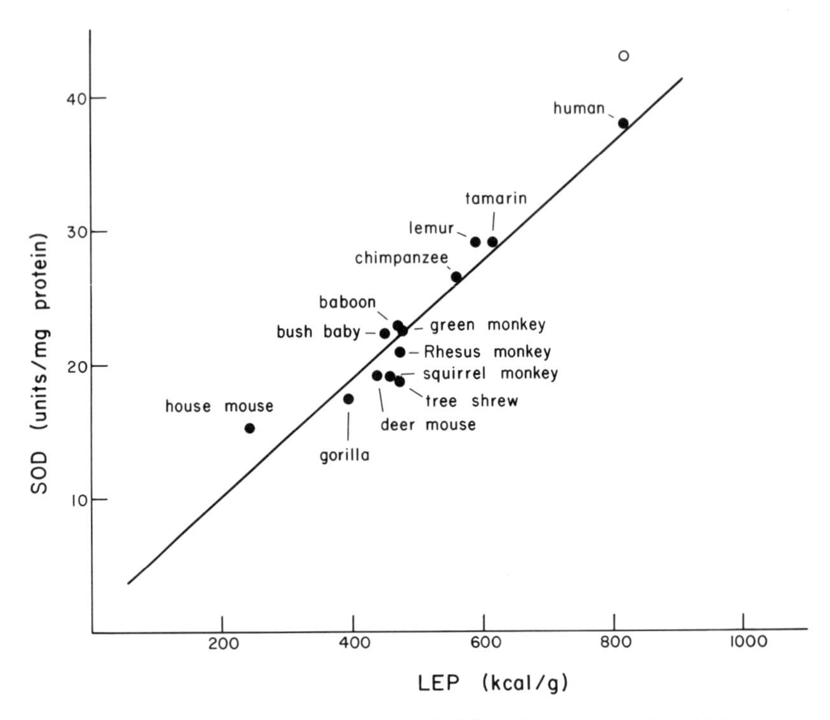

Figure 22. Correlation of superoxide dismutase (SOD) activity as a function of lifespan energy potential (LEP) in primate species. Data taken from Tolmasoff *et al.* (1980) and Cutler (in press).

LSP. From such a correlation it is not possible to determine if, with increasing LSP, SMR decreases and/or SOD increases. By noting how SMR correlates with LSP, it is found that the SMR value decreases for most species with SOD remaining constant, but for those species with unusually different LEP values, both low (such as gorilla) or high (such as human), the SOD value is seen to change. In either case, the end result is the same: longer-lived species were found to have, in proportion to their LSP, higher levels of protection offered by SOD per unit of SMR. SOD levels in tissues were not found to vary significantly with age for human and Rhesus (Tolmasoff *et al.*, 1980). These results suggested that potential antioxidant longevity determinants should be evaluated by three different procedures. These are by plotting the tissue levels of the antioxidant as a function of (a) LSP, (b) LEP, and (c) by plotting the level of antioxidants per SMR vs. LSP. In this way it can be determined if the increased protection arose by a decreased SMR or by an increased SOD or by both of these mechanisms. By plotting SOD vs. LEP, as shown in Fig. 22, it is evident that most species increased their tissue levels of SOD as well as have

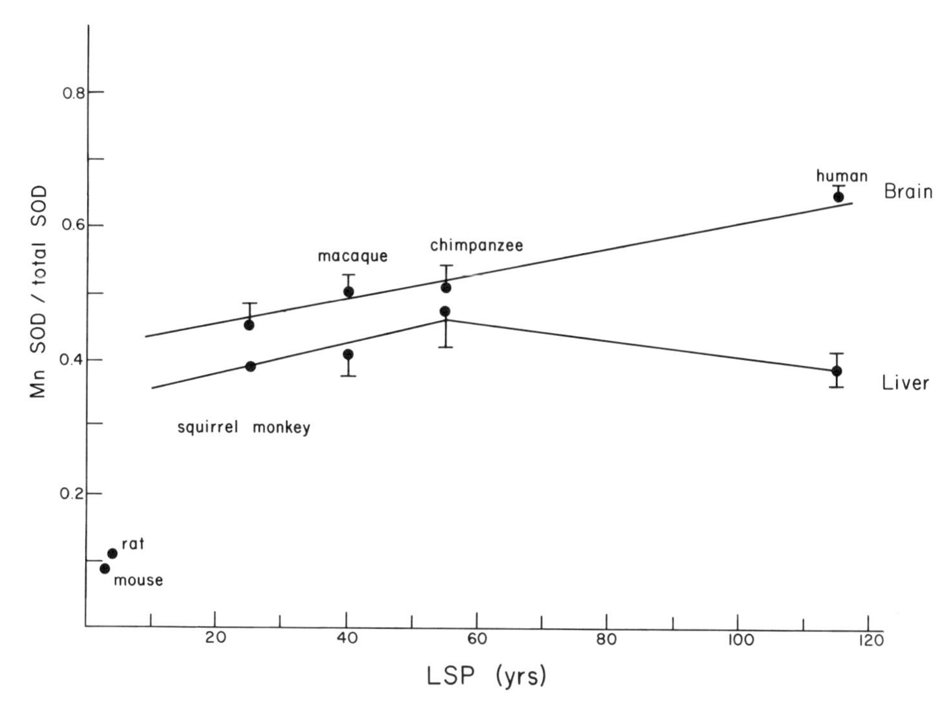

Figure 23. Ratio of MN-SOD in total SOD activity in primates as a function of lifespan potential (LSP) for liver and brain tissues.

different SMRs. The total SOD activity measured was a mixture of the concentration of the CuZn-SOD and the Mn-SOD. The Cu-SOD is thought to be mostly located in the cytoplasm of the cell and the Mn-SOD in the mitochondria. On analyzing the relative contribution of each type of SOD (see Fig. 23), it was found that primates overall appear to have a higher ratio (as compared to rodents) of Mn-SOD per total SOD, and that with increased LSP, the ratio of Mn-SOD per total SOD increased for brain but not for liver.

It should be noted that the linearity of SOD per SMR vs. LSP or SOD vs. LEP that was found was somewhat surprising, considering that the prediction being tested is only that longer-lived species would be expected to have higher levels of total tissue antioxidant activity, not necessarily a correlation with each specific antioxidant making up that total.

Considering that SMR reflects net rate of production of free radicals in a given tissue, it follows that

$$\text{LEP} \propto [\text{total tissue antioxidant capacity (AO)}]$$

where

$$\text{LEP} = K(\text{F}_1\text{AO}_1 + \text{F}_2\text{AO}_2 + \cdots + \text{F}_n\text{AO}_n)$$

and where

$$\sum_{i=1}^{n} \text{F}_i = 1$$

giving

$$\text{LEP} = K \sum_{i=1}^{n} \text{F}_i\text{AO}_i$$

The experimental result we found for SOD was

$$\text{SOD} = k\text{LEP}$$

This relationship implies the following: (a) F_is may be similar in different primate species for each antioxidant; and (b) the fraction of SMR producing toxic active oxygen species may also be similar in different primate species.

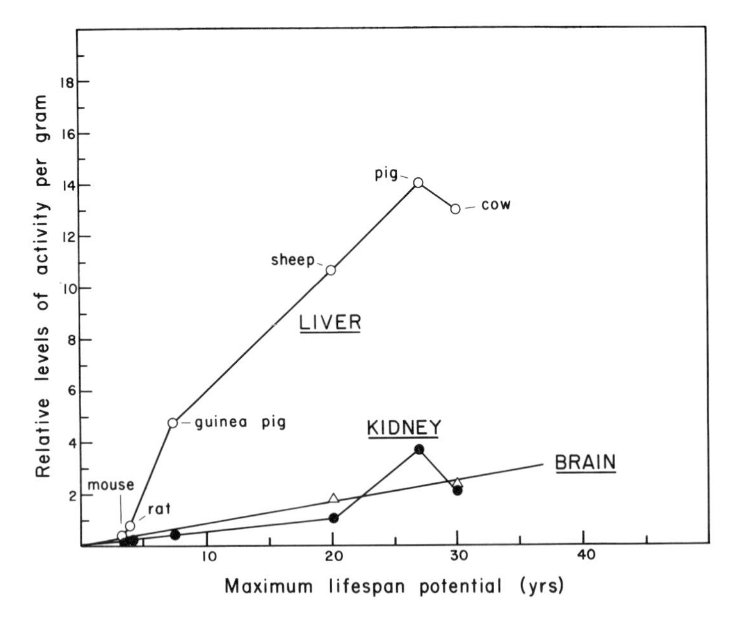

Figure 24. Catalase activity per specific metabolic rate (SMR) as a function of lifespan potential (LSP) in liver, kidney, and brain for mammalian species. Taken from Cutler (in press).

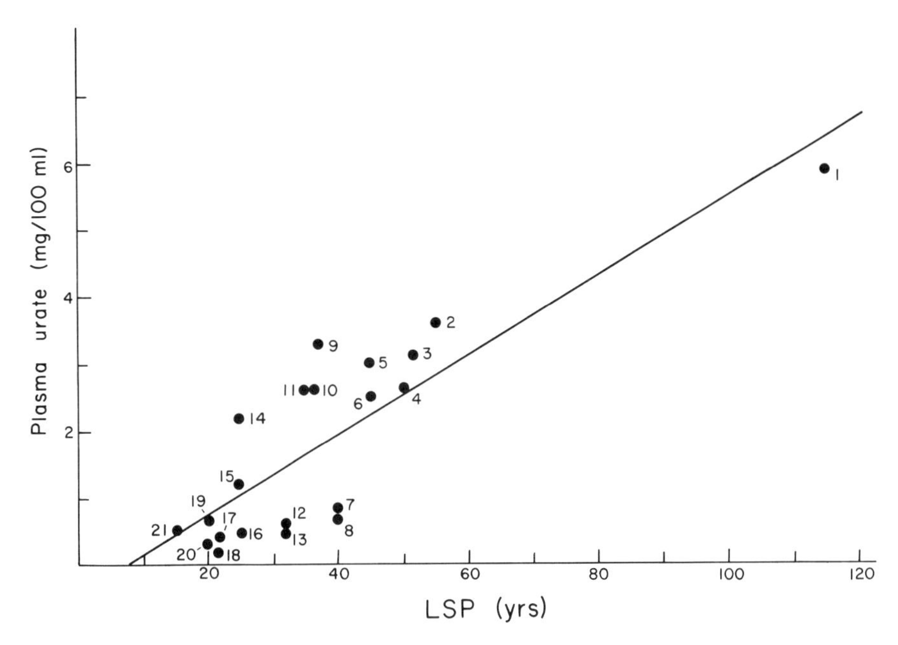

Figure 25. Plasma urate levels in primates as a function of LSP. Species represented are as follows:
1, human; 2, chimpanzee; 3, orangutan; 4, gorilla; 5, gibbon; 6, capuchin; 7, macaque; 8, baboon; 9, spider monkey; 10, Siamang gibbon; 11, wooly monkey; 12, langur; 13, grivet; 14, tamarin; 15, squirrel monkey; 16, night monkey; 17, potto; 18, patas; 19, galago; 20, howler monkey; 21, tree shrew. Correlation coefficient line is $r = 0.85$ where $P \leq 0.001$. Taken in part from Cutler (in press).

Finally, it should be noted that the amino acid sequence of SOD appears similar in most mammalian species. Thus, the increase in SOD per SMR with LSP is likely to be the result of higher enzyme concentrations due to changes occurring in regulatory genes.

2. Catalase: Tissue levels of catalse per SMR generally increase with LSP as shown in Fig. 24. Values of catalase activity for the primate species are not yet available.

3. Uric: Uric acid or urate, the sodium salt, has commonly been thought of as being a waste product of purine degradation (Wyngaarden and Kelly, 1976; Seegmiller, 1979). However, it has recently been shown to have strong antioxidant properties (Ames *et al.,* 1981). To test for its possible role in determining longevity, plasma urate concentrations were determined in a number of primate and nonprimate mammalian species. As shown in Fig. 25, an excellent correlation is found between plasma urate levels and LSP. On the other hand, a rather poor correlation is found beween plasma urate levels and LEP

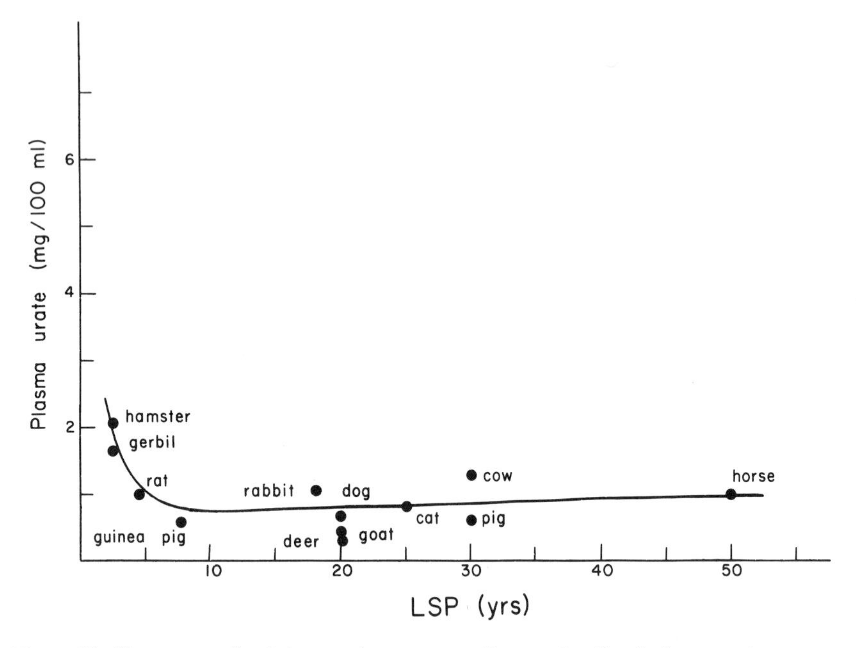

Figure 26. Plasma urate levels in nonprimate mammalian species. Species known to have strong histochemical reactions for uricase are rat, guinea pig, dog, and squirrel. Taken in part from Cutler (in press).

values for the primates. However, plasma urate levels were found to be fairly constant for nonprimate species, as shown in Fig. 26, and these species do have about the same LEP values.

This comparative analysis of plasma urate levels is not likely to be influenced by different ages of the species, for urate is found to be remarkably constant after the age of sexual maturity (Mikkelsen *et al.,* 1965). We have also found that tissue levels of urate (for example in the brain of the primate species) vary in proportion to plasma levels, but the levels run about 1/3 lower in terms of mg/100 g tissue. One of the problems of correlating antioxidants like urate, whose levels are determined both by endogenous factors as well as by diet, is the great variability between individuals of the same species. Thus, the high correlation coefficient that was found was quite surprising and strongly suggests that urate is playing some role in determining species' longevity.

The fact that urate levels correlate better with LSP than with LEP suggests that its effect may not entirely be as an antioxidant. Support for this latter view is found by the fact that urate levels may also be reflecting the tissue levels of some type of neurological stimulants that are operating in primates, particularly the human and New World monkeys. For example, a number of central nervous system stimulants like the methylated xanthines have structures similar to uric acid and are known to affect cAMP levels. Also, there is the interesting observation that many distinguished men frequently suffer from gout, a disease often related to high serum urate levels. Examples are Alexander the Great, Isaac Newton, Charles Darwin, William Harvey, Benjamin Franklin, Martin Luther, John Calvin, John Wesley, Cardinal Wolsey, and Francis Bacon. Finally, there are a number of reports finding a highly significant correlation of the level of serum uric acid to intelligence, achievement, and the need for achievement (Solyom *et al.,* 1968; Mueller *et al.,* 1970; Keehner, 1979; Park and Asaka, 1980). Thus, it appears possible that uric acid may serve both as an antioxidant and as a neural stimulator during the evolution of primate species. Both of these effects would have been important to achieve a higher learned-behavior trait.

It is also important to mention that uric acid levels are in part determined by the tissue levels of the enzyme urate oxidase or uricase, which acts to further degrade urate, and renal excretion (Roch-Ramel *et al.,* 1976). In general, the tissue levels of uricase are lower in the species having the higher serum uric acid levels. The enzyme is completely missing in the great apes and human. This correlation is consistent with urate levels being determined by changes occurring in regulatory genes. When uricase acts on the urate substrate, hydrogen peroxide is produced. Thus, another possible advantage of losing the uricase enzyme is, in addition to increasing the levels of an antioxidant or a neurostimulator, there is a decrease in the amount of hydrogen peroxide in a cell.

4. Ascorbate: Ascorbate is a well-known antioxidant, but its possible important role in this capacity has been controversial (Lewin, 1976; Chatterjee, 1978; Naito, 1979). Plasma ascorbate levels in primates and nonprimate mammals appear to be at about the the same levels and independent of LSP, as shown in Fig. 27. Ascorbate tissue levels decrease dramatically with increasing age (Kirk, 1962), so this comparative analysis was made using young adults when possible. However, if ascorbate per SMR vs. LSP is analyzed, an increase in this ratio is found up to the LSP of about 50 years (Fig. 28). The important observation here is that the human does not appear to have any higher levels of ascorbate or ascorbate per SMR than chimpanzee or other shorter-lived primates. Also, plasma ascorbate levels do not change as a function of LEP, as shown in Fig. 29.

The comparative results of ascorbate levels in tissues, however, is more clear. Ascorbate levels in the adrenal gland (Table XI), eye lens (Table XII), and liver (Table XIII) indicate that shorter-lived species have the higher levels. Levels of ascorbate in brain tissues indicate no correlation with either LSP or LEP (Table XIV). Ascorbate levels in cerebrospinal fluid of different species appear to decrease with increasing LSP or LEP, as shown in Table XV. However, on comparing ascorbate levels in a number of different tissues between

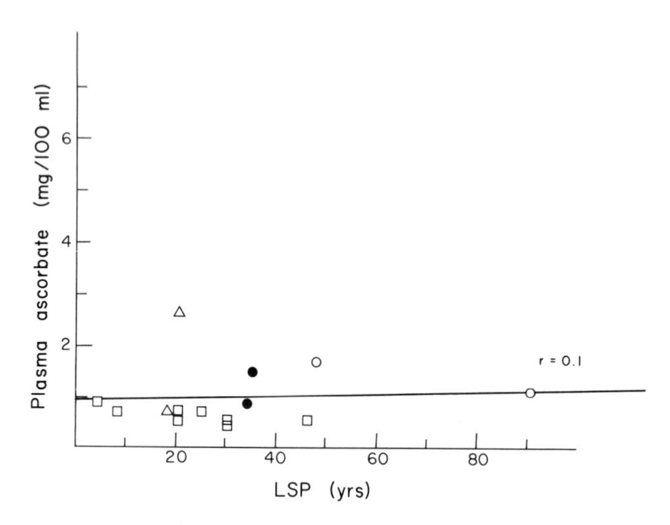

Figure 27. Plasma ascorbate levels in primate and nonprimate species as a function of lifespan potential (LSP). (○) Hominidae; (●) Old World monkeys; (△) New World monkeys; (□) nonprimate mammals. Taken in part from Cutler (in press).

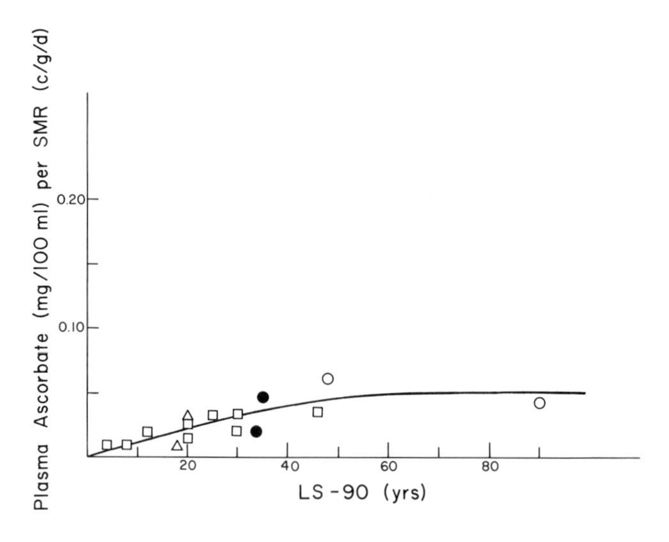

Figure 28. Plasma ascorbate levels per specific metabolic rate as a function of lifespan potential in primate and nonprimate species. Symbols are same as for Fig. 27. Taken in part from Cutler (in press).

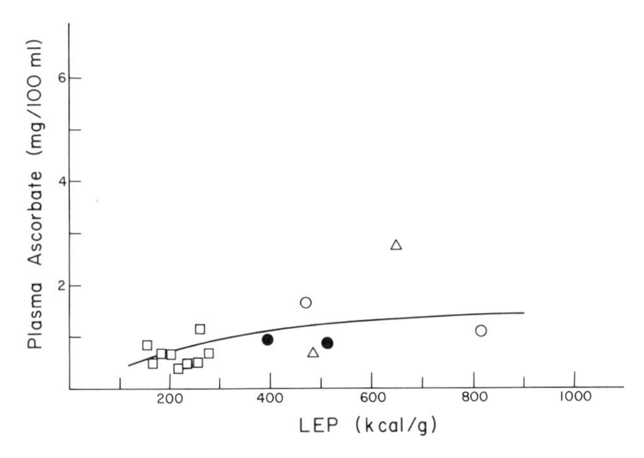

Figure 29. Plasma ascorbate levels as a function of lifespan energy potential (LEP) in primate and nonprimate species. Symbols are same as for Fig. 27. Taken in part from Cutler (in press).

Table XI. Ascorbate Levels in Adrenal Gland of Different Species[a]

Species	LSP (yr)	LEP (kcal/g)	Ascorbate (mg/100 g)	LGO
Human	90	815	44.8	No
Macaca	40	512	112.4	No
Guinea pig	8	204	90.8	No
Rat	4	151	392.0	Yes

[a]LGO is presence of the enzyme L-gulonolactone oxidase. Ascorbate data from Altman and Dittmer (1961, 1974).

Table XII. Ascorbate Levels in Eye Lens of Different Species[a]

Species	LSP (yr)	LEP (kcal/g)	Young adult (mg/100 ml)	Old adult (mg/100 ml)	LGO
Human	90	815	30	20	No
Cow	30	164	35	30	Yes
Rabbit	12	278	20	9	Yes
Rat	4	151	22	7	Yes

[a]LGO is presence of the enzyme L-gulonolactone oxidase. Ascorbate data from Altman and Dittmer (1961, 1974).

Table XIII. Ascorbate Levels in Liver of Different Species[a]

Species	LSP (yr)	LEP (kcal/g)	Ascorbate (mg/100 g)	LGO
Human	90	815	13.5	No
Macaca	35	512	14.9	No
Cow	30	164	26.1	Yes
Pig	30	219	11.2	Yes
Chicken	30	—	36.8	Yes
Rabbit	18	—	27.0	Yes
Guinea pig	8	204	16.6	No
Rat	4	151	32.9	Yes

[a]Ascorbate data from Altman and Dittmer (1961, 1974).

Table XIV. Ascorbic Acid Concentration in Whole Brain Tissues of Different Species with Age

Species	LSP (yr)	Mg % wet weight[a]	
		Young	Adult
Mouse	3	41	27
Rat	4	12	8
Guinea pig	8	15	6
Rabbit	12	39	20
Cow	30	40	20

[a]Nonlifespan data taken from Kirk (1962).

Table XV. Ascorbate Levels in Cerebrospinal Fluid of Different Species[a]

Species	LSP (yr)	LEP (kcal/g)	Ascorbate (mg/100 ml)	LGO
Human	90	815	2.47	No
Horse	49	233	1.70	Yes
Macaca	40	512	2.30	No
Cat	30	457	3.80	Yes
Dog	20	255	6.60	Yes

[a]LGO is presence of the enzyme L-gulonolactone oxidase. Ascorbate data from Altman and Dittmer (1961, 1974).

Table XVI. Ascorbate Levels in Tissues of Human and Macaque per Specific Metabolic Rate[a]

Tissue	Homo (mg/100 g per SMR)	Macaca (mg/100 g per SMR)	Homo macaca
Plasma	0.431	0.020	2.1
Whole blood	0.524	0.0328	1.5
White blood cells	1.07	0.290	3.6
Red blood cells	0.0443	0.0354	1.2
Cerebrospinal fluid	0.0995	0.0543	1.8
Aqueous humor	0.0427	0.0241	1.7
Whole brain	1.74	0.189	9.2
Pancreas	0.745	0.373	1.9
Adrenal gland	1.80	2.65	0.67
Liver	0.544	0.348	1.5
Kidney	0.401	0.513	0.78
Spleen	0.532	0.472	1.1
Testes	0.120	0.007	17.0

[a]Ascorbate data from Altman and Dittmer (1961, 1974).

human and the Macaca species (Rhesus), we find that a few human tissues are significantly higher, such as white blood cells, brain, and testes (Table XVI). Thus, more detailed and extensive comparative analyses of ascorbate are needed for the primate species, but so far it appears that ascorbate has not played an important role in determining the unusually high LSP or LEP value for humans.

During the evolution of the primate species, the ability to synthesize ascorbate was lost due to the loss of the enzyme L-gulonolactone oxidase (Chatterjee *et al.*, 1975; Jukes and King, 1975). As with urate oxidase, this enzyme also produces hydrogen peroxide in the synthesis of ascorbate. Thus, its loss would have had the advantage of decreasing the cellular levels of active oxygen species related to ascorbate synthesis. Also, ascorbate may have serious harmful side effects when in the presence of O_2 and metal ions like Fe^{2+}. Under these conditions, the ascorbate free radical is formed (Odumosu, 1972; Robinson and Rickheimer, 1975; Sharma and Krisna Murthi, 1976; Stich *et al.*, 1979; Gutteridge *et al.*, 1981; Ohno and Myoga, 1981).

Until we have more comparative data, it is not possible to evaluate the importance of the role ascorbate may have played in the evolution of the primate species, but it is at least clear that the loss of the ability to synthesize ascorbate cannot be assumed to be a disavantage and may have instead been important to the evolution of human longevity.

5. Carotenoids: β-carotene is known to be an excellent singlet oxygen scavenger, and recently it has been shown to protect membranes against lipid peroxidation (Peto *et al.*, 1981, Krinsky, 1982; Krinsky and Deneke, 1982). On

evaluating the carotenoids as potential longevity determinants, the following observations have been made.

It is known that mammalian species can be divided into three classes in terms of their relative tissue levels of the carotenoids (Karrer and Jucker, 1950; Goodwin, 1954, 1962; Moore, 1957; Isler, 1971). The first class is the human, which accumulates carotene and xanthrophylls in most tissues of the body at high levels. The second class contains species that accumulate primarily the carotenes. These are cow, horse, deer, and buffalo. The third class contains species that do not accumulate any appreciable carotene or xanthrophylls (rat, rabbits, guinea pig).

What is immediately distinctive about these three classes is that humans in the first class are the longest-lived, species in the second class are intermediate in lifespan, and the various rodent species in the third class are generally shorter lived.

Some of the characteristics of the carotenoids that suggest their possible involvement as longevity determinants are:

a. Microorganisms respond to light for photoprotection by increased synthesis of carotenoids (Krinsky, 1971).
b. Retinoids are, besides the steroid hormones, one of the few defined substances that govern growth and differentiation.
c. Retinoids act to stabilize normal differentiation and protect against some carcinogens (Sporn and Newton, 1979; Lotan, 1980; Newberne and Rogers, 1981; Hill and Grubbs, 1982).
d. β-carotene is the most efficient known quencher of singlet oxygen ($'O_2$) and is also found to protect against the peroxidation of membranes (Foote *et al.*, 1970).
e. An inverse correlation exists between plasma levels of β-carotene and retinol in human and the risks to obtain certain types of cancers (Peto *et al.*, 1981).
f. Down's syndrome patients have abnormally low retinol levels, higher aging rates and higher risk to some types of cancer (Palmer, 1977).
g. Humans accumulate carotene and xanthrophyll in tissues unselectively. Intermediate-lifespan mammals only accumulate carotene. Short-lifespan species accumulate neither carotene nor xanthrophyl.
h. Levels of carotenoid distribution in tissues are inversely related to intestinal carotenase activity: high in rats, intermediate in cows, and extremely low activity in human (Thompson *et al.*, 1949; Goodman, 1962; Olson and Hayashi, 1965).

Plasma levels of retinol and carotene taken from the literature are shown in Table XVII. These data suggest that a possible correlation does exist with LSP. Levels of serum carotenoid concentrations (carotene + xanthrophyll) in primates and nonprimate mammals are shown in Fig. 30, where an excellent

Table XVII. Plasma Levels of Retinol and Carotene vs. MLP[a]

Species	MLP	Retinol (μg/ 100 ml)	β-carotene (μg/100 ml)
Human	115	54	200
Horse	49	—	100
Baboon	40	37	17
Rhesus	40	36	—
Cow	30	24	36
Pig	30	20	—
Sheep	25	10	32
Deer	20	48	0.36
Dog	20	0.3	—
Rat	4	40	0

[a]Nonlifespan data taken from Altman and Dittmer (1961).

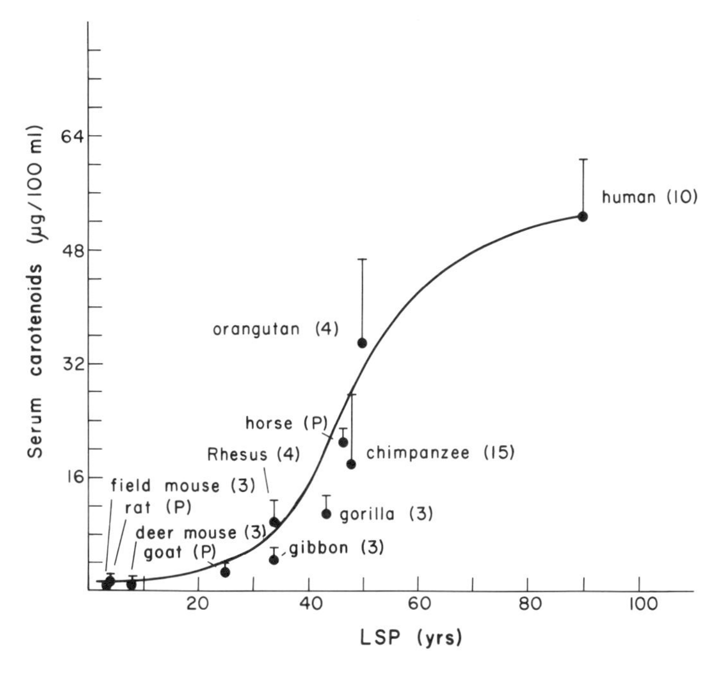

Figure 30. Serum carotenoid concentrations as a function of lifespan potential (LSP) in mammalian species. Numbers in parentheses are numbers of individual animals used in each determination. Taken in part from Cutler (in press).

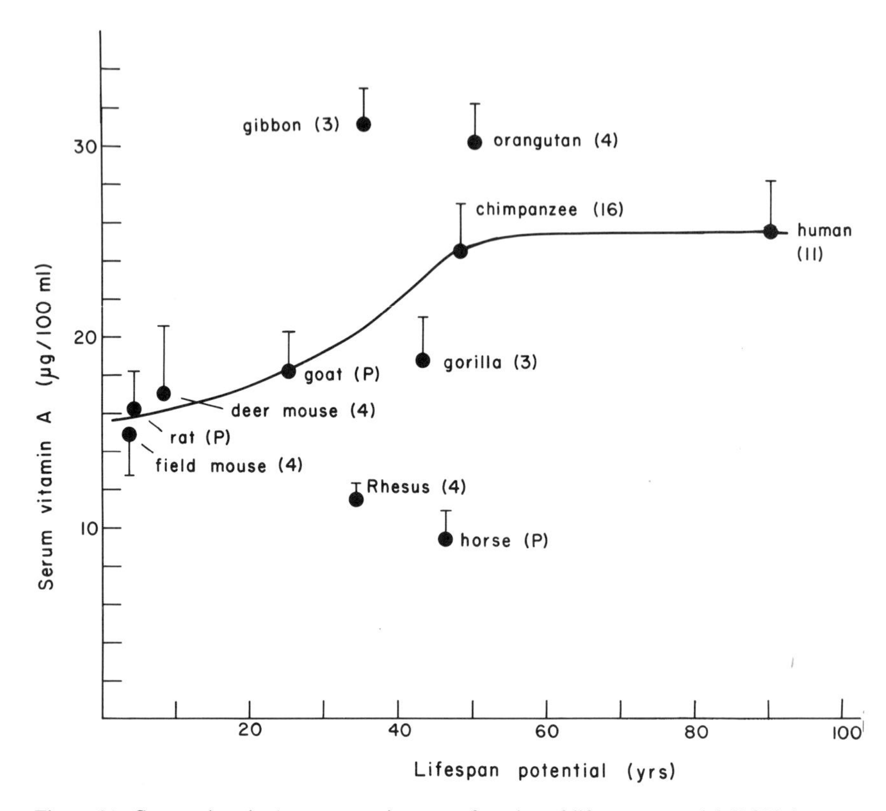

Figure 31. Serum vitamin A concentrations as a function of lifespan potential (LSP) in mammalian species. Numbers in parentheses are numbers of individual animals used in each determination. Taken in part from Cutler (in press).

correlation with LSP is found. On the other hand, Fig. 31 shows that serum vitamin A levels have no significant correlation with LSP. An excellent correlation of serum levels of carotene with LEP values of these species was also found. Considering the good correlation of carotenoids with both LSP and LEP, it could not be determined if the carotenoids were predominantly acting as antioxidants. Tissue levels of the carotenoids and vitamin A have also been determined as a function of LSP and LEP, and in general the correlations are similar to those found for the serum. Plasma levels of vitamin A and carotene in human blood change little from 10 to 80 years of age (Leitner *et al.,* 1960).

These results strongly support the possible role of the carotenoids as a longevity determinant, but not vitamin A. Thus, like urate, which was thought to have no function, the carotenoids may be more important than simply a precursor to vitamin A synthesis.

In addition, like urate and ascorbate, where tissue levels are determined in part by urate oxidase and L-gulonolactone oxidase, respectively, the levels of the carotenoids are also determined in part by the levels of the intestinal enzyme (a dioxygenase) β-carotene-15-15′-dioxygenase. Species such as humans, which have high tissue levels of the carotenoids, have low levels of the carotene dioxygenase enzyme. Thus, carotene levels are determined by the levels of an enzyme, a regulatory gene function. In addition, it is interesting that all three of these potential longevity determinants, urate, ascorbate, and carotenoids, have been affected along the evolutionary ancestral–descendant sequence leading to the humans by a loss of activity of an enzyme, a change that has a high probability of occurrence because of the many different ways it can happen.

6. α-tocopherol: α-tocopherol or vitamin E is becoming well accepted as an important tissue antioxidant, protecting the lipid or membrane components of a cell (Machlin, 1980). Table XVIII shows the plasma levels of α-tocopherol

Table XVIII. Plasma Levels of α-Tocopherol in Different Species[a]

Species (common name)	LSP (yr)	SMR (cal/g per d)	LEP (kcal/g)	α-tocopherol (mg/100 ml)	$\frac{\alpha\text{-tocopherol}}{\text{SMR}}$
Homo (human)	90	24.8	815	1.2	0.0483
Equus (horse)	46	13.9	233	0.25	0.0179
Cebus (capuchin)	42	52.2	804	0.50	0.00957
Papio (baboon)	35	30.9	394	0.73	0.0236
Macaque (Rhesus)	34	41.3	512	0.56	0.0135
Bos (cow)	30	15	164	0.40	0.0266
Sus (pig)	30	20	219	0.16	0.008
Aotus (night monkey)	20	72.7	530	0.53	0.00729
Ovis (sheep)	20	25.6	186	0.020	0.000781
Canis (dog)	20	35	255	0.41	0.0117
Rattus (rat)	4	104	152	0.31	0.00298
Mus (mouse)	3.5	182	232	0.75	0.00412

[a]LSP vs. α-tocopherol, $r = 0.554$; LEP vs. α-tocopherol, $r = 0.661$; LSP vs. $\frac{\alpha\text{-tocopherol}}{\text{SMR}}$, $r = 0.864$. Non-lifespan data taken from Altman and Dittmer (1961) and Benirschke *et al.* (1978).

Table XIX. Plasma Vitamin E Levels Are Species-Specific [a]

Species (common name)	LSP (yr)	LEP (kcal/g)	Plasma vitamin E levels (mg/100 ml) Vit. E + safflower oil	Vit. E + coconut oil
Cebus (capuchin)	42	804	0.420 ± 0.1	0.880 ± 0.079
Macaque (crab-eating monkey)	34	512	0.367 ± 0.047	0.530 ± 0.121

[a]Resultant plasma vitamin E levels after 50 weeks of diet. With species over a wide range of identical diets, Cebus always had higher plasma vitamin E levels and was least affected by vitamin E-deficient diets. Non-lifespan data from Ausman and Hayes (1974).

in different mammalian species as a function of LSP and LEP. The best correlations are found with LEP and α-tocopherol per SMR vs. LSP, indicating that the α-tocopherol is likely to play an important role as a longevity determinant, and in doing so is acting as an antioxidant. Plasma levels of α-tocopherol as a function of age in human do not change significantly from 10 to 80 years of age (Leitner *et al.,* 1960). Unfortunately, few tissue studies of α-tocopherol have been made, as with most of the antioxidants.

One study was made, however, that has an important bearing on these correlations of antioxidants that are part of the food an animal eats. In Table XIX, plasma levels of α-tocopherol in two primates are shown, Cebus and the crab-eating monkey, under different diets but where both species at any one time had the same diet. Under both diets Debus always had the higher plasma α-tocopherol level. Thus, not only does the longer-lived monkey with a higher LEP value also have a higher α-tocopherol level under identical diets than the shorter-lived species with the lower LEP value, but this higher α-tocopherol level is maintained under different dietary conditions.

Table XX. Whole Blood Glutathione Levels

Species	LSP (yr)	LEP (kcal/g)	Total[a]	Oxidized[a]	Reduced[a]
Human	90	815	36.8	31	4
Horse	49	152	60	50	—
Baboon	35	512	49	—	—
Cow	30	153	46	40	6
Pig	30	219	36	—	—
Sheep	25	186	26	—	—
Dog	20	268	31	29	—
Rabbit	12	257	45	35	—
Guinea pig	8	204	127	—	—
Rat	4	152	120	—	—
Mouse	3	232	102	—	—

[a]Units: mg/100 ml. Nonlifespan data from Altman and Dittmer (1961, 1974).

Table XXI. Glutathione Concentration in Tissues[a]

Species (common name)	LSP (yr)	LEP (kcal/g)	Relative values				
			Brain	Heart	Liver	Kidney	Spleen
Human	90	815	12.8 ± 3.45 n = 5	11.3 ± 1.95 n = 4	13.0 ± 9.41 n = 4	7.42 ±2.89 n = 4	6.2 n = 1
Baboon	35	394	17.4 ± 2.68 n = 4	15.7 ± 4.59 n = 2	36.8 ±13.0 n = 4	8.66 ±6.35 n = 3	7.66 ±6.78 n = 3
Pig-tailed macaque	34	517	17.5 ± 1.41 n = 2	13.0 ± 0.981 n = 3	42.6 ± 2.46 n = 3	4.30 ±1.13 n = 2	13.1 ±4.38 n = 2
Deer mouse	8	440	54.5 ± 2.1 n = 2	—	—	—	—
Field mouse	3	232	49.0 ± 1.4 n = 2	—	—	—	—

[a]n is number of different individuals measured. LSP vs. GSH, r = −0.803 (brain); LEP vs. GSH, r = −0.636 (brain). Taken from Cutler, submitted for publication.

7. Glutathione: Glutathione is probably one of the most abundant antioxidants in mammalian tissue (Prins and Loos, 1969; Aries and Jakoby, 1976; Reddy *et al.,* 1980; Younes and Siegers, 1981; Meister, 1982). On examining whole blood levels of glutathione as a function of LSP or LEP, however, an inverse function appears to exist (Table XX); that is, human blood levels of glutathione appear to be unusually low, not high as one might expect. Similar results are found in tissues, as shown in Table XXI. Here, it is seen that glutathione levels in brain, heart, liver, kidney, and spleen of the human are lower than other primate species. However, if the level of glutathione per SMR is calculated for brain as function of LSP, a positive correlation coefficient of r = 0.713 is found. In the eye lens (as shown in Table XXII) where glutatione

Table XXII. Concentration of Free and Protein-Bound Glutathione in Mammalian Lens

Species	LSP (yr)	GSH[a] (μmole/g wet wt.)	Mixed disulfide[a] (μmole/g wet wt.)
Human	90	2.20	1.20
Cow	30	14.0	0.20
Rabbit	12	12.0	1.30
Rat	4	7.9	0.96

[a]Nonlifespan data taken from Reddy *et al.,* (1980).

is thought to be extremely important in protecting against cataracts in aging, again no evidence is found that human eye lens has an unusually high level of this antioxidant. This evidence suggests that for some reason glutathione is not an important antioxidant contributing to the longevity differences among the various primate species, and in particular to human longevity.

Glutathione concentration is found to decrease with increasing age in blood and tissues of mice and humans (Sass *et al.*, 1964; Hazelton and Lang, 1980; Stohs *et al.*, 1980, 1982). Because longer-lived species appear to have lower blood and tissue levels, this age-dependent decrease may not be detrimental, as has been assumed, but could be adaptive, where if it did not decrease, lifespan would be shorter.

8. Glutathione peroxidase: Glutathione peroxidase is thought to be one of the most important protective enzymes against the accumulation of lipid peroxides (Flohé, 1982; Tappel *et al.*, 1982). However, like glutathione, this enzyme also appears to decrease in concentration as LSP or LEP values increase. This is shown in Table XXIII for liver and Table XXIV for liver and brain for primate and nonprimate species. Blood selenium and glutathione peroxidase are shown as a function of LSP and LEP in Table XXV. Here, no clear correlation is seen with selenium, and glutathione peroxidase activity is seen to decrease with LSP and LEP, as with the other tissues. Of interest is

Table XXIII. Glutathione Peroxidase Activity in Liver

Species	LSP (yr)	Se GPX[a]	Non-Se GPX[a]	Total GPX activity	Non-Se GPX (% of total)
Hamster	3	26.0 ± 2.8	19.1 ± 1.3 $n = 3$	45.1	43
Rat	4	19.6 ± 4.5	8.4 ± 1.4 $n = 3$	28.0	35
Guinea pig	8	0	7.3 ± 0.7 $n = 2$	7.3	100
Sheep	20	3.8 ± 0.9	17.6 ± 2.7 $n = 3$	21.4	81
Pig	30	2.5 ± 5	6.0 ± 2.4 $n = 3$	8.5	67
Chicken	30	1.5 ± 0.02	3.4 ± 0.2 $n = 2$	4.9	70
Human	90	1.3 ± 0.32	8.4 ± 1.2 $n = 6$	9.7	84

[a]Se GPX and non-Se GPX 105,000 × g activity measured as units/mg protein. One unit of activity is 1 μmole NADPH oxidized per min. Non-Se GPX is glutathione S-transferase. n is number of animals (adult). Se GPX vs. MLP, $r = -0.489$. Non-Se GPX vs. MLP, $r = -0.269$. Total GPX vs. MLP, $r = -0.469$. Nonlifespan data from Lawrence and Burk (1978).

Table XXIV. Glutathione Peroxidase Activity in Liver and Brain

Species	n	LSP (yr)	LEP (kcal/g)	Liver[a]	Brain[a]
Mouse	5	3	232	1140	23 ± 3
Rat	5	4	152	153 ± 23	5
Guinea pig	5	6	206	57	14 ± 4
Rabbit	5	12	257	381	20 ± 5
Dog	2	20	268	ND	3
Cow	5	30	153	ND	5 ± 1

[a]Activity of cytosol fraction: mean ± SD. Nanomoles GSH oxidized/min per mg protein. n is number of animals used for determination. LSP vs. GPX: $r = -0.315$, liver; $r = -0.551$, brain. LEP vs. GPX: $r = 0.476$, liver; $r = 0.412$, brain. Nonlifespan data taken from De Marchena *et al.*, 1974.

that the glutathione peroxidase/selenium ratio decreases markedly with increased LSP or LEP.

9. Glutathione S-transferase: Glutathione S-transferases are important enzymes for detoxification reactions (Jakoby, 1978). These enzymes are also known as the non-Se glutathione peroxidases. Figure 32 shows the decreasing levels of this enzyme in liver tissues in three cellular fractions with increasing age. Although glutatione S-transferase does decrease with LSP, the percent a tissue has of this enzyme in relation to total tissue glutathione peroxidase activity increases significantly with LSP—particularly so if expressed per SMR, as shown in Table XXVI.

On examining a wide number of tissues for their relative glutathione S-

Table XXV. Blood Selenium and Glutathione Peroxidase Activity

Species	LSP (yr)	LEP (kcal/g)	Se[a] (ppm)	GPX[a]	GPX/Se
Rat (6 male, 3 female adults)	4	152	0.0775 ± 0.060 $n = 4$	120 ± 100 $n = 4$	1529 ± 161
Sheep (all adult females)	20	186	0.115 ± 0.102 $n = 6$	152 ± 143 $n = 6$	128 ± 73.6
Rhesus (all adult females)	34	512	0.72 ± 0.054 $n = 5$	19.8 ± 18 $n = 5$	257 ± 71.9
Human (2 male, 8 female adults)	90	815	0.10 $n = 10$	19.0 $n = 10$	190

[a]Se and GPX data taken from Butler *et al.* (1982). Glutathione peroxidase activity as nanomoles NADPH oxidized/min per mg hemoglobin. Se vs. LS, $r = 0.254$. Se vs. LEP, $r = -0.010$. GPX vs. LS, $r = -0.720$. GPX vs. LEP, $r = -0.893$. GPX/Se vs. LS, $r = -0.812$. GPX/Se vs. LEP, $r = -0.928$.

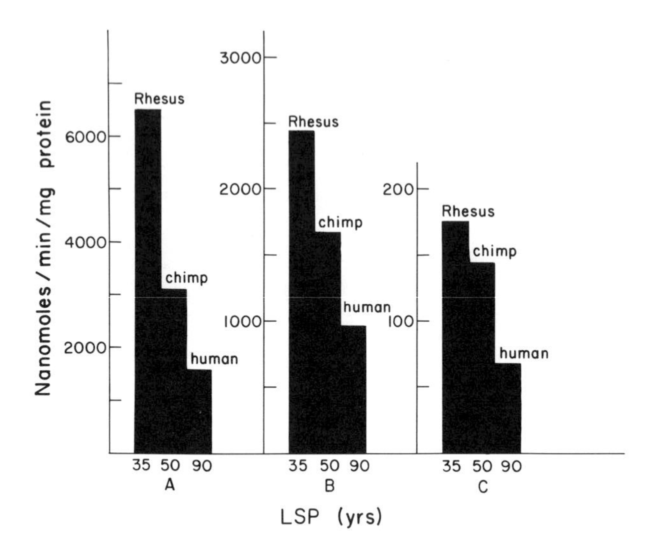

Figure 32. Glutathione S-transferase activity in liver as a function of LSP in primates. (A) 100,000 × *g* supernatant, (B) 9000 × *g* supernatant, (C) microsomes. Data taken in part from Summer and Grien (1981).

transferase activity in human versus rat (Table XXVII), it is seen that human tissues are not significantly higher, except for muscle. Perhaps in this tissue a LSP or LEP correlation exists, but as yet there are not sufficient data to answer this question. Epoxide hydratase also appears to be lower in human as compared to baboon (Pacifici *et al.,* 1981a).

Table XXVI. Percent Glutathione S-Transferase to Total Glutathione Peroxidase Activity in Liver[a]

Species	LSP (yr)	SMR (cal/g per d)	LEP (kcal/g)	Non-Se GPX (% of total GPX)	% Non-Se GPX / SMR
Hamster	3	118	129	43	0.364
Rat	4	104	152	35	0.336
Guinea pig	8	69.8	294	100	1.43
Sheep	20	25.6	186	81	3.16
Pig	30	20	219	67	3.36
Human	90	24.8	815	84	3.38

[a]LSP vs. % Non-Se GPX, r = 0.131. LEP vs. % Non-Se GPX, r = 0.393. LSP vs. % Non-Se GPX/SMR, r = 0.701. Nonlifespan data from Lawrence and Burk (1978).

Table XXVII. Glutathione S-Transferase Activity in Human and Rat

Tissue	Human[a]	Rat[a]	Human/rat
Kidney	127	86	1.47
Liver	119	708	0.167
Adrenal	112	42	2.66
Muscle	103	10	10.3
Pancreas	102	42	2.42
Testes	97	301	0.322
Brain	91	66	1.37
Ovary	80	58	1.37
Stomach	51	40	1.27
Heart	46	26	1.15
Spleen	38	31	1.22

[a]1-chloro-2,4-dinitrobenzene as substrate. Activity expressed as nmoles of conjugate produced per min per mg protein. Nonlifespan data from Baars *et al.* (1981).

The negative correlations found for glutathione, glutathione peroxidase, and glutathione S-transferase with LSP are similar to the results of Schwartz and co-workers, showing that the mutagenic activities of carcinogens decrease with increasing LSP and that the P-450 and P-455 enzymes are shown to decrease in time (Schwartz and Perantoni, 1975). These results suggest that the reactions involving glutathione may have harmful toxic side effects that have been reduced with the evolution of longer LSPs.

There is evidence that mixed-function oxidation systems produce active oxygen products capable of inactivating proteins (Fucci *et al.*, 1983) and that dietary restriction (Martin *et al.*, 1980) and adding DHEA to the diet, reduces the activity of the mixed-function oxidation system. Thus, the decrease in activity of this detoxification system may be part of the mechanism of how these two methods act to increase lifespan.

10. Ceruloplasmin: Ceruloplasmin has been suggested to be the major antioxidant in blood plasma (Denko, 1979; Plonka and Metodiewa, 1980; Frieden, 1980; Gutteridge *et al.*, 1980). An examination of the relation of ceruloplasmin with LSP and LEP (Table XXVIII) for primate species shows a positive but not highly significant correlation. Looking at nonprimate species (Tables XXIX and XXX) the human does not appear to have unusually high levels of ceruloplasmin, copper, or ceruloplasmin oxidase activity. Thus, the overall correlation with LSP and LEP is weak, and it appears that ceruloplasmin does not play a major role as an antioxidant in determining the unusually high LSP and LEP value for humans.

11. Dehydroepiandrosterone: DHEA has been suggested to be an antiox-

Table XXVIII. Plasma Ceruloplasmin Levels in Primates versus LSP

Species	LSP (yr)	LEP (kcal/g)	CP (A_{530} mμ)[a]	CP (per mg protein)[a]
Human	90	815	0.550	0.0785
Capuchin	42	804	0.490	0.0671
Spider monkey	30	524	0.600	0.0705
Night monkey	20	530	0.630	0.0875
Howler monkey	20	371	0.215	0.0405
Squirrel monkey	18	485	0.125	0.0189

[a]Measured as ρ-phenylenediamine oxidase activity. LSP vs. CP, $r = 0.445$. LEP vs. Cp, $r = 0.511$. Nonlifespan data from Seal (1964).

Table XXIX. Plasma Ceruloplasmin Concentration in Different Species

Species	LSP (yr)	Ceruloplasmin[a] (mg/100 ml)	Ceruloplasmin-bound[a] copper (μg/100 ml)
Human	90	35.6	121
Cow	30	20.3	69
Sheep	25	26.5	90
Pig	25	35.3	120
Dog	20	17.4	59
Rat	4	35.9	122

[a]Nonlifespan data from Evans and Wiederanders (1967).

Table XXX. Plasma and Erythrocyte Copper and Ceruloplasmin Oxidase Activity in Different Species

Species	LSP (yr)	Plasma copper[a] (μg/100 ml)	Erythrocyte copper[a] (μg/100 ml)	Ceruloplasmin[a] oxidase activity
Human	90	129	111	0.268
		(95–172)	(92–130)	(0.141–0.400)
Cow	30	91	81	0.052
		(28–153)	(42–102)	(0.010–0.210)
Sheep	25	115	82	0.114
		(70–198)	(45–126)	(0.040–0.270)
Pig	25	215	89	0.375
		(148–267)	(76–100)	(0.162–0.511)
Dog	20	67	82	0.033
		(51–108)	(67–100)	(0.010–0.087)
Rat	4	123	96	0.228
		(99–144)	(67–115)	(0.178–0.267)

[a]Nonlifespan data from Evans and Wiederanders (1967). Parentheses indicate normal range.

idant (Demopoulos, personal communication), but its greatest interest to longevity studies is in its remarkable age-dependent decline in human serum (De Peretti and Forest, 1976). In addition, the work of Schwartz and co-workers has indicated that DHEA may be a protectant against carcinogenic agents, perhaps acting as an antioxidant (Schwartz and Perantoni, 1975; Schwartz, 1979).

On examining the serum levels of DHEA and DHEA-S (the sulfate derivative), we find no LSP or LEP correlation for DHEA (Fig. 33a), but humans do have an unusually high DHEA-S level in relation to other primates and nonprimate mammals (Fig. 33b). Thus, DHEA-S may be an important longevity determinant for the human.

12. Choline: Choline appears to have antioxidant properties. The pathological effects of choline deficiency can be prevented or reduced by feeding these animals synthetic antioxidants such as BHT, BHA or vitamin A (Newberne *et al.,* 1969; Spratt and Kratzing, 1971). In addition to its potential antioxidant properties, choline is the precursor to the synthesis of acetylcholine. A deficiency of this neurotransmitter is present in brain in senile dementia, Alzheimer's disease, and Down's syndrome (Boyd *et al.,* 1977). On examining the relative levels of choline in the plasma of different species, we find that humans appear to have higher levels in relation to shorter-lived species (Table XXXI). However, because of the severe limitation of data for other species, the importance of choline as a longevity determinant remains unclear.

13. Glucose: Glucose has been reported to scavenge the hydroxyl radical at physiological concentrations (Sagone *et al.,* 1983). It is also known that lower glucose levels frequently aggravate the symptoms of diabetes. Thus, it is possible that high glucose levels may act as an adaptive protective mechanism for the beta cells, which appear to be unusually sensitive to damage by free radicals.

Tables XXXII and XXXIII show serum levels of glucose for primate and nonprimate mammals, respectively. These data indicate no correlation of glucose levels with LSP or LEP, but a significant correlation of glucose levels per SMR with LSP. This correlation is illustrated in Fig. 34. From this figure, it is seen that high plasma levels of glucose do not likely play a role in the evolution of longevity of the human but may be important for shorter-lived species. It is most surprising, however, that plasma glucose levels do not correlate inversely more strongly with SMR, although there is some correlation in this direction for the smallest species.

A summary of the antioxidants that were found to be potentially most important in determining human longevity are (a) superoxide dismutase, particularly the Mn-SOD, (b) catalase, (c) urate, (d) α-tocopherol, and (e) the carotenoids. Antioxidants that may play an important role in determining lon-

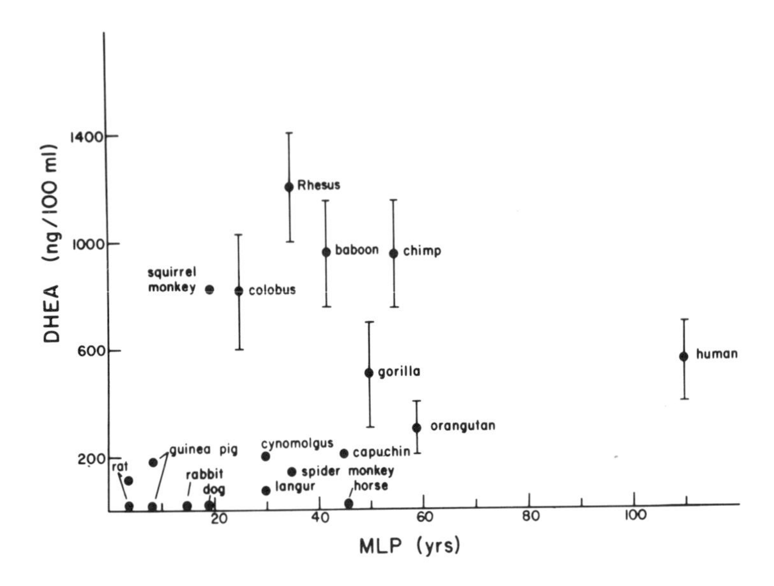

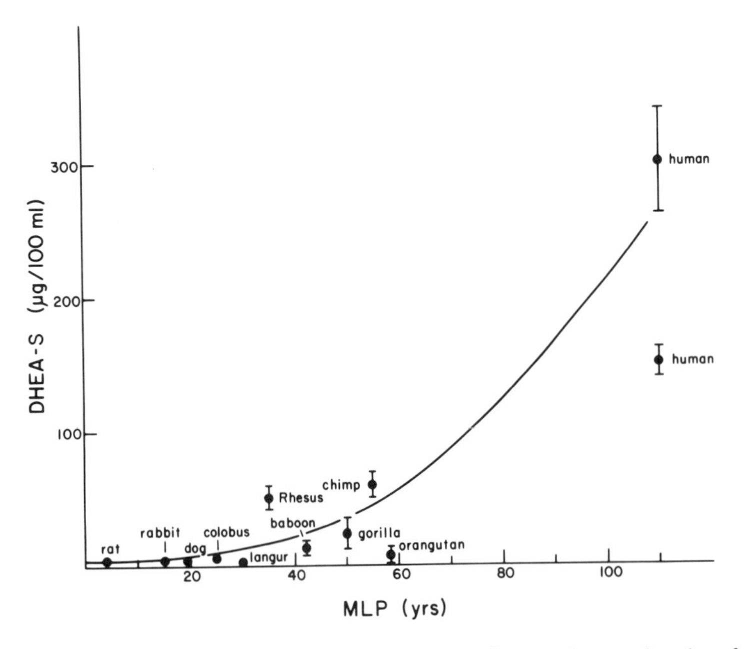

Figure 33. Serum dehydroepiandrosterone levels in mammalian species as a function of LSP. Figure 33a shows LSP correlation with DHEA and 33b with the sulfate derivative, DHEA-S. Data taken in part from Snipes *et al.* (1969), Townsley and Pepe (1981), Cutler *et al.* (1978), and Schwartz (1979).

Table XXXI. Choline Plasma Levels versus MLP[a]

Species	MLP (yr)	Choline (mg/100 ml)
Human	115	30.5
Horse	49	15.0
Cow	30	21.0
Pig	30	2.60
Sheep	25	10.0
Dog	20	12.5
Rabbit	18	0.50
Rat	4	0.15

[a]Nonlifespan data from Altman and Dittmer (1961).

Table XXXII. Serum Glucose Levels in Primates

Species (common name)	LS-90 (yr)	SMR (cal/g per d)	LEP (kcal/g)	Plasma glucose[a] (mg/100 ml)	Plasma glucose per SMR
Human	90	24.8	815	82 ± 8.09 $n = 5$	3.31
Chimpanzee	48	26.8	469	95 ± 5.13 $n = 3$	3.54
Gorilla	43	19.7	309	83 $n = 1$	4.21
Gibbon	35	44.6	569	57 $n = 1$	1.27
Baboon	35	30.9	394	85 ± 12.2 $n = 2$	2.75
Macaque	34	41.3	512	72 ± 20.5 $n = 9$	1.74
Langur	30	35.5	388	106 $n = 1$	2.98
Owl monkey	20	72.7	530	113 $n = 1$	1.55
Squirrel monkey	18	73.9	485	79 ± 4.59 $n = 2$	1.06
Bush baby	17	67.6	419	98 $n = 1$	1.44

[a]n is number of individuals measured. Nonlifespan data from Benirschke *et al.* (1978).

Table XXXIII. Serum Glucose Levels in Nonprimate Mammals[a]

Species	LS-90	SMR	LEP	Plasma glucose[b] (mg/100 ml)	Plasma glucose per SMR
1 Horse	46	13.9	233	83 $n = 1$	5.97
2 Cow	30	15	164	65 $n = 1$	4.33
3 Pig	30	20	219	78 $n = 2$	3.90
4 Cat	25	50.1	457	133 ± 26.8 $n = 3$	2.65
5 Goat	25	30.4	277	66 $n = 2$	2.17
6 Sheep	20	25.6	186	63 $n = 1$	2.46
7 Dog	20	35	255	93 ± 12.3 $n = 10$	2.65
8 Deer	20	24.6	179	101 $n = 1$	4.10
9 Rabbit	12	58.7	257	120 ± 28.1 $n = 3$	2.04
10 Guinea pig	8	69.8	204	110 ± 21.2 $n = 2$	1.57
11 Gerbil	6	138	302	94 $n = 1$	0.681
12 Rat	4	104	152	105 ± 32.8 $n = 9$	1.00
13 Mouse	3.5	182	232	112 ± 38.5 $n = 5$	0.615
14 Hamster	3.0	118	129	118 ± 29.9 $n = 9$	1.00

[a] n is the number of individuals measured.
[b] Nonlifespan data taken from Benirschke *et al.* (1978).

gevity (but not the unusual longevity of the human) are (a) ascorbate, (b) ceruloplasmin, and (c) glucose. Agents that may not be antioxidants but may play an important role in determining human longevity are choline and the sulfate derivative of dehydroepiandrosterone. The most surprising aspect of this survey was the negative correlation of LSP and LEP for glutathione and the related enzymes, glutathione peroxidase, and glutathione S-transferase. This negative correlation suggests that metabolic pathways associated with glutathione utilization, as the mixed-function oxidase system, are somehow detrimental to the organism in terms of evolving longer lifespans.

An important difficulty with the type of approach just demonstrated in determining if antioxidants are longevity determinant processes is that we probably do not know all the important antioxidants to investigate. For example, a potentially important but unknown antioxidant that is heat-labile and found in all tissues of the rat is still unidentified.

One method used to get around this problem is to measure the rate of autoxidation of whole tissue homogenates. Crude tissue homogenates are incubated at 37°C in air, and the amount of thiobarbituric acid-reacting material produced is measured as a function of time of incubation. This test, called the TBA test for malonaldehyde, is a reasonably good assay for the relative amounts of lipid peroxides produced by the autoxidation process (Ohkawa *et al.*, 1979). The tissue constituents subject to autoxidation would include the unsaturated lipids and also the nucleic acids, and prostaglandin endoperoxides.

Typical results of these experiments are shown in Fig. 35 for whole brain homogenates for a number of primate species and two rodent species. Similar results were also found for kidney. The results show a homogenate of human tissue autoxidizes very little, but there is a progressive rank order in tissue sensitivity to autoxidation with decreasing LSP. This correlation is seen more clearly in Fig. 36, showing the rate of autoxidation for 14 species. The corre-

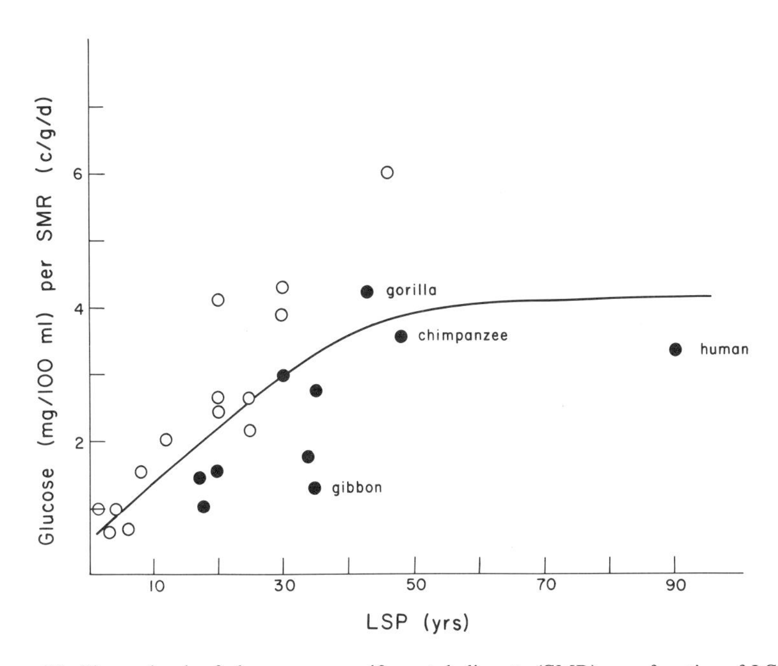

Figure 34. Plasma levels of glucose per specific metabolic rate (SMR) as a function of LSP in different mammalian species. Primate (●); nonprimate mammals (○). Data taken from Tables XXII and XXXIII.

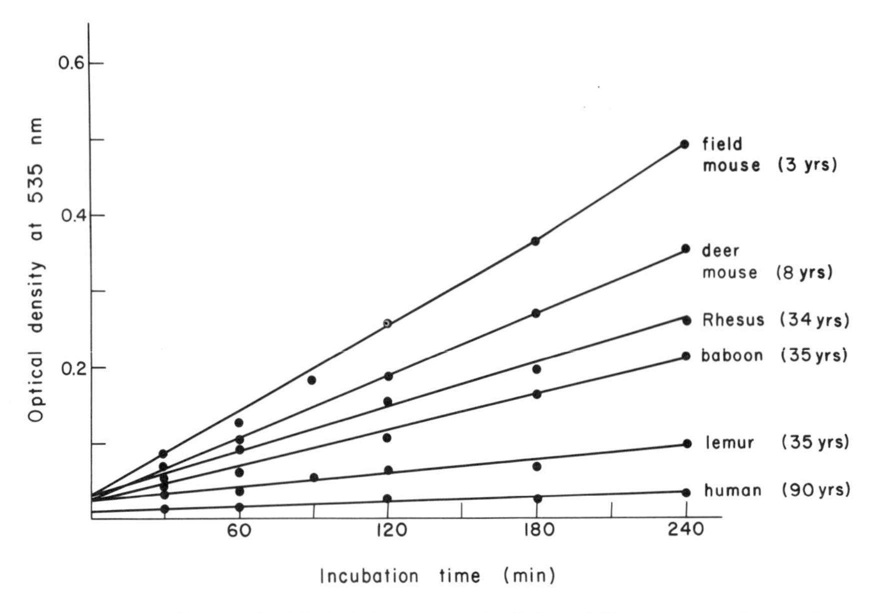

Figure 35. Autoxidation of whole brain homogenate in air from different mammalian species as a function of lifespan potential (LSP). Data taken in part from Cutler (in press).

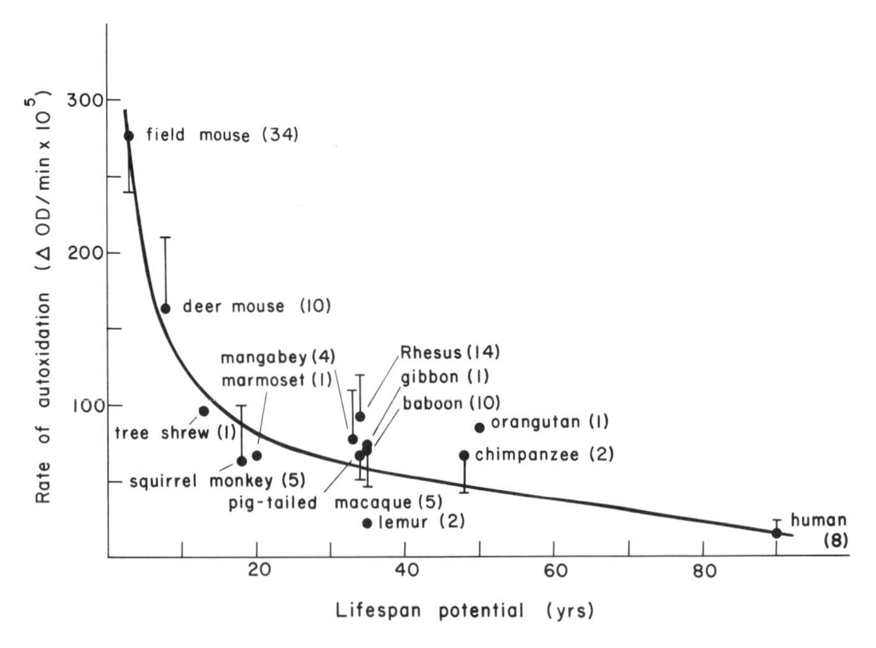

Figure 36. Rate of autoxidation of whole brain homogenate in air from different mammalian species as a function of lifespan potential (LSP). Numbers in parentheses represent number of individuals used in each determination. Data taken in part from Cutler (in press).

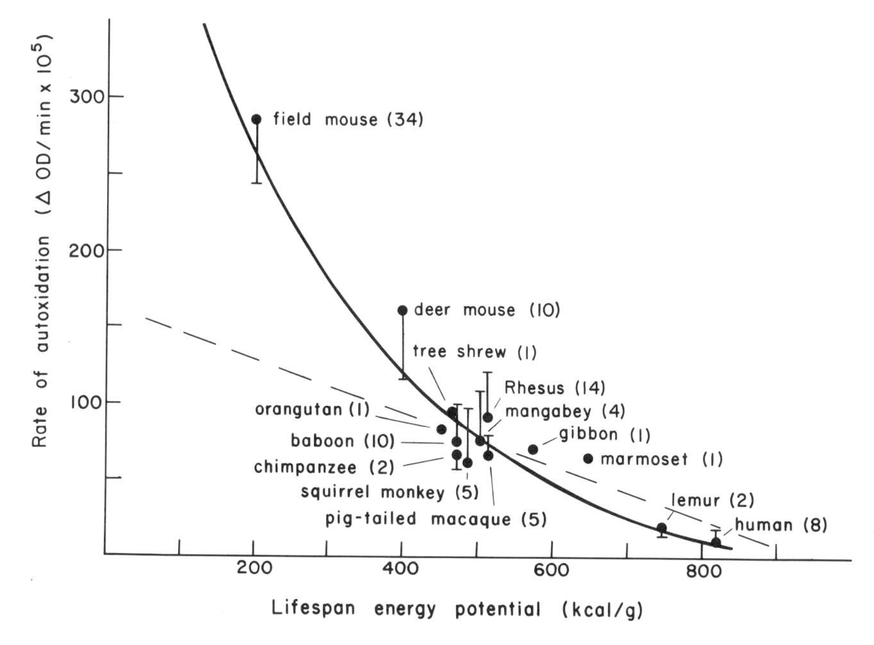

Figure 37. Rate of autoxidation of whole brain homogenate in air from different mammalian species as a function of lifespan energy potential (LEP). Numbers in parentheses represent number of individuals used in each determination. Data taken in part from Cutler (in press).

lation is significant, but on plotting these same data against LEP, as shown in Fig. 37, the correlation is even more significant.

Other experiments have shown that, if the autoxidation is taken to completion for the primate species, a common plateau value is eventually reached. Thus, the amount of peroxidizable material in these tissue homogenates appears to be similar. Other control experiments have also been done, indicating that the rate of autoxidation is dependent on the antioxidant levels in the homogenate (Bieri and Anderson, 1960; Barber and Bernheim, 1967). Thus, addition of α-tocopherol, for instance, is effective in decreasing the autoxidation rate.

The rate of autoxidation therefore seems to be reflecting the net tissue levels of antioxidants, and if this is confirmed by further experiments of this nature, then the hypothesis that LEP value is proportional to the net tissue level of antioxidants is supported and, most importantly, we will then have good evidence that antioxidants are potentially important determinants of human longevity.

Similar results to these autoxidation experiments can be found in the literature if LSP values are assigned to the animals used in the experiments. For example, serum taken from longer-lived species has been found to have a greater inhibition to the autoxidation of a given tissue fraction (Table XXXIV). Also, when erythrocytes were exposed to H_2O_2 vapor, the rate of autoxidation and the plateau value reactions are found to correlate with LSP (Fig. 38). These experiments suggest that the membrane of the red blood cells from longer-lived species may be less susceptible to peroxidation, independent of their antioxidant defense system. More work along this line is also clearly needed.

There is evidence that species do differ in the composition of their membranes, which could affect their innate susceptibility to lipid peroxidation. For example, as shown in Table XXXV, humans have lower level of 16:0, 18:0, and 20:0 folic acids in myelin, but a higher level of 22:4 and 22:6, as compared to shorter-lived species. In this example, one would predict that myelin from longer-lived species would be more susceptible to lipid peroxidation. Clearly, total lipid composition of brain and other tissues needs to be measured as a function of LSP and LEP to determine if compensation does play a role in determining LSP.

The amount of lipid peroxidation in a tissue *in vivo* reflects both the level of endogenous protection a tissue has against the accumulation of such peroxides and the potential tissue damage that is present (Tappel, 1980; Yagi, 1982). Measurement of the level of peroxides in tissues would therefore represent another means of determining if longer-lived species have lower levels of toxic by-products of metabolism throughout most of their lifespan. Lipid peroxide levels in plasma of different mammalian species, are shown in Fig. 39 (determined by the TBA test—not allowed to autoxidize). These data indicate that shorter-lived species such as the mouse have high plasma levels of lipid

Table XXXIV. Serum Lipoxidase-Inhibiting Activity of Serum of Different Species[a]

		Percent inhibition activity		
Species	LSP (yr)	O[b]	1.2[b]	2.0[b]
Hamster	3	0	5	10
Rat	4	0	10	20
Rabbit	12	0	20	35
Human	90	0	70	100

[a]Nonlifespan data taken from Placer and Slabochova (1961).
[b]Serum concentration measured as weight/volume. Inhibition activity of serum (as compared to human) of the rate of autoxidation of rat liver homogenate using the 2-thiobarbituric assay with different serums being added to the homogenate.

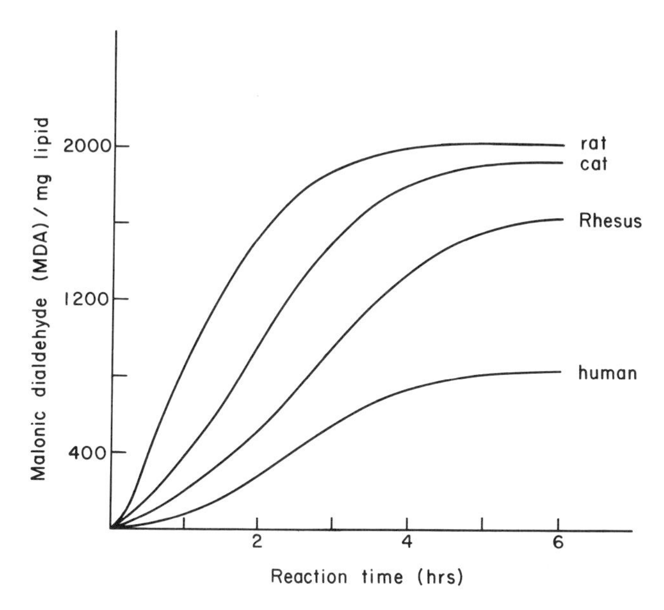

Figure 38. Lipid peroxidation of erythrocytes exposed to hydrogen peroxide as a function of lifespan potential (LSP) of mammalian species. LSPs of rat, cat, Rhesus, and human are 4, 20, 34, and 90 years, respectively. Taken in part from Kurian and Iyer (1976).

Table XXXV. Acyl Group Composition of Ethanolamine Phosphoglycerides of Myelin Isolated from Different Species of the Mammalian Brain (% wt.)[a]

Fatty acids	Human (90)	Rhesus (35)	Ox (30)	Rat (4)	Mouse (3)
16:0	2.9	3.4	4.2	4.4	7.1
18:0	6.0	7.3	7.3	9.0	12.0
18:1	41.3	35.8	46.2	43.8	33.0
20:1	8.1	10.7	13.2	18.9	14.5
20:3	—	2.8	0.8	—	1.0
20:4	8.7	11.3	7.5	7.7	12.9
22:4	21.6	17.8	12.4	4.8	9.9
22:6	5.8	3.1	2.7	3.7	6.9
24:4	1.3	3.9	—	—	—

[a]Values in parentheses are LSP (yr). Nonlifespan data taken from Sun and Sun (1982).

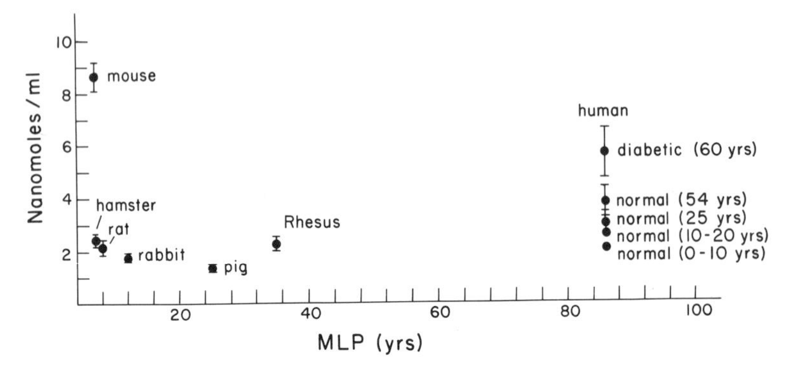

Figure 39. Plasma levels of lipid peroxides as a function of lifespan potential (LSP) in mammalian species. Taken in part from Suematsu *et al.* (1977), Sato *et al.* (1979), Nakakimura *et al.* (1980, 1981).

peroxides, but that the human is not unusually low in relationship to other medium lifespan species. Nevertheless, other data suggests that the human appears to have the highest level of antioxidant protection against lipid peroxidation in serum in comparison to other species (Vidlakova *et al.*, 1972). It is interesting to note that human diabetics generally have higher tissue levels of lipid peroxides than normal nondiabetic persons. Also, it should be cautioned that the method of assay here is the TBA test, which is not necessarily specific for lipid peroxidation and may be measuring some peroxides of arachidonic acid-related compounds produced during the assay procedure (Hayaishi and Shimizu, 1982).

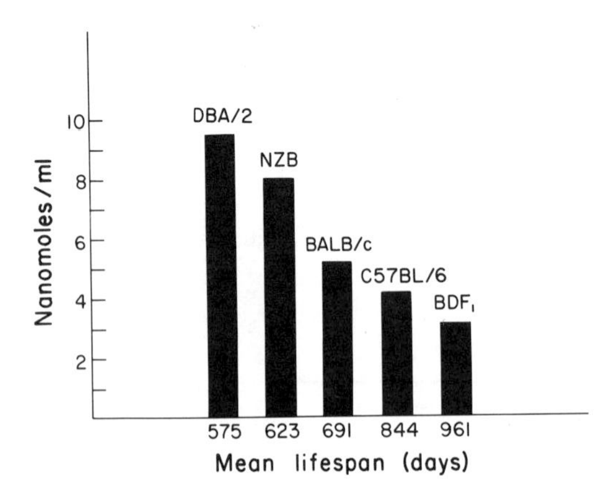

Figure 40. Plasma levels of lipid peroxides as a function of lifespan potential (LSP) in different mouse strains. All animals were 5 weeks of age. Data taken in part from Myers (1978), Nakakimura *et al.* (1978).

Measurement of lipid peroxides in the serum of a number of inbred mouse strains having different LSPs, using the TBA test, shows a perfect rank order of serum lipid peroxide levels with LSP, as shown in Fig. 40. These data suggest that a basis of strain differences in longevity may be related to a deficiency in their antioxidant protection system. This concept is supported by data indicating that the shorter-lived mouse strains are the most susceptible to lifespan extension by feeding them antioxidants or by putting them on a calorie-restricted diet (Harman, 1982).

On examining the serum lipid peroxide levels using the TBA assay of different species, mainly primates, we found that in general, longer-lived species have lower lipid peroxide levels (Fig. 41). Most interesting, however, was the remarkably good correlation found between serum lipid peroxides and LEP

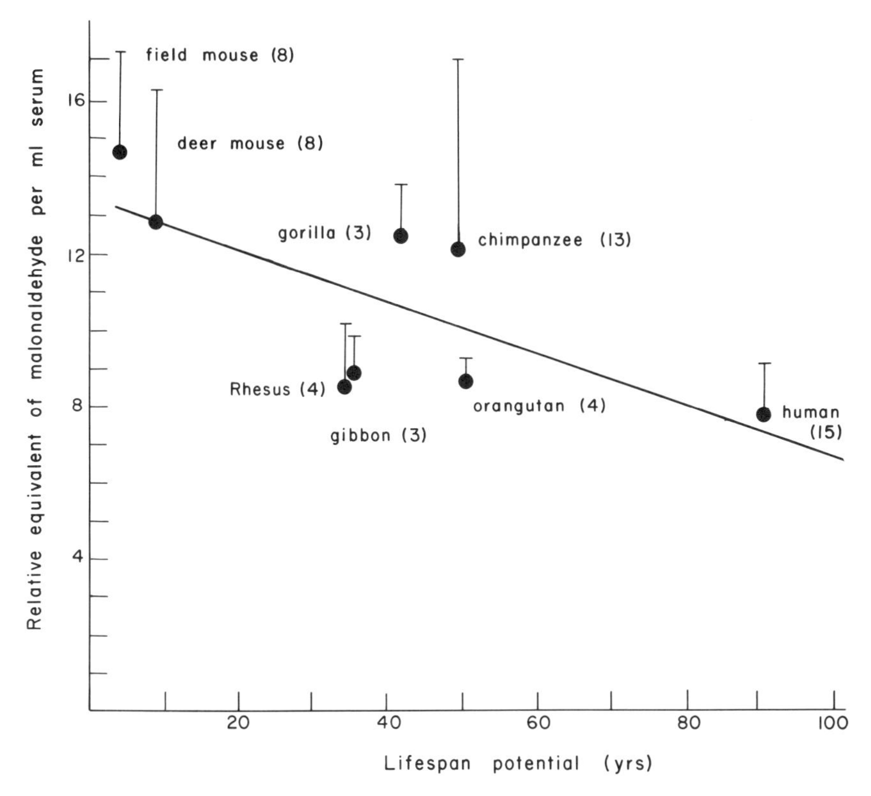

Figure 41. Serum lipid peroxide levels in mammalian species as a function of lifespan potential (LSP). TBA assay used. Correlation coefficient $r = -0.70$. Data taken in part from Cutler (in press).

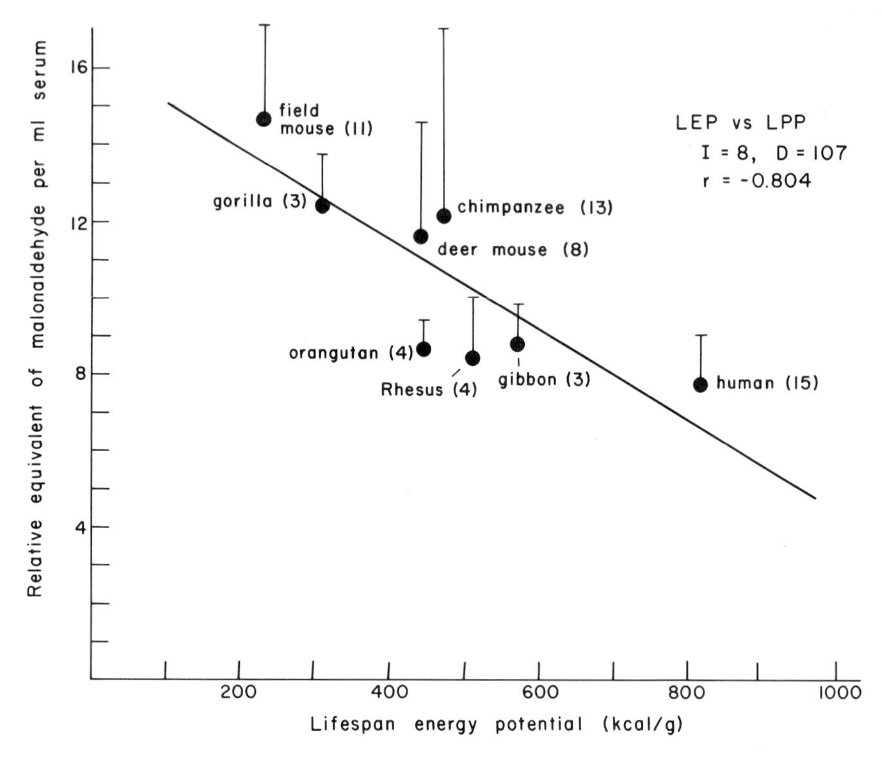

Figure 42. Serum lipid peroxide level in mammalian species as a function of lifespan energy potential (LEP). Same data as in Fig. 40. Correlation coefficient $r = -0.80$. Data taken in part from Cutler (in press).

values of the different species (Fig. 42). Nevertheless, a close examination of the data indicates that human lipid peroxide levels are not significantly lower than orangutan or chimpanzee and thus may not play an important role in determining human longevity.

In addition to the DNA repair, detoxification and antioxidants that may be potential longevity determinants, a few others will be mentioned briefly. These are as follows:

1. Cholesterol: Cholesterol is well known to be an important risk factor for arteriosclerosis. Although a positive correlation exists in the plasma level of cholesterol with the onset frequency of this disease, this does not prove that it is a causative agent of this disease. Instead, the correlation may be due to an adaptive-protective response, such that the damage to the arterial walls would be even more severe if cholesterol were not deposited at the organism's lesion

site. Because cholesterol is a well-known membrane-stabilizing agent and is not readily oxidized, it may help to dilute out highly unsaturated fatty reagents in membranes and reduce the rate of propagation of the lipid peroxidation chain reaction (Nagatsuka and Nakazuwa, 1982). It is also known that, although cholesterol is not a good antioxidant, it does increase the antioxidant efficiency of α-tocopherol (Fukuzawa *et al.*, 1981).

With this as a basis, it was of interest to see if average serum levels of cholesterol of different species correlated with LSP and LEP values. Table XXXVI summarizes such data for primate and nonprimate species. Human levels of cholesterol do not appear higher than other species, although cholesterol level per SMR does show a significant correlation, particularly for the nonprimate mammalian species (Fig. 43). However, data from another paper, where free and ester derivatives of cholesterol were analyzed, are shown in Table XXXVII, indicating that the human has an unusually high fraction of the ester derivative. Calculation of the total cholesterol per SMR vs. LSP from these data results in a more impressive positive correlation.

Table XXXVI. Plasma Cholesterol

Species	LSP (yr)	SMR (cal/g per d)	LEP (kcal/g)	Plasma cholesterol[a] (mg/100 ml)	SMR
1 Human	90	24.8	815	195 ± 48.4	7.86
2 Chimpanzee	48	26.8	469	268 ± 31.2	9.44
3 Orangutan	50	24.5	447	214 ± 9.19	8.73
4 Gorilla	43	19.7	309	333 ± 52.2	16.9
5 Baboon	35	30.9	394	100 ± 15.5	3.23
6 Macaque	34	41.3	512	120 ± 8.42	2.51
				$n = 5$	
7 Langur	30	35.5	388	173 ± 34.2	4.87
8 Grivet	25	43.4	394	222 ± 13.5	5.11
9 Mangabey	25	—	—	154 ± 32.3	—
10 Squirrel monkey	18	73.9	485	177 ± 31.1	2.39
11 Pig	30	20	219	117 ± 15.4	5.85
12 Cat	25	50.1	457	120 ± 32.8	2.39
				$n = 14$	
13 Deer	20	24.6	179	55 ± 8.7	2.23
				$n = 5$	
14 Dog	20	35	255	183 ± 30.2	5.22
				$n = 7$	
15 Rabbit	12	58.7	257	27 ± 27.2	0.459
16 Guinea pig	8	69.8	204	44	0.630
				$n = 1$	
17 Rat	4	104	152	52 ± 3.25	0.500

[a]Nonlifespan data from Benirschke *et al.* (1978).

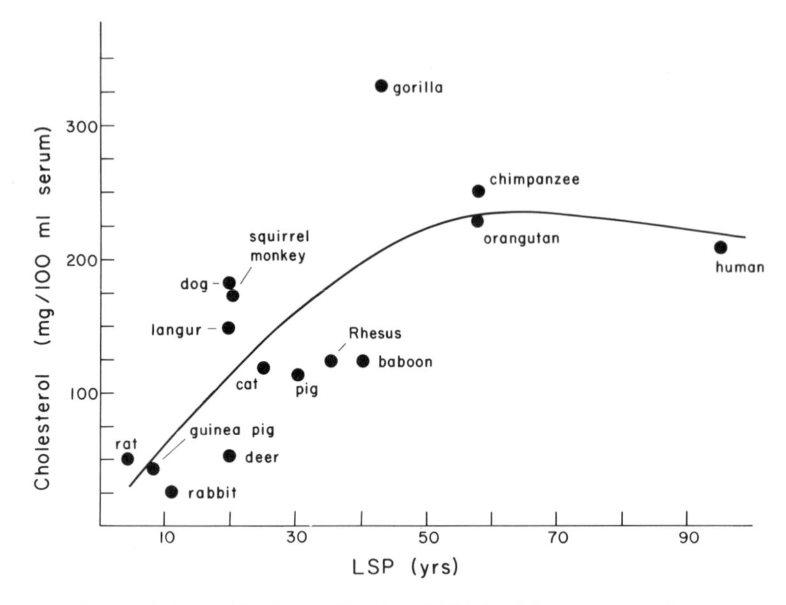

Figure 43. Serum cholesterol levels as a function of LSP for different mammalian species. Data taken from Table XXXVI.

Thus, like urate, high plasma levels of cholesterol may indeed lead to disease. On the other hand, drugs which lower urate or cholesterol to unusually low values may remove from the serum an important protective agent against membrane damage. Support for this concept comes from data suggesting that people with unusually low cholesterol levels have high levels of colon cancer (Kark *et al.*, 1981b).

Table XXXVII. Serum Cholesterol Levels of Different Species[a]

Species	LSP (yr)	SMR (cal/g per d)	LEP (kcal/g)	Free[b]	Ester[b]	Total[b]	Ester/Total	Total/SMR
Human	90	24.8	815	60	166	226	73.5	9.13
Pig	30	20.0	219	14.7	55.4	70.1	79.0	3.50
Dog	20	35	255	31.1	98.3	129.4	76.0	3.69
Rabbit	12	58.7	257	11.8	27.4	39.2	69.9	0.667
Guinea pig	8.0	69.8	204	10.8	32.3	43.1	74.9	0.617
Rat	4.0	104	152	20.6	41.5	70.1	69.9	0.674

[a]Nonlifespan data taken from Swell *et al.* (1960).
[b]Serum cholesterol, mg. Data represents mean of 5 to 6 animals.

2. Tissue renewal: An obvious potential determinant of longevity is tissue renewal, where one might guess that higher levels of tissue maintenance would be provided by high removal rates to prevent the accumulation of abnormal or damaged cells. On reviewing the literature on rate of cell turnover in different tissues as a function of LSP (Altman and Dittmer, 1972), it is found that actually the reverse occurs; that is, longer-lived species (or probably larger species) have slower rates of cell turnover. It therefore appears that rate of cell turnover in different tissues may relate inversely to SMR.

These data do not necessarily argue against turnover being an important protective mechanism. It may be that longer-lived and/or larger animals do not require such rapid cell turnover because of other superior protective processes. Also, it could be aruged that cell division is a dangerous act to a cell in terms of its ability to maintain its proper state of differentiation. The longer it can stretch out the time interval between division, the higher the effective stability of the differentiated state would be in terms of time of maintenance.

3. Protein turnover: Similarly to cell turnover, it could be argued that protein turnover might be beneficial to a cell by reducing the rate of accumulation of altered defective proteins. Indeed, the mechanism responsible for the accumulation of abnormal proteins was postulated to be due to the well-known slowing down of protein synthesis and degradation that occurs with increasing age. However, it was found that rate of protein turnover of plasma (Table XXXVIII) or for the whole body (Table XXXIX) is an inverse function of body size and not LSP. Thus, net rate of protein turnover does not appear to play an important role as a protective process in determining species longevity. Instead, the decrease in the rate of protein synthesis with body size represents a beneficial mechanism by reducing the metabolic energy necessary to sustain a high synthesis rate. In this regard, the age-dependent decrease in RNA and protein synthesis may be adaptive as well.

Table XXXVIII. Serum Protein Turnover as Function of LSP[a]

Species	LSP (yr)	Albumin half-life	γ-Globulin half-life
Human	90	15	13.1
Rhesus	35	—	6.6
Cow	30	20.7	21.2
Dog	20	8.2	6.6
Rabbit	12	5.7	8.0
Guinea pig	8	—	4.6
Mouse	3	1.2	1.9

[a]Nonlifespan data from Spector (1974).

Table XXXIX. Rate of Total Protein Turnover as a Function of LSP[a]

Species	LSP (yr)	Body wt. (g)	Protein turnover: g/kg per day	g/kg$^{0.75}$ per day: A	g/kg$^{0.75}$ per day: B
Mouse	3	0.04	43.4	19.5	
Rat	4	0.51	20.5	17.4	
Rabbit	6	3.6	18.0	24.8	15.0
Dog	15	10.2	12.1	21.5	
Sheep	20	67.0	5.3	15.9	
Cow	30	628.0	3.7	18.7	16.1
Human	95	77.0	5.7	16.7	12.5

[a]Nonlifespan data from Waterlow and Jackson (1981).

8. GENETIC REGULATION OF LONGEVITY DETERMINANTS

8.1. Compensative Nature of Tissue Antioxidant Levels

The major conclusion of this comparison of tissue levels of antioxidants with LSP and LEP is that a number of antioxidants do appear important as longevity determinants. However, an argument used against such a conclusion is the wide variations in the tissue concentrations of an antioxidant such as urate or β-carotene between individuals of the same species. Yet, in spite of this wide variation, no similar range is evident in aging rates. Also, some humans are known to have unusually low serum levels of urate or β-carotene with no obvious symptoms such as accelerated aging.

An example of the wide distribution of the antioxidants, vitamins E, A, and C and carotene, in the serum of humans are shown in Fig. 44. This represents the normal range, which can vary tenfold without apparent ill effects. Thus, the question naturally arises as to how such wide variation of antioxidants can exist and still not affect the general health status or aging if indeed these antioxidants are important as longevity determinants.

It was previously pointed out that there is very little evidence for a large heterogeneity in aging rates among different human individuals or in other species. It appears that the aging rate of each individual is under a tight genetic regulatory mechanism such that lifespan differences due to aging do not vary between individuals any more, and perhaps even less, than one sees in the normal body weight or height.

Another argument against antioxidants (particularly those that are part of the diet) being longevity determinants is that antioxidants have not been able to significantly extend LSP or even mean lifespan of mice or rats when added to their diet in high concentrations (Harman, 1981, 1982). The little lifespan extension effect that is seen is in the mean lifespan, and even here this effect is usually only found when unusually short-lived mouse strains are used. If antioxidants are really playing an important role in aging, then lifespan extension of both mean and maximum, should be at least 50%–100%, not the typical 5%–10%.

One of the best examples of an experiment to determine the effects of vitamin E on longevity is shown in Fig. 45, for mice. Here, the rate of accumulation of lipofuscin pigment was shown to decrease in the animals fed the vitamin E-enriched diet. This important control argues that the vitamin E was getting into the cells. However, in spite of the significant reduction of the rate of accumulation of lipofuscin pigment, neither mean nor maximum lifespan was affected as compared to the controls. In addition to lifespan, vitamin E-enriched diet was found to have no effect on the aging rate of connective tissue (Blackett and Hall, 1980).

Studies on the mortality of individuals who have taken higher than average levels of vitamins A and C for many years and, in general, have undertaken what might be called a healthful lifestyle, did not show any significant increase in mean or maximum lifespan potential (Enstrom and Pauling, 1982).

Finally, a study was made on how serum concentrations of vitamins A, C, E, and carotene were affected by taking one of these vitamins as a supplement. The general finding was that, under a wide range of dosage of one vitamin, the

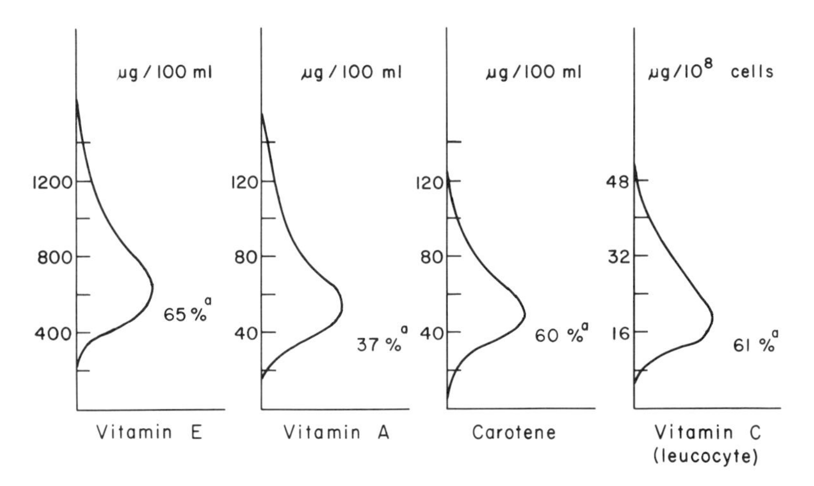

Figure 44. Range of vitamin antioxidants in human plasma. Data taken in part from Kelleher and Losowsky (1978).

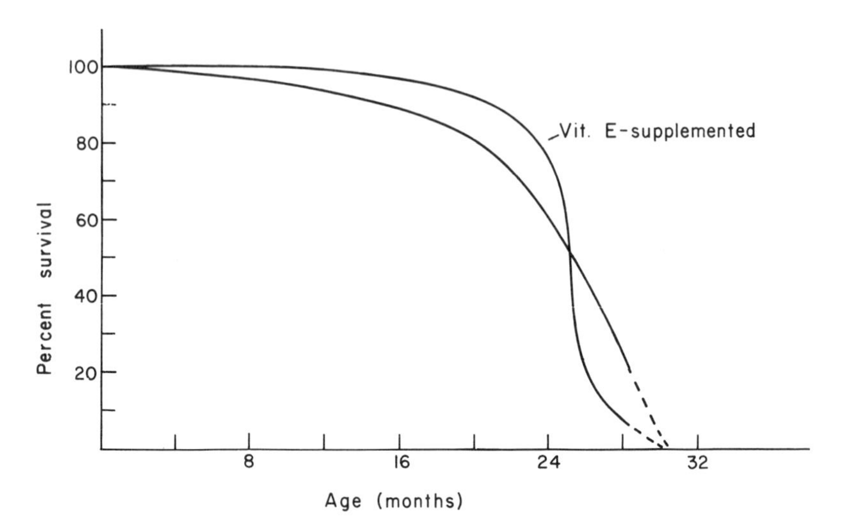

Figure 45. Effect of vitamin E-supplemented diet on lifespan of mice. Data derived from Blackett and Hall (1981).

serum concentrations of the others were affected very little (Urbach *et al.*, 1952). Thus, there appears to be a compensatory or regulatory effect maintaining a constant antioxidant level of protection with these vitamins in the serum over a moderate dosage range.

All of these data argue against the importance of dietary intake of antioxidants as well as endogenously synthesized antioxidants in affecting longevity and general pathology. Thus, although there is ample evidence that extreme deficiencies of antioxidants such as vitamin E do result in symptoms similar to accelerated aging, there is no evidence that additional vitamin E supplementation of the diet above the normal range slows down the aging process.

These arguments against antioxidants being potential longevity determinants can be answered by the new concept that all longevity determinant processes including the antioxidants act in a compensatory manner to maintain a constant net level of protection and thus a constant LSP when a deficiency or excess of one occurs. Thus, over a wide range of dietary intakes of antioxidants or pro-oxidants, exposure to toxins, and of exercise, the net levels of protection the tissue receives from antioxidants is maintained by compensatory antioxidant processes. In this way, like so many other critical biological processes in an organism that are known to be maintained by compensatory processes, the maintenance of LSP is also under a compensation process that is genetically determined. Since it is the net level of antioxidant protection that determines LSP, then the ability to maintain this level through compensatory processes forms a critical aspect of the genetic regulatory basis of species' longevity.

Some of the major experimental evidence supporting this compensational model follow:

1. Vitamin E: When rats are placed on a long-term vitamin E-deficient diet, glutathione peroxidase, glutathione reductase, superoxide dismutase, and catalase levels are increased, according to what tissues is being examined (Chow *et al.*, 1973; Chow, 1977, 1979; Gross, 1979; Chen *et al.*, 1980; Lee *et al.*, 1981; Sklan *et al.*, 1981). Typical data are shown in Table XL.

One of the difficulties of such studies is that compensatory processes are only likely to work over a moderate dietary range of intake. Beyond a certain point, such as extreme deficiencies, pathological consequences begin to be observed. For example, the level of glutathione peroxidase in liver decreases with decreased levels of Vitamin E if the level of vitamin E is less than 25 IU, but above the level of vitamin E, an increase in glutathione production occurs with a decrease in vitamin E (Chen *et al.*, 1980).

Vitamin E deficiency also increases the level of glutathione S-transferase (Reddy *et al.*, 1981) and xanthine oxidase, which is likely but not yet proven to result in an increase of serum and tissue levels of urate (Dinning, 1953; Catignani and Dinning, 1971; Catignani *et al.*, 1974; Folkers, 1974; Masugi and Nakamura, 1976). Typical data showing the increase of xanthine oxidase in vitamin deficient animals are shown in Tables XLI and XLII.

Rats on a vitamin E-deficient diet have higher serum cholesterol levels, and on a vitamin E-supplemented diet have lower levels (Chen *et al.*, 1972). Human subjects on a vitamin E-supplemented diet tend to have lower levels of vitamin A (Garrett-Laster *et al.*, 1981) and also reduced calcium and phosphorus in blood plasma and bone (Murphy *et al.*, 1981). Unfortunately, it is not known what the long-term effects might be of high levels of vitamin E

Table XL. Effect of Long-Term α-Tocopherol Deficiency in the Rat on Other Antioxidants and Related Enzymes[a]

	Fraction of control						
Tissue	α-Tocopherol	Glutathione peroxidase	Glutathione reductase	G-6-P-D	SOD	Catalase	Ascorbate
RBC	0.72	0.80	0.82	1.38	1.40	1.10	—
Plasma	0.045	0.70	NS	—	—	—	0.61
Heart	—	NS	1.26	1.78	1.42	1.37	—
Liver	0.77	0.69	NS	1.63	1.41	NS	0.77
Kidney	—	0.89	NS	NS	NS	NS	—
Testes	—	1.35	1.33	1.26	—	1.52	—
Muscles	—	1.21	1.31	3.64	1.75	NS	—

[a]Weanling Sprague Dawley rats on 100 IU vitamin E per kg diet for 12 months. Adapted from Chen *et al.* (1980).

Table XLI. Possible Compensation for Vitamin E-Deficient Rats by Increased Urate Synthesis[a]

	Xanthine oxidase[b]	Xanthine dehydrogenase[b]
Control	0.5 ± 0.5	28.3 ± 9.5
Vitamin E-deficient	21.3 ± 5.7	168.0 ± 58.1

[a]Adapted from Folkers (1974).
[b]Enzyme activities are Δ OD per ml. Coenzyme Q (C_0O_{10} or $H_6C_0O_4$) can bring back normal levels of the xanthine oxidase and xanthine dehydrogenase enzyme activities.

dietary supplementation on other tissue antioxidants. These experiments are presently being undertaken in our laboratory.

Evidence that compensatory processes can operate to reduce the effects of a vitamin E-deficient diet are that cystein, methionine, quinones (C_0Q_{10}), synthetic antioxidants, vitamin C, and selenium can prevent the pathological effects of long-term vitamin E-deficient diets (Folkers, 1974).

2. Selenium/glutathione peroxidase/glutathione: Selenium is required as a cofactor for glutathione peroxidase, and thus it is expected that, in selenium-deficient animals, this enzyme is decreased in activity. Animals placed on a selenium-deficient diet show an increase of catalase, superoxide dismutase (Combs, 1981), glutathione (Hill and Burk, 1982), glutathione peroxidase (Gross, 1979), and glutathione S-transferase (Stone and Dratz, 1980).

Lead and iron, which are known to be toxic and to increase the oxidative stress of cells, have been shown to increase levels of glutathione peroxidase, catalase, and vitamin E (Macdougall, 1972; Hsu, 1981).

It is also of interest that, by utilization of agents known to deplete cellular levels of glutathione, no increase of lipid peroxidation by the lower glutathione level was seen until a certain threshold was reached (Younes and Siegres, 1981). This is evidence that, over a limited range of glutathione depletion, compensation processes do operate to prevent lipid peroxidation.

Table XLII. Possible Compensation by Other Antioxidants in Vitamin E-Deficient Rats[a]

	Xanthine oxidase	SOD (units/g)	Gluthione peroxidase (Δ A_{340}/min per g)	Catalase (units/g)
Control:	0.55 ± 0.06	704 ± 31	34.6 ± 2.2	930 ± 55
Test:	0.84 ± 0.07	764 ± 11	32.6 ± 1.6	1010 ± 98
Change:	52.7% (+)	8.5% (+)	5.78% (−)	(8.6%)

[a]Values are for liver; deficiencies were for 16 weeks. Adapted from Masugi and Nakamura (1976).

3. Ascorbate and Detoxification Systems: Guinea pigs placed on an ascorbate-deficient diet show a decrease in cytochrome P-450 enzyme and an increae in ceruloplasmin in serum and copper in the liver (Omaye and Turnbull, 1980). On the other hand, animals placed on both vitamin E- and Se-deficient diets showed an increase of liver expoxide hydratase activity (Reddy *et al.*, 1982). Animals placed on under-nutrition conditions show a general lowering of the microsomal mixed functional oxidases (Kalamegham *et al.*, 1981). This response may be important to the lifespan extension effects of dietary restriction.

4. Uric acid: Individuals with hypouricemia or xanthinurea have extremely low plasma urate levels, frequently much less than 2 mg/100 ml, but no accelerated aging changes in these individuals have been observed (Wyngaarden and Kelley, 1976; Seegmiller, 1979). It would be important to determine tissue levels of antioxidants in these individuals with unusually low levels of urate. There are numerous reports showing an association of hypouricemia with malignant neoplasms and other diseases, but it is not clear what the cause and effect relationship might be (Weinstein *et al.*, 1965; Bennett *et al.*, 1972; Ramsdell and Kelley, 1973; Dwosh *et al.*, 1977; Mitnick and Beck, 1979). If urate were an important antioxidant, and compensation processes were operating, then these people would be expected to have higher than normal levels of other antioxidants.

On the other hand, hyperuricemia is associated with gout, but no direct connection has been made between high serum urate levels and the symptoms of this disease (Stetten, 1958). Also, hyperuricemia has been identified as a risk factor for coronary heart disease, but these data do not differentiate whether this association is causative or adaptive (Jacobs, 1976).

Plasma urate levels can be increased by feeding animals a diet with a high composition of nucleic acids such as yeast RNA (Wyngaarden and Kelley, 1976). There is some experimental evidence that lifespan (Gardner, 1963), resistance to radiation (Rounds, 1961), carcinogens (Dunning and Curtis, 1958), and viral infections (O'Dell *et al.*, 1958) are decreased when animals are fed diets which enhance plasma urate concentration, but these effects are small.

In addition to vitamin E deficiency increasing the levels of xanthine oxidase, it has been found that copper deficiency results in higher serum levels of cholesterol and urate (Klevay, 1980). In another study, copper-deficient diets resulted in an increase of glutathione peroxidase and superoxide dismutase (Russanov and Kassabova, 1982). Interestingly, plasma levels of DHEA are low or absent in patients with high plasma levels of urate or with gout, suggesting a compensation mechanism (Sparagana and Phillips, 1972).

Finally, it should be noted that urate is derived from the endogenous synthesis and dietary intake of purines (Wyngaarden and Kelley, 1976; Roch-

Ramel and Peters, 1978). However, plasma and tissue levels of urate do not appear to depend largely on dietary composition but rather on hereditary factors such as the rate of endogenous synthesis or degradation of urate and the extent plasma urate is returned to the plasma by the kidney. A similar situation exists for ascorbate, where in addition to the endogenous synthesis and dietary intake levels, the resultant plasma and tissue levels are determined by a number of endogenous factors such as relative levels of degradation, absorption, and excretion (Lewin, 1976; Chatterjee, 1978; Sato and Udenfriend, 1978). Thus, although plasma and tissue levels of urate and ascorbate are influenced by diet, species-specific levels exist that are largely independent of this variable. Plasma urate remains constant throughout lifespan, beginning with the young adult (Wyngaarden and Kelley, 1976), but plasma and tissue ascorbate levels frequently decrease dramatically with increasing age (Kirk, 1962).

5. Superoxide dismutase: As noted before, selenium, vitamin E, and copper deficiency result in increased levels of superoxide dismutase activity. Also, diets deficient in vitamin A increase levels of both superoxide dismutase and glutathione peroxidase (Sklan *et al.*, 1981).

Rats fed corn oil for two weeks show an increase of glutathione peroxidase, SOD, and catalase (Iritani and Ikeda, 1982). Noradrenaline has been shown to increase the levels of SOD in brown adipose tissue (Petrovic *et al.*, 1981) as well as thyroid hormone (Petrovic *et al.*, 1982), so this suggests an oxygen stress produced by this hormone. SOD has been shown to be induced by a number of free radical-enhancing agents as well as by exercise (Reznick *et al.*, 1982).

6. Cholesterol: Drugs that decrease plasma level of cholesterol such as clofibrite also increase hepatic peroxisomes and decrease levels of SOD and glutathione peroxidase (Ciriola *et al.*, 1982). The possible regulatory involvement of these factors with one another is apparent from this study.

High levels of cholesterol in the diet act to increase liver catalase (Fuentes, 1978) and vitamin E (Rubenstein *et al.*, 1968). Consistent with this observation, is that low serum levels of cholesterol are associated in human with low serum levels of vitamin A, and these patients have higher probabilities of colon cancer (Kirk *et al.*, 1981).

7. Compensation in Disease: Diabetic patients have been found to have unusually high levels of glutathione peroxidase but lower levels of SOD in red blood cells and other tissues (Matkovics *et al.*, 1982b). Children with cystic fibrosis have been found to have remarkably high levels of SOD and catalase in their red blood cells (Matkovics *et al.*, 1982a). β-thalassemia patients have unusually high levels of catalase and glutathione peroxidase in their red blood cells (Gerli *et al.*, 1982). And in patients with Down's syndrome, where Cu–Zn SOD is higher than normal due to a gene dosage effect, glutathione per-

oxidase activity was found to be increased in red blood cells and Mn-SOD to be decreased in the platelets (Kedziora *et al.,* 1982; Sinet, 1982).

When taken together, these results suggest that, when tissues are placed under an oxidative stress, an increased level of antioxidants is induced. This increase of oxidative stress can be accomplished by directly increasing the production of active oxygen species (such as by drugs or high oxygen tension) or by the reduction of one of the enzymes or substances that make up the tissues' net antioxidant defense system.

These data provide convincing evidence that antioxidant compensation processes do occur, although it is not known how extensive such compensational processes are and how effective they are. Nevertheless, these data do provide an explanation of why the aging rates of humans and other species can be maintained in spite of a wide range of intake of dietary antioxidants. Thus, the failure of past experiments to enhance lifespan of experimental animals by feeding them antioxidants can be explained on the basis of the antioxidant compensatory process preventing a significant net gain of antioxidant protection in the tissues. If this conclusion is correct, then the ability of antioxidants to enhance lifespan has not really been tested yet until a means can be devised to block the antioxidant compensatory process such that a substantial net increase in antioxidant protection can be realized.

The existence of an antioxidant compensation system also has important implications in explaining the past difficulties in defining the minimum daily requirement for the antioxidant vitamins. For example, people's vitamin E requirements are known to depend on their intake of many other antioxidants as well as their overall body oxidation stress. Such factors are (a) unsaturated fatty acids, (b) selenium, (c) copper, (d) iron, (e) zinc, (f) carotenoids, (g) ascorbate, (h) nucleic acids, and (i) exercise. In addition to these, a person's innate ability to compensate against a deficiency of vitamin E would also play an important role. Thus, it is not surprising that pathologies for vitamin E deficiency are rare in humans and that there is controversy in determining what a minimum daily requirement for the antioxidant vitamins might be (Draper, 1980).

An antioxidant is not expected to be able to compensate equally well for any of the other antioxidants. Those that best compensate for one another are predicted to be those with overlapping protective functions. Thus, the antioxidants expected to work best in a compensatory manner are predicted to be those having a common parallel and/or serial function. This is illustrated in Fig. 46.

Finally, it is important to discuss what the consequences might be if a deficiency in an antioxidant exists and compensation for this antioxidant was not possible. It is first necessary to realize that there are a number of different

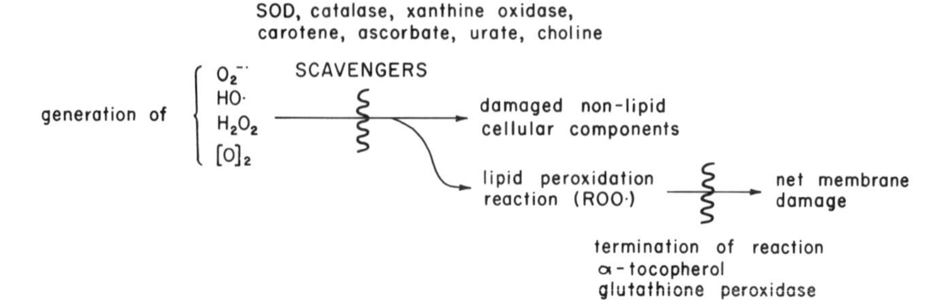

Figure 46. Compensation model of antioxidant longevity determinants. Antioxidants having overlapping properties and operating in serial or parallel defense as scavengers or chain terminators will compensate one another. Data taken in part from Cutler (in press).

antioxidants in tissues; but let us say that for illustrative purposes, there are ten antioxidants most important in determining human longevity. If each of these contributed an equal fraction of defense, then the complete loss of one could reduce the human LEP value from 814 to 734 kcal/g and lifespan from 90 to 81 years. It is unlikely that a loss of 9 years would be evident in anyone that has such a deficiency.

8.2. A Model for the Regulation of the Oxidizing–Reductive Environment of a Cell

We now have some evidence supporting the role of antioxidants as being importantly involved as longevity determinants. Some antioxidants have been identified as having an important role in determining the unusual longevity of humans, while others (although still important in determining longevity) do not have this distinction. Many of the antioxidants are likely to be involved in a compensatory/homeostatic-like regulatory system designed to maintain a species' characteristic levels of reactive oxygen species in the cell over a wide range of endogenous antioxidant synthesis rates, dietary intake, and metabolic products of active oxygen species. This mean tissue level of active oxygen species, of which each differentiated cell type probably has a characteristic level, is proposed to govern the rate of dysdifferentiation of the cell and thus the aging rate of the organism.

From this, it follows that inadequacies of endogenous synthesis of antioxidants or other components of the antioxidant compensatory system could be involved in producing specific types of disease processes, some of which may result in accelerated aging syndromes. Such may be the case for a number of inbred mice strains (see Fig. 40) as well as a number of inherited human diseases such as ataxia telangiectasia, Bloom's syndrome, and collagen diseases (Emeritt and Cerutti, 1981, 1982).

Species having different LSPs would be predicted to have unique and characteristic set points to maintain a given mean level of active oxygen species within their cells about which the compensational regulation of antioxidant levels would operate. The level of this set point (level of active oxygen species) would be inversely related to a species' LSP. The lower the set point, the lower would be the mean value of active oxygen species in a cell and the greater the LSP. This set point would be maintained by governing the species' net antioxidant protection per SMR, which is by definition directly proportional to LSP.

What are the biochemical mechanisms for maintaining this set point, or, more specifically, what are the mechanisms maintaining SOD, catalase, α-tocopherol, urate, or β-carotene so high in humans as compared to the shorter-lived species? And more generally, what are the genetic regulatory mechanisms that govern and maintain the proper levels of all tissue antioxidants? An answer to this question is fundamental to an understanding of the biological mechanisms determining human longevity. Also, an understanding of this regulatory process may lead to means of intervention and the development of therapeutic methods to correct a number of human hereditary diseases as well as to further prolong the healthy years of human lifespan.

A model is presented in this section that may be involved in the regulation of tissue levels of antioxidants. It is based on the properties of three substances that are unusually sensitive to the effects of active oxygen species. On this capacity they may prove to be effective detectors of endogenous levels of active oxygen species. These three substances are (a) arachidonic acid, an essential unsaturated fatty acid (20:4) that is extremely sensitive to lipid peroxidation and is the only precursor to prostaglandin, prostacycline, thromboxane, and the leukotrienes; (b) prostaglandin cycloxygenase, the first enzyme in the conversion of arachidonic acid to the prostaglandin endoperoxide (PGG_2)—cycloxygenase is unusual because it is sensitive to activation by low levels of hydroperoxides and inactivation by high levels of hydroperoxides and number of active oxygen species; and (c) guanylate cyclase, which is the enzyme catalyzing the synthesis of cGMP and is very sensitive to activation by a wide number of different active oxygen species (Mittal and Murad, 1977; Murad *et al.*, 1978, 1979; Kuehl and Egan, 1980; Karmali, 1980; Gemsa *et al.*, 1982; Iida *et al.*, 1982; Pekoe *et al.*, 1982).

The guanylate cyclase–arachidonate–cycloxygenase model (GAC model) is shown diagrammed in Fig. 47. The basis of the model, as already noted, is the unusual sensitivity of three constituents to their oxidation–reduction environment and the ability of these three constituents, when altered by their oxidation–reduction environment, to affect changes in the intracellular level of the cGMP/cAMP ratio and thus the protective rates of the cell.

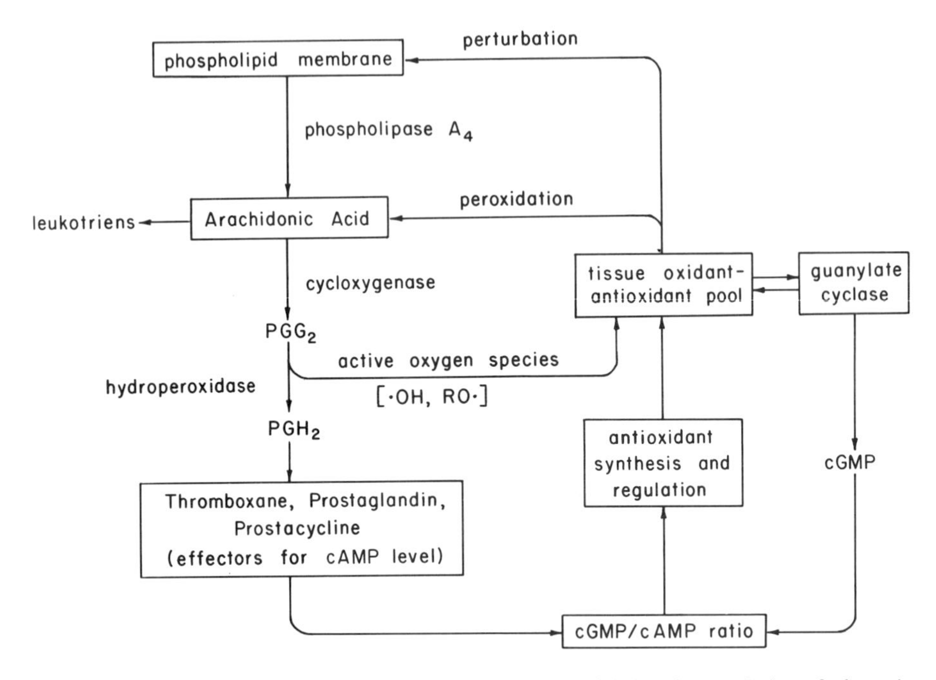

Figure 47. Guanylate cyclase-arachidonate-cycloxygenase model for the regulation of the oxidation–reduction potential of a cell. Data taken in part from Cutler (in press).

The amount of arachidonic acid produced and/or the activity of cycloxygenase determines the rate of PGH_2 synthesis, which then determines the rate of prostaglandin synthesis and in turn the cellular levels of cAMP. Active oxygen species will peroxidize arachidonic acid, resulting in a decrease in the available arachidonic acid for synthesis of PGH_2 and lower cellular levels of cAMP. Active oxygen species also act to inactivate cycloxygenase and lower cAMP. On the other hand, antioxidants will tend to lower the levels of the active oxygen species and thus increase the overall arachidonic acid availability and/or increase the activity of cycloxygenase, resulting in higher rate of PGH_2 synthesis and in turn a higher cellular level of cAMP. Important to this scheme is that, in the synthesis of PGG_2 from arachidonic acid, an active oxygen species is produced which adds to the cell's tissue oxidation–reduction load. Thus, a self-producing negative feedback process is present which acts to stabilize the overall prostaglandin synthesis pathway by being sensitive to the intercellular levels of active oxygen species.

In contrast to cycloxygenase, guanylate cyclase is activated by active oxygen species. An increase of guanylate cyclase activity results in an increase of cellular levels of cGMP. Thus, the overall result is a "push-pull effect," where if the intercellular oxidative environment increases above the correct set point of the cells, for example, then cGMP will increase and cAMP will decrease, resulting in a large net increase in the cGMP/cAMP ratio. Conversely, if antioxidant levels are in excess (a lower than normal level of active oxygen species), then this push-pull effect will result in a net decrease in the cGMP/cAMP ratio.

The rate of antioxidant synthesis is proposed to be determined by the intracellular levels of cGMP to cAMP. It is of course unlikely that the details of this model are correct, but the essential feature is the concept of identifying certain substances or enzymes that can act to detect endogenous levels of active oxygen species which can in turn govern gene activity.

One of the most interesting features of the GAC model is that it suggests methods of how to trick the regulatory system to react as if active oxygen species are too low or too high and thus be able to enhance or inhibit endogenous antioxidant synthesis by nontoxic means. For example, the inhibition of cyclooxygenase by indomethacine or aspirin or many of the other nonsteroid antiinflammatory agents is seen to mimic the condition of high intercellular levels of active oxygen species. Thus, the resultant decrease in prostaglandin and cAMP synthesis would increase the cGMP/cAMP ratio and be predicted to stimulate increased antioxidant synthesis. This leads to one of the key predictions of the model, which is that many antiinflammatory agents act not only by reducing the level of prostaglandin endoperoxides and related active oxygen species, but also by enhancing higher levels of tissue antioxidants.

Another important feature of the model is the consequences predicted by the endogenous addition of large amounts of antioxidants. In this case, the effect would be a decrease in the cGMP/cAMP, resulting in a decrease in endogenously synthesized antioxidants. Thus, as we have observed, the model can account for the fact that feeding of high levels of antioxidants in the diet would depress endogenous synthesis and therefore not result in a net gain of antioxidant protection.

Some of the experimental data supporting the GAC model are as follows:

1. The activity of guanylate cyclase is known to be increased by its interactions with free radicals and a number of mutagenic agents (Murad *et al.*, 1978, 1979). Cigarette smoke is found to activate guanylate cyclase and to result in an increase of guanosine 3′,5′-monophosphate in tissue, the active component of the smoke being nitric oxide (Arnold *et al.*, 1977). Estrogens and progesterone increase guanylate cyclase activity, and this may be related to their possible promoter effect for cancer (Vesely and Hill, 1980). In fact, there are substantial data linking an increase of cyclic GMP to cell injury and to the increase of detoxification of chemical carcinogens as well as natural toxins (DeRubertis and Craven, 1980).

Adenylate cyclase is activated in rat brain tissue by lipid peroxidation reactions (Baba *et al.*, 1981), so perhaps both adenylate and guanylate play a role in detecting cell injury. Also of interest is the observation that plasma levels of insulin and glucagon are correlated with hepatocyte levels of guanylate cyclase, suggesting some type of regulatory linkage (Earp, 1980). This may be important in view of the fact that the β cells of the islets of Langerhans appear to be unusually sensitive to the toxic effects of free radicals (Crouch *et al.*, 1981). The concept that changes in the oxidation–reduction state of a cell may be linked to guanylate cyclase activity in the cell, and thus cGMP levels, is supported by the reversible inactivation of guanylate cyclase by disulfides (Brandwein *et al.*, 1981).

It has been observed that, when the nematode *Caenorhabditis elegans* was exposed to cGMP in the growth media, a significant increase in longevity was observed (33 to 42 days) (Willett *et al.*, 1980). The levels of cGMP used appeared to be nontoxic and did not alter normal development. Possibly related to the effects of altering the intercellular levels of cGMP/cAMP ratio is the evidence that addition of cAMP to cell cultures or injecting it into mice enhances their radiation resistance by a yet undiscovered mechanism (Lehnert, 1975a,b, 1979a,b; Dubravsky *et al.*, 1978). Guanylate cyclase activity has also been found to increase with age in the skeletal muscle of the Rhesus monkey (Beatty *et al.*, 1981).

To determine if guanylate cyclase might play an important role in the detection and genetic regulation of LDPs in the cell, the activity of soluble guanylate cyclase was measured in the brain tissue of primate species. The results showed an inverse correlation of guanylate cyclase activity with LSP (Tolmasoff *et al.*, 1980; Cutler, 1980). This would be the expected result if the GAC model was correct, where the ratio of cGMP/cAMP would be proportional to the active oxygen set point.

2. Catecholamines have antioxidant properties and stimulate prostaglandin synthesis. This stimulation can be prevented by monamine oxidase inhibitors (Seregi *et al.*, 1983). Catecholamines are degraded by monamine oxidation, producing H_2O_2—which in turn activates cycloxygenase. These reactions are known to occur, for example, in brain tissue mitochondria. Much evidence suggests that a key factor in determining intercellular levels of antoxidants may be H_2O_2, and so the ability of H_2O_2 to modulate the activity of cycloxygenase may be important in this respect. The difficulty here, however, is that if H_2O_2 increases prostaglandin synthesis, and if guanylate cyclase were not affected, the ratio of cGMP/cAMP would decrease and less antioxidant would be synthesized. Perhaps the activation of guanylate cyclase by H_2O_2 occurs to an even greater extent to produce more cGMP, or that, as with other active species, high levels of H_2O_2 inactivate cycloxygenase.

3. Uric acid, as well as other antioxidants such as epinephrine, phenol, tryptophane, and hydroquinone stimulate prostaglandin synthesis (Ogino *et al.*,

1979; Deby *et al.*, 1981; Bourgain *et al.*, 1982). Because uric acid is known to be an antioxidant (Ames *et al.*, 1981), this result is consistent with urate protecting cycloxygenase from being inactivated. The result is that urate increases prostaglandin synthesis, resulting in a decreased cGMP/cAMP ratio and less antioxidant synthesis. Unusually high levels of uric acid, however, may lower the synthesis rate of the other antioxidants to the point that free radical pathology appears. This would be consistent with the enhancement of arterial thromboformation by uric acid (Bourgain *et al.*, 1982) and some of the symptoms associated with gout.

4. Prostaglandin endoperoxides (PGG_2 and PGH_s) and hydroperoxides and fatty acid hydroperoxides have been found to enhance the activity of guanylate cyclase (Graff *et al.*, 1978). Thus, the active oxygen species are able to decrease or increase activity of cycloxygenase, depending on their concentration and to increase the activity of guanylate cyclase.

5. Prostaglandins, particularly the E series as PGE_2, are elevated in cancerous cells (Goodwin *et al.*, 1980; Karmali, 1980). In addition, it is known that cancer cells have a high innate rate of dysdifferentiation (metastases), reflecting an instability in maintaining a stable differentiated state. This characteristic is consistent with cancer cells being normally low in antioxidant levels such as catalase and superoxide dismutase.

Now, according to the GAC model, an unusually high level of prostaglandin synthesis would be characteristic of cells having a higher than normal level of antioxidants, but in fact the cancerous cell has a deficiency in antioxidant protection. This implies that the cell is trying to overcome the antioxidant deficiency but that there is a defect somewhere in the feedback control system where the high levels of prostaglandin synthesis are not able to stimulate sufficient antioxidant synthesis. Thus, a common defect in cancer cells would be predicted to involve the antioxidant regulatory system of the cell. Similar regulatory defects might be involved in Hodgkin's disease, which is also characterized by high levels of prostaglandin synthesis.

6. Enhanced levels of prostaglandin have been shown to both decrease and increase levels of DNA excision repair (Egg *et al.*, 1978).

7. Tumor promotors such as 12-0-tetradecanoylphorbol-13-acetate (TPA) enhance prostaglandin synthesis through phospholipase A_2 activation. Consequently, it has been found that inhibition of phospholipase A_2 protects cells from TPA promotion. One mechanism of how tumor promotors function is by increasing intercellular levels of active oxygen species. Thus, scavengers against active oxygen species such as OH have also been shown to inhibit the effects of tumor promotors (Emerit and Cerutti, 1982; Novogrodsky *et al.*, 1982; Borek and Troll, 1983).

One of the interesting effects of the tumor promotor phorbol 12-myrestate 13-acetate (PMA) is that it enhances a clastogenic factor similar to that found in the chromosomal breakage disorder of ataxia telangiectasia and Bloom's syndrome. In these disorders, SOD can decrease the levels of this clastogenic factor and the severity of the disease (Emerit and Cerutti, 1981, 1982).

These results suggest that tumor promotors interfere with the normal operation of the antioxidant regulatory system of a cell. The tumor promotors TPA or PMA appear to interfere with the antioxidant regulatory system such that the cell reacts as if there is excess antioxidant protection. This is achieved by activating the phospholipase A_4 enzyme, as already described, which is one effect of TPA or PMA, resulting in an increase of prostaglandin synthesis and thus a decrease of antioxidant synthesis. The end result is that the cell responds by lowering its normal levels of antioxidant protection, thus increasing the lifetime of the toxic clastogenic factor. This clastogenic factor would be normally produced in cells but rendered inactive by SOD before it has a chance to cause cell damage.

Tumor promotors are also known to produce a wide spectrum of other effects on cells, such as changes in membrane and gene expression. These results suggest that tumor promotors act to enhance the alteration of the proper differentiated state of cells. This property of tumor promotors would also be consistent with their lowering normal antoxidant defense levels within a cell, and, in addition, would support the general hypothesis that stability of the differentiated state is, in part, determined by the antioxidant levels. Consistent with this view is the finding that α-tocopherol has also been shown to reduce the effects of the tumor promotor TPA on dysdifferentiation of cells (Ohuchi and Levine, 1980), as well as SOD (Borek and Troll, 1983).

These effects of tumor promotors are consistent and support the hypothesis that aging is a result of dysdifferentiation, that cancer is a special case of dysdifferentiation, and that aging rate is determined in part by antioxidant levels in a cell.

Finally, these data support the concept that ataxia telangiectasia and Bloom's syndrome represent human inherited diseases involving improper regulation of antioxidant levels in cells. From this, it would be predicted that deficiency in antioxidant levels and/or the ability of the cells to respond in a compensational manner to oxidative stress are likely to be important in a number of other human inherited diseases.

8. Feeding vitamin E to animals affects prostaglandin levels (PGE_2 and OGF_{2a}) in an inverse manner (Hope *et al.,* 1975; Horwitt, 1976; Machlin, 1978); that is, α-tocopherol inhibits prostaglandin synthesis, and thus has been found to lower platelet aggregation. However, prostaglandin inhibitors like indomethacin or aspirin also reduce platelet aggregation by inhibition of cycloxygenase activity and thus prostaglandin synthesis. These results suggest that such inhibitors of prostaglandin synthesis have α-tocopherol-like properties. This idea was confirmed by noting that aspirin prevented anemia and

thrombocythemia in vitamin E-deficient animals (Machlin, 1978). However, other effects of vitamin E deficiency were not reduced. These data support the concept of partial compensation among the different antioxidants and that inhibition of prostaglandin synthesis by drugs results in a further enhancement of other antioxidants in vitamin E-deficient animals. It is important to note that higher than normal levels of vitamin E exert a suppressive effect on prostaglandin synthesis and lower than normal levels enhance protaglandin synthesis. The first effect is likely to be by way of protecting the cycloxygenase from inactivation and the second effect by enhancement of the lipoxygenation of arachidonic acid (Goetzl, 1980; Valentovic *et al.,* 1982).

9. A key prediction of the GAC model is that inhibition of prostaglandin synthesis should increase tissue levels of antioxidants. Thus anti-inflammatory agents like indomethacin or aspirin would be predicted to decrease a tissue's susceptibility to lipid peroxidation. Evidence that this may occur is found where lipid peroxidation levels were measured in the serum of rabbits 6 hr after the intravenous injection of aspirin (50 mg/kg) (Hayaishi and Shimizu, 1982). The result was a decrease in serum lipid peroxidation level by twofold, as measured by the thiobarbituric acid reaction. However, this result might also be related more directly to a decrease in prostaglandin endoperoxides. The answer could be obtained by direct measurement of the antioxidant levels in the tissues of animals given aspirin or other anti-inflammatory agents. This work is now in progress, in our laboratory.

10. The effects of aspirin on lifespan of *Drosophila melanogaster* have been studied, and an increase of lifespan (both mean and maximum lifespan) of over 20% was found (Hochschild, 1972). It is interesting that similar lifespan extension was observed by the addition of corticosterol to the diet of flies (Hochschild, 1971). Aspirin given to mice, however, did not significantly increase lifespan and, at one concentration, actually shortened lifespan (Hochschild, 1973). This result may indicate the critical dosage effect of aspirin, a phenomenon similar to its anticoagulation effects. Both of these agents, aspirin and corticosteroids, are known not only to be membrane stabilizers but also to be anti-inflammatory agents. Thus, the lifespan-increasing effects could have been related to an increased level of tissue antioxidants.

11. Finally, it is of interest to note that rats placed on a diet deficient in essential fatty acids show a remarkable decrease in glutathione peroxidase activity (Jensen and Clausen, 1981), suggesting a possible role of essential fatty acids in regulating tissue levels of this enzyme.

In conclusion, evidence has been presented indicating that, within a certain range, antioxidants have sufficient overlap in protective properties to provide effective compensation if a deficiency in an antioxidant exists. Such a compensatory action of antioxidants accounts for the ability of an organism to maintain its characteristic longevity potential over a wide range of levels of antioxidants in the diet, endogenous synthesis levels of antioxidants, and exposure to oxidative stress.

A model to account for the maintenance of the general level of antioxidant protection (not the antioxidant compensation mechanism of individual antioxidants) has been presented, based on the sensitivity of arachidonic acid to lipid peroxidation, cycloxygenase to inactivation and guanylate cyclase to activation by active oxygen species. This model has far-reaching predictions, suggesting a new class of genetic diseases of which diabetes, Hodgkin's disease, Bloom's syndrome, and ataxia telangiectasia are examples, as well as some of the shorter-lived mouse strains like NZB. In addition, certain types of cancer may also involve defects in the postulated antioxidant regulatory system. In all these dysfunctions, it is suggested that an impairment of the proper regulation of compensation and/or overall levels of antioxidants is involved. This model provides a rational basis for the treatment of these dysfunctions and points to some of the genetic and biochemical processes that may have been affected during the evolution of human longevity.

9. NEOTENY, AGING, AND LONGEVITY

The biological nature of aging was postulated to be pleiotropic in nature, being the result of normal by-products of the living process; that is, aging is a result of evolutionarily nonselected endogenous properties of an organism. For the mammalian species, two classes of pleiotropic aging processes were postulated to exist. One is associated with energy metabolism and has been called the continuously acting biosenescent processes (CABPs). The second is associated with developmental processes and has been called the developmentally linked biosenescent processes (DLBPs). It is not known which of these two classes of aging processes might play the most important role in causing aging. This would be determined in part by how many and how effective each was in causing aging and how effective defense systems acting against them can evolve.

The DLBPs were postulated to exist largely on the basis of the following reasons: (a) Medawar (1957), Williams (1957), and Hamilton (1966) have proposed that aging is caused by a special class of pleiotropic-acting genes whose expression is linked to development; and (b) there is much experimental data indicating that developmentally linked hormonal processes such as those involved in the sexual maturation of the Pacific coast salmon can act to accelerate the aging process (Cutler, 1972, 1976a,b, 1982b).

The inadequacy of the proposal suggested by Medawar, Williams, and Hamilton was that they failed to consider species differences in longevity and how differences in species' longevity could have evolved relative to their aging hypothesis. In fact, because of the apparent high complexity of this type of aging process, which involves essentially all biological aspects of the organism, the development of any effective antiaging method to extend lifespan was pointed out to be impossible (Wallace, 1967, 1975).

Being convinced that developmentally linked biosenescent processes do exist, I suggested that the means which evolved to counter their aging effects was retardation of development; that is, to postpone the time when DLBPs become expressed (Cutler, 1972, 1976a,b). The major experimental evidence supporting this idea is the high correlation found between the rate of development of an organism from embryo to the mature adult stage of life and its lifespan potential. Other evidence is the finding that experimentally induced retardation of developmental rate, prevention of sexual maturity, or removal of the pituitary gland in the young adult all increase lifespan.

The hypothesis that one of the major longevity determinants is a retardation in development has a number of far-reaching ramifications. This concerns the relationship of the evolution of LSP with other characteristics of the species which are known to be influenced by the process called neoteny or paedomorphosis. Neoteny refers to the retardation of fetal or juvenile traits in the adult by the retardation of the developmental process and appears to have played a major role in the evolution of morphological and behavioral characteristics in mammalian species, particularly in the evolution of the hominid species. For example, the chimpanzee fetus is remarkably similar in morphology to an adult human. In this sense, it has been said humans are like a primate fetus which has become sexually mature. The major proponents of neoteny are: Bolk, 1926; Haldane, 1932; Portman, 1945; Keith, 1949; Schultz, 1950; Abbie, 1958; de Beer, 1958; Gould, 1977.

On comparing the morphology of the premature versus adult fossils of the hominid species, it was found that, indeed, the evolutionary stages do have a large neotenous component; that is, both retardation of development and neotenous traits increase steadily with hominid evolution. The unusual high degree of neotenous evolution found for the hominids has resulted in humans having the most neotenous traits of the mammalian species. Some of the outstanding physical neotenous traits of the human, as compared to other primates, are (a) large brain size, (b) small teeth, (c) late eruption of teeth, (d) gracile skeleton, (e) thin nails, (f) relative hairlessness of the body, and (g) shape of the skull—flatness of face and lack of heavy brow ridges. Some of the outstanding functional neotenous traits of the human are (a) rapid growth of the brain in the third year of life, (b) prolonged immaturity, (c) prolonged growth period, (d) late development of reproductive maturity (Montagu, 1982).

Although neoteny can explain many of the characteristic features of humans which appeared during hominid evolution, it cannot explain all of them. Some of the nonneotenous human traits are (a) overbite, (b) erect posture, (c) head hair, and (d) large penis. To account for all human characteristics, the process of mosaic evolution or heterochronic evolution had to be evoked. This type of evolution involves a combination of both acceleration and retardation of morphological and physiological functions during development (Gould, 1977).

An extremely important implication of the heterochronic evolution model is that no new features evolved; instead, evolution of the hominid species involved a quantitative and not a qualitative modification of features common to all mammalian species during the embryonic and later stages of development. Thus, again we arrive at the conclusion that the difference between humans and the apes is essentially quantitative in nature, the difference being alterations in the rate and duration of common features of ontogenetic development. It is evident, then, that the heterochronic model of hominid evolution is consistent with other evolutionary and comparative molecular biological evidence, suggesting a common genetic basis for the appearance of most species' characteristics (Cutler, 1976, 1976; Wilson, 1976; Zuckerkandl, 1976b; Gould, 1979). This involves changes occurring in regulatory gene expression, resulting in quantitative changes in the expression of structural genes. These changes in turn result in quantitative changes in developmental rate and morphology.

This model indicates that retardation of development gave rise to most of the morphological characteristics of the hominids and the innate ability to live out the extended neotenous developmental period; that is, the innate aging rate of the organism needed to be reduced in order for the organism to remain free from the detrimental effects of aging and to carry out the retarded developmental program. It cannot be automatically assumed that retardation of development, which implies greater longevity, can be accomplished without somehow also affecting the aging rate of the organism.

The fascinating aspect in all this is that the very act of slowing down development, which was necessary to evolve many of the hominid's characteristics necessary for their evolutionary success, may in itself have played a direct role in decreasing aging rate, by postponement in time of the DLBPs.

It would be of interest to determine whether or not retardation of development also affects the CABPs or the antiaging processes acting against the CABPs. There is evidence that DNA repair levels are at a similar level and unusually high during embryonic development for different mammalian species and only adjust to the species' characteristic levels during later development after birth (Cutler, 1976a). Thus, retardation of development or neoteny might result in the retention of higher DNA repair levels into later stages of development as well as the retention of other defense processes normally found to be higher in fetal stages of development. On the other hand, the age-dependent loss of neotenous traits may include the LDPs, explaining the loss of DNA repair, SOD, decrease in protein synthesis rates, and other changes previously thought to be the result of aging. Consistent with this line of thought, developmental genes are known to be associated with the major histocompatibility complex, where some LDP genes appear to be located (Cutler, 1982b). Are the neoteny genes located here? It would be important to determine what changes, that occur with age, identify with a loss of neoteny.

Another point of interest is that Mongoloids have the most neotenous characteristics of the human races, such as larger brains, broader skulls, flatness of the nose, and being less hairy. The human female of all races is also more neotenous, having a more delicately constructed skeleton, larger brain relative to body size, less hair, more delicate skin and smaller body size. Also, in spite of having a smaller body size, the human female has a specific metabolic rate of some 5%–7% lower than the equivalent body weight for males. Consistent with these findings, we also find data indicating that the Mongoloids have longer lifespans on the average and higher IQs, and the females of all races are longer lived (Montagu, 1982). Finally, it appears that in general for the primate species, the degree of neotenous traits that have evolved correlate well with their LSPs. This is consistent with species having higher LSPs also having higher ratios of learned-to-instinctive behavior.

In addition to the evolution of morphological changes, many behavioral traits can be explained by a neotenous model of hominid evolution; that is, the evolution of the hominid species leading to the human has also resulted in the retention of childlike behavioral patterns into the adult stages of life (Portmann, 1945; Starck, 1960; Lorenz, 1971; Montagu, 1982). Some outstanding human neotenous traits of behavior are (a) curiosity, (b) imagination, (c) playfulness, (d) open mindedness, (e) sense of humor and laughter, (f) willingness to work hard at making sense of it all, (g) flexibility, (h) educability, (i) willingness to experiment, (j) receptiveness to new ideas, and (k) eagerness to learn. These qualities are characteristic of young premature mammals but are rapidly lost as development progresses. In those species where neotenous traits have evolved further, these behavioral traits are retained for longer periods of time and into deeper stages of development.

This list of neotenous traits of human behavior consists of some of the most desirable human characteristics and certainly some of the most important in contributing to his evolutionary success. An interesting point in this regard is that persons today with the most outstanding neotenous traits are frequently long-lived and have higher than normal intelligence. Examples are A. Einstein and I. Newton. In fact, many outstanding scientists are frequently criticized for their childlike behavior. The two characteristics, plasticity and educability, are similar to the features which I have previously termed learned behavior. Learned behavior was proposed to be a new trait and a driving force behind all mammalian species' evolution as compared to instinctive behavior, where increased longevity coevolved to make this new trait possible. Maximum plasticity and educability occur in the premature phase of human growth. This is also the phase of life that has been increased excessively in primate evolution (see Fig. 10). Thus, retardation of development also appears to have provided not only the material basis for learned behavior but also the time to achieve it and the postponement of aging to make it all possible.

In summary, the heterchronic model of primate evolution provides a rather simple explanation for what appears to be an extremely complex phenomenon. Neoteny accounts for many aspects of morphology and behavior and also provides at the same time a mechanism for dealing with the DLBPs. Thus, some of the major characteristics of the human appear to be governed by regulatory genes controlling developmental rates. There is also some evidence indicating that only a few mutations are required to affect developmental rate (Cutler, 1975, 1976a,b; Gould, 1977). This is consistent with the prediction that longevity evolved by relatively few genetic alterations occurring in regulatory genes (Cutler, 1975).

An interesting question suggested by the neoteny model is, what would happen if a juvenile chimpanzee were arrested in further development but allowed to grow to its normal size? Would it develop into a hominidlike creature and would its lifespan potential and intelligence also be increased? If the human does go through the same structural and biochemical stages of brain development as the chimpanzee—but more slowly—then the degree of intelligence might largely be a matter of time spent at a critical stage of development; that is, the height of intelligence a species reaches might be related not only to the morphological and biochemical makeup of the brain but also how long the animal stays at a given stage of brain development. Moreover, if it were possible to further retard human development to further increase neotenous characteristics then, in addition to resulting in a larger brain size, a higher intellect might also evolve as the result of more time being spent in a childlike developmental stage.

One of the most important aspects of the neoteny model concerns differentiating the processes of development from those of aging. Aging has been roughly defined as those processes which lead to increased probability of death. Although this definition is clearly correct, the aging processes are usually thought of as being the result of a loss in the innate ability to function (Cutler, 1976a,b). According to the neoteny model, many of the most desirable of human characteristics arise by retardation of development and thus the prolongation of childlike properties. However, this is a result of retardation of development. Development would still be expected to continue but at a slower rate throughout the extended lifespan. This retardation is not limited to morphological changes but also applies to behavioral changes. Thus, the neotenous traits of morphology, behavior, and intellect would all be expected to change with increasing chronological age. After the age of sexual maturation, many of the childlike properties are diminished—and become even further diminished with increasing age. Indeed all of the behavioral neotenous human traits expressed in the premature years appear to be slowly lost with increasing age. *Growing older chronically therefore results in changes that might be erroneously thought to be due to biological aging processes.* For example, it is well known that learning a language is a relatively simple task for a child and quite difficult for an adult. Is it due to biological aging or some other factor? This increased difficulty in learning a new language may not be the result of a biological aging process but, instead, may be related to the loss of retarded childlike neotenous traits as the developmental program proceeds. From this, the question then arises of how many behavioral as well as morphological and physiological age-dependent characteristics are due to the neotenous program of development and thus not due to biological aging? The results of classic studies concerning the age of achievement, showing different optimum ages for different types of performance (Lehman, 1953, 1960), may be largely due to the loss of neotenous traits in the human rather than to biological aging processes.

Relative to this question is the now classical novel by Aldous Huxley entitled *After Many a Summer Dies the Swan.* This novel is based on the idea that, if human aging can be prolonged sufficiently, the continued development of the person would lead to a person slowly developing both behavioral and physical characteristics of our hominoid ancestors. Does a person continue to develop after sexual maturation is reached—to stages like his hominoid ancestors—or does the developmental or genetic program end at this point? The catch may be that, in order to increase longevity, you must retard the rate of development. Thus, in effect it would be impossible to ever live beyond the point where apelike qualities appear.

A key question in all of this is whether the genetic program of development continues after sexual maturity is reached or is stopped at this stage. There are some data indicating that gene expression in mouse liver does continue in an ordered manner after the age of sexual maturation (Cutler, 1972).

10. IMPLICATIONS: PRESENT AND FUTURE

10.1. Nonlifespan Implications

The long-term objective in investigating the biological basis of human longevity is to help lay the scientific foundation to develop more effective means to treat the undesirable physical and behavioral aspects associated with aging. So far, the most important development of this effort is the possibility that, in spite of the vast complexity of the aging process, separate longevity determinant processes may exist common to all mammalian species and under the control of a relatively few regulatory genes. Some of these longevity determinant processes appear to be antioxidants, where higher tissue levels are found in the longer-lived species. Another development of this effort is the concept that aging is dysdifferentiation and that longevity determinants act primarily to stabilize the differentiated state of cells. The dysdifferentiation model of aging has been successful in accounting for a wide variety of age-dependent changes in animals and provides a rational basis for searching for longevity determinant processes. A paper describing how this hypothesis of aging and longevity might be further tested has been published (Cutler, 1982b).

A number of other interesting aspects, apart from the aging and longevity aspects, have developed from our studies. Some of these are as follows:

1. The possible identification of a new class of human inherited diseases that are represented by deficiencies in regulation and/or maintainance of proper tissue levels of antioxidants.

2. The possible identification of a new class of mouse inherited diseases, represented by strains having shorter lifespans, higher frequencies of various types of diseases due to an innate deficiency in regulating and/or maintaining proper tissue levels of antioxidants.

3. An explanation of the genetic and biochemical basis of cancer arises naturally from the hypothesis of longevity and aging and accounts for the age-dependent association of the incidence of cancer (Cutler, 1982a,b). It is argued that, although cancer incidence is related to the environmental source of carcinogens (Doll, 1978, 1980; Weisburger *et al.*, 1977), even without this source, cancer is a normal result of the aging process, a dysdifferentiation (Ono and Cutler, 1978; Cutler, 1982a,b); that is, the endogenous cause of cancer is the same as those processes causing transformation of cells to a cancerous form. Such a transformation would normally result in a benign tumor if normal antioxidant or defense levels in the cell were maintained. However, if a dysdifferentiation event occurs in these benign cells, resulting in an unusually low antioxidant level state, then their metastasizing potential is dramatically increased and the tumor becomes malignant. This model suggests that a treatment increasing antioxidant levels in tumors would decrease their spontanous transformation to the highly metastatic state. The model also suggests that high tissue levels of antioxidant may be an important preventive treatment for cancer. Recent evidence has indicated that the tumor promoter action mechanism involves free radicals and that superoxide dismutase is protective against the effects of tumor promoters. Such evidence is consistent with the above-mentioned model which implies that aging and cancer have similar causative mechanisms (Emerit and Cerutti, 1982; Borek and Troll, 1983).

4. A new approach to the experimental evaluation of toxic environmental hazards to humans is suggested using short-lived experimental animals (Tol-

masoff *et al.*, 1980; Cutler, 1982b). If longevity of species is determined in part by their innate sensitivity to endogenous produced free radicals, hydroperoxides, aldehydes, and other similar agents, then it would be expected that a species' sensitivity to exogenous sources of the same type of toxic agents would be related to its LSP. Shorter-lived species such as the mouse or rat would be expected to be unusually sensitive to these toxic agents as compared to humans. Thus, on testing mutagens, carcinogens, and other toxic agents using mice and rats or any other experimental animals with lifespan potentials shorter than humans, it would be important to take their LSP differences into consideration in the extrapolation of toxicological data from mice to humans (Cutler, 1982b).

5. The compensation model for LDPs provides an explanation for the difficulties found in defining minimum daily doses for the antioxidant vitamins. This model also suggests that large doses of antioxidants might lower levels of other antioxidants to dangerous levels.

6. Beta-carotene may be important as a longevity determinant, a role possibly as important as vitamin A. Because of this and the nontoxic nature of beta-carotene, this substance may be the preferred dietary supplement rather than vitamin A, as is now commonly used.

7. Urate may be an important longevity determinant. Therefore, high levels of plasma urate may not necessarily be of pathologic nature, and in the treatment of gout, urate levels should not be dramatically lowered.

8. The guanylate–arachidonic acid–cycloxygenase (GAC) model for regulation of cell oxidation–reduction potential provides a mechanism of how anti-inflammatory agents operate, to enhance tissue levels of antioxidants. This model also suggests means to enhance natural endogenous levels of antioxidants. Such means may have a valuable application by increasing normal resistance to persons working in unusually high environmental levels of radiation and other mutagenic agents and perhaps helping their recovery after a hazardous exposure.

10.2. Lifespan Extension

The LSP of humans is about 100 years, and the longest authentic lifespan recorded is 113 years (Comfort, 1978). No good evidence exists that any person has ever lived longer. In addition, there is no evidence that any individual of any species has lived longer than what would be expected according to the LSP for their species. For example, although intermittent fasting and/or dietary restriction lengthens mean lifespan and LSP of mice and rats, it does so only with reference to *ad libitum*-fed controls. Evidence has already been discussed indicating that *ad libitum* feeding probably shortens normal lifespan, and that dietary restriction and/or fasting simply returns the animal back to normal

(Cutler, 1982b). Similarly, antioxidants, membrane stabilizers, and other nutrients and drugs have all failed to extend the lifespan of an animal beyond the normal LSP it would expect to have in the wild.

These results are explained by the compensatory model for LDPs, where, for example, supplementation of antioxidants to the diet would not likely result in a net increase in the tissue of antioxidant protection. Thus, although there are many claims that lifespan extension has been accomplished using experimental animals, both in the scientific literature and in the public press, careful examination of the data has shown that the experiments are flawed and/or that abnormally short lifespan strains are brought up to a normal lifespan. Finally, it is extremely important to recognize that procedures such as calorie restriction, with reference to *ad libitum* controls, that do increase lifespan of mice and rats (the shortest-lived of all mammals) may not work at all for longer-lived species like the human.

One possible answer to the question of whether it will ever be possible to extend human LSP beyond 100 years will depend on determining if the hypothesis of longevity and aging (as described in this chapter) is correct. If it is true that specific longevity determinants do exist, then to further increase LSP using the same mechanisms that had previously evolved in hominid evolution, it is necessary to learn how to manipulate the LSP-regulatory genes to enhance higher levels of their expression. Although it appears possible to be able to enhance the longevity determinants associated with the energy metabolic by-products (CABPs) without changing basic human characteristics, the more serious problem is how the developmentally linked by-products can be handled (DLBPs) without affecting neotenous traits? One approach in dealing with the problem is to identify the most causative DLPBs related to aging and somehow remove or neutralize them after the adult stage of sexual maturity has been reached. Although this seems to be a formidable task, there is evidence that hormones or related growth factors, when reduced by the removal of the pituitary gland, do result in slowing down the aging rate and even rejuvenating some aspects of the organism (Bilder and Denckla, 1977; Parker *et al.*, 1978; Denckla, 1978; Scott *et al.*, 1979; Everitt, 1980; Everitt *et al.*, 1980; Miller *et al.*, 1980).

These results clearly point out that no life extension method or technology is at hand, and that we are just beginning to evaluate the possibility of this achievement. Nevertheless, the possibility that relatively few regulatory genes appear to be involved in governing the level of expression of the proposed longevity determinant processes does offer a scientific basis whereby human LSP may truly be increased in the not too distant future. Because of this possibility, it is important to evaluate the possible impact such an increase might have.

One misconception is that lifespan extension will result in dramatically increasing the population of the country and the percent of people living

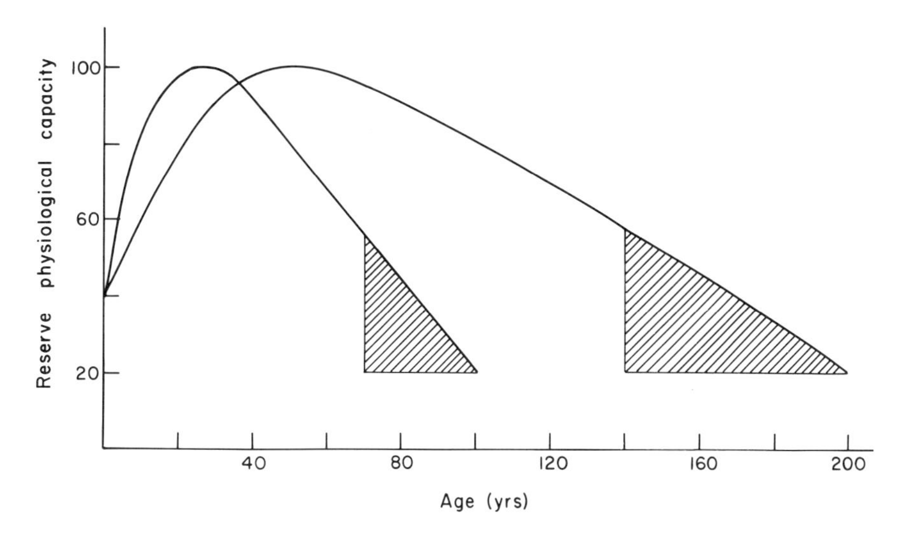

Figure 48. Reserve capacity of physiological functions. Curves represent age-dependent decline in the average of the reserve capacity of physiological functions for a typical human individual today having an LSP of 100 years and a projection to a human having an LSP of 200 years. The hatched area represents the length of time spent beyond the age of 70 or 140 years, respectively. It is apparent that doubling LSP doubles the number of years spent in both an optimum and a declining health state.

beyond their most productive years. Figure 48 illustrates how the general decline in reserve capacity of physiological functions would be if LSP could be doubled from 100 to 200 years using the same mechanisms that have evolved in the past evolutionary increase of LSP. It is clear that if LSP is extended along the same lines that it has evolved, then more years would be spent in declining health; that is, one would have both a longer period of youth and a longer period of old age.

How this doubling in LSP might affect the survival curves of a human population is shown in Fig. 49. The dotted line represents the exponentially declining survival curve typical for a human wild-type population where LSP is 100 years. The squaring of this survival curve is seen where LSP still remains at 100 years. This squaring of the survival curve occurred over the past 400 years or so by decreasing exogenous environmental hazards. The result was going from a population where few individuals lived into old age to a population where today over 10% of the people are 65 years or more of age.

Now let us predict what might happen if, instead of decreasing environmental hazards, that LSP is now increased and the intensity of the environmental hazards is kept constant (the probability of death is maintained at what it now is for a 30-year-old male in the United States). The result of this situ-

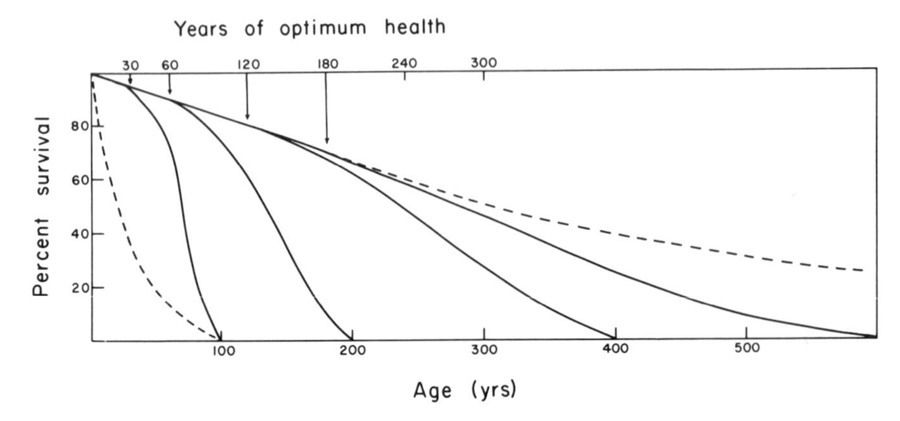

Figure 49. Percent survival curves for humans in the past, present, and projected future. Dotted lines represent the two exponentially declining curves where the smallest fraction of senescent individuals occurs. For these curves, number of years of optimum health is shown where the curve breaks from the exponential. Taken from Cutler (1978).

ation is shown as a family of curves in Fig. 49. As LSP increases, an increased fraction of people will die from exogenous rather than endogenous (aging) causes. The shape of the survival curve begins to return to an exponentially declining type, characteristic of the primitive human populations. Thus, it becomes clear that an increase in LSP actually decreases the percent in the population of those with declining health due to aging. In fact, essentially all problems of biological age would effectively be eliminated if LSP would increase to about 600 years. By increasing LSP and keeping environmental hazards constant, human populations would return to the natural conditions found in most species living in the wild where aging is essentially nonexistent. Thus, individuals making up a population do not need to become biologically immortal to minimize the effects of their biological aging.

If lifespan extension requires a decrease in development rate, what changes might this bring in relation to stages of development? Development rates of different stages of development of the primate species are all remarkably correlated with LSP. This is illustrated in Fig. 50. Thus, if LSP is increased from 100 to 200 years along the same lines as it has previously evolved, it is expected that all stages of life would be equally extended. This retardation of development would of course also affect the neotenous characteristics of humans, and a neoteny model of the morphological proportion of a human having a LSP of 200 years compared to 100 years is shown in Table XLIII and Fig. 51.

One outstanding feature of such a retardation of development to achieve a LSP of 200 years is that the human would clearly have a longer childlike

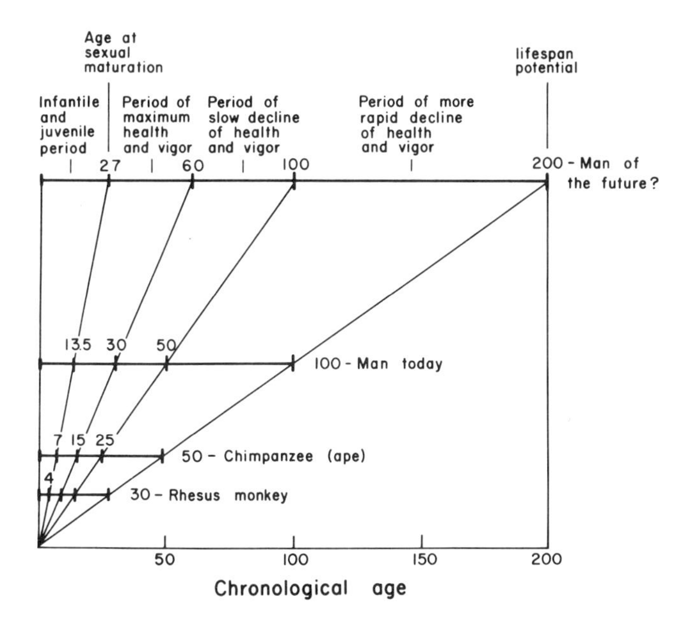

Figure 50. Relative length of time spent in various stages of development as a function of LSP. Projection is made, assuming a linear extrapolation from existing living species to a human having a LSP of 200 years. Taken from Cutler (1978).

period of development. On reflecting how essential childlike or neotenous behavior has been in human evolutionary success, perhaps further changes along the same lines would represent an improvement of humanlike qualities. But if an increase of neotenous characteristics and LSP is an important basis of human evolutionary success, why have those traits not continued to evolve? About 100,000 to 50,000 years ago, LSP appeared to suddenly stop at a stage

Table XLIII. Model of Continued Evolution of Human Longevity[a]

MLP, yr	100	150	200
Brain capacity, cm^3	1412	3396	5688
EQ	7.8	15.2	22.6
$N_c \times 10^{-8}$	90	168	240
Body wt., kg	63.5	81	97
SMR, cal/g per d	23.6	21.8	20.8
LEP, kcal/g	861	1198	1523
ASM, yr	14–17	22–27	30–37
% Brain wt./body wt., g/g	2.2	4.2	5.8

[a]The assumption is made that these parameters continued to evolve along the same equation as they have during the past 2.5 million years of hominid evolution.

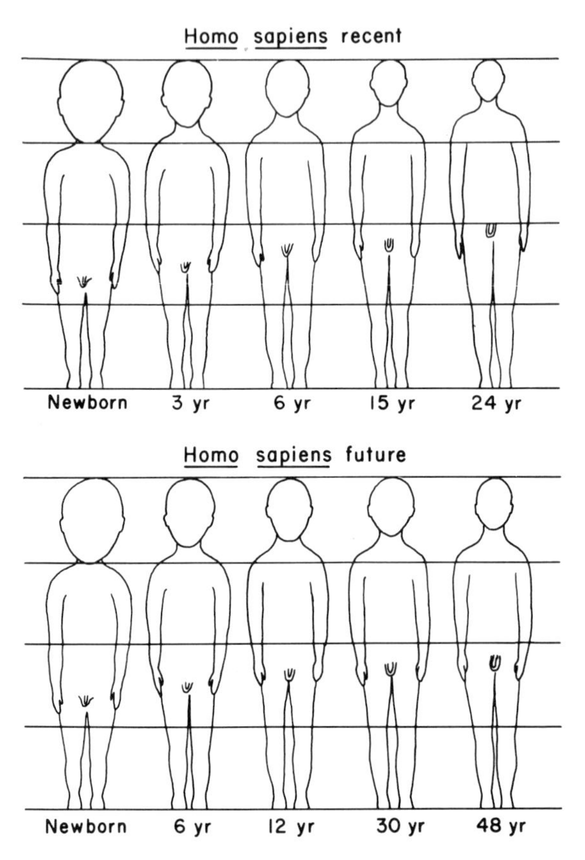

Figure 51. Neotenous traits of human with LSP of 100 years and for human if LSP were increased to 200 years by further retardation of development. Taken from Cutler (1978).

when the rate of increase was the greatest (see Fig. 17). One possible answer to this question is that selection for the most fit in terms of reproduction no longer occurred. This would be due to the loss of small inbreeding populations, where one or a few dominant males were responsible for reproduction, to where reproduction potential became largely independent of physical or intellectual fitness.

Many arguments have been advanced as to the cultural disadvantages that would occur if LSP could be increased (Vetch, 1978). To many the more acceptable model of lifespan extension would be a further squaring of the curve, resulting in increasing old age longer where LSP remains constant. However, I believe a more careful analysis will show the advantages of a society made up of individuals living longer, healthier lifespans will far outweigh the possible disadvantages socially, culturally, and economically (Cutler and Kalish, 1977; Cutler, 1978, 1981a).

ACKNOWLEDGMENTS

I thank George Roth for helpful comments on the manuscript and Don Ingram for his encouragement and interesting discussions. The assistance of Edith Cutler is gratefully acknowledged for laboratory assistance, drawings, and typing of the manuscript. Her help was made possible by support from the Glenn Foundation for Medical Research.

REFERENCES

Abbie, A. A., 1958, Timing in human evolution, *Proc. Linn. Soc. New South Wales* **83:**197.

Acsádi, G., and Nemeskéri, J., 1970, *History of Human Lifespan and Mortality,* Akadémiai Kiadó, Budapest.

Adachi, L., 1973, The metabolism and control mechanism of human hair follicles, *Curr. Probl. Dermatol.* **5:**37.

Albritten, E. C., ed., 1952, *Standard Values in Blood,* Fed. Am. Soc. Exp. Biol., Bethesda, Maryland.

Altman, P., and Dittmer, D., eds., 1961, *Blood and Other Body Fluids,* Fed. Am. Soc. Exp. Biol., Bethesda, Maryland.

Altman, P., and Dittmer, D., eds., 1962, *Biological Handbook. Growth,* Fed. Am. Soc. Exp. Biol., Bethesda, Maryland.

Altman, P., and Dittmer, D., eds., 1972, *Biology Data Book,* Fed. Am. Soc. Exp. Biol., Bethesda, Maryland.

Altman, P., and Dittmer, D., eds., 1974, *Biology Data Book,* 2nd Ed., Fed. Am. Soc. Exp. Biol., Bethesda, Maryland.

Ames, B. N., Cathcart, R., Schwiers, E., and Hochstein, P., 1981, Uric acid: An antioxidant defense in humans against oxidant- and radical-caused aging and cancer. A hypothesis, *Proc. Natl. Acad. Sci. USA* **78:**6858.

Aries, I. M., and Jakoby, W. B., eds., 1976, *Glutathione, Metabolism and Function,* Raven Press, New York.

Arnold, W. P., Aldred, R., and Murad, F., 1977, Cigarette smoke activates guanylate cyclase and increased guanosine 3′,5′-monophosphate in tissues, *Science* **198:**934.

Ausman, L. M., and Hayes, K. C., 1974, Vitamin E deficiency anemia in Old and New World monkeys, *Am. J. Clin. Nutr.* **27:**1141.

Baars. A. J., Mukhtar, H., Zoetemelk, C. E. M., Jansen, M., and Breimer, D. D., 1981, Glutathione S-transferase activity in rat and human tissues and organs, *Comp. Biochem. Physiol. Part C* **70:**285.

Baba, A., Lee, E., Ohta, A., Tatsuno, T., and Iwata, H., 1981, Activation of adenylated cyclase of rat brain by lipid peroxidation, *J. Biol. Chem.* **256:**3679.

Banroques, J., Gregori, C., and Schapira, F., 1980, Aging of aldolase and pyruvate kinase in eye lens, in: *Ageing of the Lens* (F. Regnault, F., O. Hockwin, and Y. Courtois, eds.), pp. 63–69, Elsevier–North Holland, Amsterdam.

Barber, A. A., and Bernheim, F., 1967, Lipid peroxidation: Its measurement, occurrence and significance in animal tissues, *Adv. Gerontol. Res.* **2:**355–403.

Barrett, J. C., Wong, A. and McLachlan, J. A., 1981, Diethylstilbestrol induces neoplastic transformation without measurable gene mutation at two loci, *Science* **212:**1402.

Bauchot, R., and Stephan, H., 1969, Encephalisation et niveau evolutif chez les simiens, *Mammalia* **33:**225.

Beatty, C. H., Bocek, R. M., Herrington, P. T., Lamy, C., and Hoskins, M. K., 1982, Aged rhesus skeletal muscle: Histochemistry and lipofuscin content, *Age* **5**:1.

de Beer, G., 1958, *Embryos and Ancestors,* Oxford Univ. Press, London.

Bennett, J. S., Bond, J., Singer, I., and Gottlieb, A. J., 1972, Hypouricemia in Hodgkin's disease, *Ann. Int. Med.* **76**:751.

Berdyshev, G. D., and Protsenko, N. A., 1972, Current theories of the mechanism of the post-spawning death of Pacific salmon, *Hydrobiological J.* **8**:85.

Bernirschke, K., Garner, F. M., and Jones, T. C., eds., 1978, *Pathology of Laboratory Animals,* Springer-Verlag, Basel.

Bieri, J. G., and Anderson, A. A., 1960, Peroxidation of lipids in tissue homogenates as related to vitamin E, *Arch. Biochem. Biophys.* **90**:105–110.

Bilder, G. E., and Denckla, W. D., 1977, Restoration of ability to reject xenografts and clear carbon after hypophysectomy of adult rats, *Mech. Ageing Dev.* **6**:153.

Blackett, A. D., and Hall, D. A., 1981, The effects of vitamin E on mouse fitness and survival, *Gerontology* **27**:133.

Blair, W., 1951, "Population Structure, Social Behavior and Environmental Relations in a Natural Population of the Beach Mouse," Contribution from the Laboratory of Vertebrate Biology, No. 48, Univ. of Michigan, Ann Arbor, Michigan.

Bolk, L., 1926, La recapitulation ontogenique comme phenomene hormonique, *Arch. Anat. Histol. Embryol.* **5**:87.

Borek, C., and Troll, W., 1983, Modifiers of free radicals inhibit in vitro the oncogenic actions of x-rays, bleomycin, and the tumor promoter 12-0-tetradecanoylphorbol 13-acetate, *Proc. Natl. Acad. Sci. USA* **80**:1304.

Botkin, D., and Miller, R., 1974, Mortality rates and survival of birds, *Am. Naturalist* **108**:181.

Bourgain, R. H., Deby, C., Deby-Dupont, G., and Andries, R., 1982, Enhancement of arterial thromboformation by uric acid, *Biochem. Pharmacol.* **31**:3011.

Bowden, D., and Jones, M. L., 1979, Aging research in nonhuman primates, in: *Aging in Nonhuman Primates* (D. M. Bowden, ed.), pp. 1–13, Van Nostrand Reinhold, New York.

Boyd, M. R., Catignani, G. L., Sasame, H. A., Mitchell, J. R., and Stiko, A. W., 1979, Acute pulmonary injury in rats by nitrofuratoin and modification by vitamin E, dietary rat and oxygen, *Am. Rev. Resp. Dis.* **120**:93.

Brandwein, H. J., Lewicki, J. A., and Murad, F., 1981, Reversible inactivation of guanylate cyclase by mixed disulfide formation, *J. Biol. Chem.* **256**:2958.

Braun, A. C., 1981, An epigenetic model for the origin of cancer *Q. Rev. Biol.* **56**:33.

Brot, N., Weissbach, L., Werth, J., and Weissbach, H., 1981, Enzymatic reduction of protein-bound methionine sulfoxide, *Proc. Natl. Acad. Acad. Sci. USA* **78**:2155.

Bush, G. L., Case, S. M., Wilson, A. C., and Patton, J. L., 1977, Rapid speciation and chromosomal evolution in mammals, *Proc. Natl. Acad. Sci. USA* **74**:3942.

Butler, J. A., Whanger, P. D., and Tripp, M. J., 1982, Blood selenium and glutathione peroxidase activity in pregnant women: Comparative assays in primates and other animals, *Am. J. Con. Nutr.* **36**:15.

Cairns, J., 1979, Differentiation and cancer, *Differentiation* **13**:67.

Cameron, I. L., and Thrasher, J. D., eds., 1971, *Cellular and Molecular Renewal in the Mammalian Body,* Academic Press, New York.

Catignani G. L., and Dinning, J. S., 1971, Role of vitamin E in the regulation of rabbit liver xanthine dehydrogenase activity, *J. Nutr.* **101**:1327.

Caughley, G., 1966, Mortality patterns in mammals, *Ecology* **47**:906.

Chatterjee, I. B., 1978, Ascorbic acid metabolism, *World Rev. Nutr. Diet.* **30**:69.

Chatterjee, I. B., Majumder, A. K., Nandi, B. K., and Subramanian, N., 1975, Synthesis and some major functions of vitamin C in animals, *Ann. N.Y. Acad. Sci.* **258**:24.

Chance, B., Sies, H., and Boveris, A., 1979, Hydroperoxide metabolism in mammalian organisms, *Physiol. Res.* **59**:527.

Chen, L. H., Liao, S., Packett, L. V., 1972, Interaction of dietary vitamin E and protein level on lipid source with serum cholesterol level in rats, *J. Nutr.* **102**:729.

Chen, L. H., Thacker, R. R., and Chow, C. K., 1980, Tissue antioxidant status and related enzymes in rats with long-term vitamin A deficiency, *Nutr. Rept. Intern.* **22**:873.

Cherry, L. M., Case, S. M. Kunkel, J. G., and Wilson, A. C., 1979, Biochemical evolution, *Ann. Rev. Biochem.* **46**:573.

Chow, C. K., 1977, Dietary vitamin E and levels of reduced glutathione, glutathione peroxidase, catalase and superoxide dismutase in rat blood, *Int. J. Vit. Nutr. Res.* **47**:268.

Chow, C. K., 1978, Effect of dietary selenium and vitamin E on the antioxidant defense system of rat erythrocytes, *Int. J. Vit. Nutr. Res.* **49**:182.

Chow, C. K., Reddy, K., and Tappel, Al L., 1973, Effect of dietary vitamin E on the activities of the glutathione peroxidase system in rat tissues, *J. Nutr.* **103**:618.

Ciriolo, M. R., Mavelli, I., Rotilio, G., Borzatta, V., Cristofari, M., and Stanzani, L., 1982, Decrease of superoxide dismutase and glutatione peroxidase in liver of rats treated with hypolipidemic drugs, *FEBS Lett.* **144**:264.

Combs, G. F., 1981, Influences of dietary vitamin E and selenium on the oxidant defense system of the chick, *Poultry Sci.* **60**:2098.

Comfort, A., 1978, *The Biology of Senescence,* Elsevier, New York.

Cone, C. D., 1971, Unified theory on the basic mechanism of normal mitotic control and oncogenesis, *J. Theor. Biol.* **30**:151.

Cone, C. D., 1974, The role of surface electrical transmembrane potential in normal and malignant mitogenesis, *Ann. N.Y. Acad. Sci.* **238**:420.

Cone, C. D., 1980, Ionically mediated induction of mitogenesis in CNS neurons, *Ann. N.Y. Acad. Sci.* **339**:115.

Cone, C. D., and Cone, C. M., 1978, Evidence of normal mitosis with complete cytokinesis in central nervous system neurons during sustained depolarization with ouabain, *Exp. Neurol.* **60**:41.

Coyle, J. T., Price, D. L., and DeLong, M. R., 1983, Alzheimer's disease: A disorder of cortical cholinergic innervation, *Science* **219**:1184.

Criss, W. E., 1971, A review of isozymes in cancer, *Cancer Res.* **31**:1523.

Crouch, R. K., Gandy, S. E., Kimsey, G., Galbraith, R. A., Galbraith, G. M. P., and Buse, M. G., 1981, The inhibition of islet superoxide dismutase by diabetologenic drugs, *Diabetes* **30**:235.

Crowley, C., and Curtis, H. J., 1963, The development of somatic mutations in mice with age, *Proc. Natl. Acad. Sci. USA* **49**:626.

Curtis, H. J., 1966, *Biological Mechanisms of Aging,* C. C. Thomas, Springfield, Illinois.

Curtis, H. J., 1971, Genetic factors in aging, *Adv. Genetics* **16**:305.

Curtis, H. J., and Miller, K., 1971, Chromosome aberrations in liver cells of guinea pigs, *J. Gerontol.* **26**:292.

Curtis, H. J., Leigh, J., and Tilley, J., 1966, Chromosome aberrations in liver cells of dogs of different ages, *J. Gerontol.* **21**:268.

Cutler, G. B., Glenn, M., Bush, M., Hodgen, G. D., Graham, C. E., and Loriaux, D. L, 1978, Adrenarche: A survey of rodents, domestic animals, and primates, *Endocrinology* **103**: 2112.

Cutler, R. G., 1972, Transcription of reiterated DNA sequence classes throughout the lifespan of the mouse, *Adv. Geront. Res.* **4**:219–312.

Cutler, R. G., 1973, Redundancy of information content in the genome of mammalian species as a protective mechanism determining aging rate, *Mech. Ageing Dev.* **2**:381.

Cutler, R. G., 1975, Evolution of human longevity and the genetic complexity governing aging rate, *Proc. Natl. Acad. Sci. USA* **72**:4664.
Cutler, R. G., 1976a, Evolution of longevity in primates, *J. Human Evol.* **5**:169.
Cutler, R. G., 1976b, Nature of aging and life maintenance processes, in: *Interdiscplinary Topics in Gerontology,* Vol. 9 (R. G. Cutler, ed.), pp. 83–133, Karger, Basel.
Cutler, R. G., 1976c, Cross-linkage hypothesis of aging: DNA adducts in chromatin as a primary aging process, in: *Aging, Carcinogenesis, and Radiation Biology* (K. C. Smith, ed.), pp. 443–492, Plenum Press, New York.
Cutler, R. G., 1978, Evolutionary biology of senescence, in: *The Biology of Aging* (J. A. Behnke, C. E. Finch, and G. B. Moment, eds.), pp. 311–360, Plenum Press, New York.
Cutler, R. G., 1979, Evolution of longevity in ungulates and carnivores, *Gerontology* **25**:69.
Cutler, R. G., 1980a, Evolution of human longevity, in: *Advances in Pathobiology 7. Aging, Cancer and Cell Membranes* (C. Borek, C. M. Feneglio, and D. W. King, eds.), pp. 43–79, Thieme-Stratton, Inc., New York.
Cutler, R. G., 1980b, Stabilization of the differentiated state of cells: Possible role of the major histocompatibility complex loci, *Gerontologist* **20**:88.
Cutler, R. G., 1981a, Life span extension, in: *Aging: Biology and Behavior* (J. L. McGaugh and S. B. Kiesler, eds.), pp. 31–76, Academic Press, New York.
Cutler, R. G., 1981b, Thyroid hormone receptors in the brain of rat as a function of age and hypophysectomy, in: *Brain Neurotransmitters and Receptors in Aging and Age-Related Disorders. Aging,* Vol. 17 (S. J. Enna, T. Samora, and B. Beerjeds, eds.), pp. 117–132, Raven Press, New York.
Cutler, R. G., 1981c, Lifespan determinant mechanisms: Identification and genetic regulation of cellular integrity maintenance processes, Abstr. XII Intern. Congr. Gerontol., Hamburg, Germany, p. 272.
Cutler, R. G., 1982c, The dysdifferentiative hypothesis of mammalian aging and longevity, in *The Aging Brain. Aging,* Vol. 20 (E. Giacobini, G. Giacobini, G. Filogamo, A. Vernadakis, eds.), pp. 1–18, Raven Press, New York.
Cutler, R. G., 1982b, Longevity is determined by specific genes: Testing the hypothesis, in: *Testing the Theories of Aging* (R. Adelman and G. Roth, eds.), pp. 25–114, CRC Press, Boca Raton, Florida.
Cutler, R. G., 1983, Superoxide dismutase, longevity and specific metabolic rate, *Gerontology* **29**:113.
Cutler, R. G., and Kalish, R. A., 1977, Prolongation of mental and physical health: Its possibility and advisability, *Gerontologist* **17**:141.
Deby, C., Deby-Dupont, G., Noel, F.-X., and Lavergne, L., 1981, In vitro and in vivo arachidonic acid conversions into biologically active derivatives are enhanced by uric acid, *Biochem. Pharm.* **30**:2243.
Deevey, E. S., 1947, Life tables for natural populations of animals, *Q. Rev. Biol.* **22**:283.
Deevey, E. S., 1960, The human population, *Sci. Am.* **203**:195.
Demopoulos, H. B., Flamm, E. S., Pietronigro, D. D., and Seligman, M. L., 1980, The free radical pathology and the microcirculation in the major central nervous system disorders, *Acta Physiol. Scand. Suppl.* **492**:91.
Denckla, W. D., 1975, A time to die, *Life Sci.* **16**:31.
Denckla, W. D., 1978, Interactions between age and the neuroendocrine and immune systems, *Fed. Proc.* **37**:1263.
Denko, C. W., 1979, Protective role of ceruloplasmin in inflammation, *Agents Actions* **9**:333, 1979.
De Peretti, E., and Forest, M. G., 1976, Unconjugated dehydroepiandrosterone plasma levels in normal subjects from birth to adolescence in human; the use of a sensitive radioimmunoassay, *J. Clin. Endocrinol. Metab.* **43**:982.

DeRubertis, F. R., and Craven, P., 1980, Cyclic nucleotides in carcinogenesis: Activation of guanylate cyclase-cyclic GMP system by chemical carcinogens, *Adv. Cyclic Nucleotide Res.* **12**:97–109.

Dilman, V. M., 1979, Hypothalamic mechanisms of ageing and of specific age pathology V. A model for the mechanism of human specific age and natural death, *Exp. Gerontol.* **14**:287.

Dinning, J. S., 1962, Nutritional requirements for blood cell formation in experimental animals, *Physiol. Rev.* **42**:169.

Dixon, M., and Webb, E. C., 1964, *Enzymes,* Academic Press, New York.

Doll, R., 1978, An epidemiological perspective of the biology of cancer, *Cancer Res.* **38**:3573.

Doll, R., 1980, The epidemiology of cancer, *Cancer* **45**:2475.

Doolittle, W. F., and Sapienza, C., 1980, Selfish genes, The phenotype paradigm and genome evolution, *Nature* **284**:601.

Dover, G., and Doolittle, R. F., 1980, Modes of genome evolution, *Nature* **288**:646.

Dover, G., Brown, S., Coen, E., Dallas, J., Strachan, T., and Trick, M., 1982, The dynamics of genome evolution and species differentiation, in: *Genome Evolution* (G. A. Dover and R. B. Flavell, eds.), pp 343–372, Academic Press, New York.

Draper, H. H., 1980, Nutrient interrelationships, in: *Vitamin E. A Comprehensive Treatise* (L. J. Machlin, ed.), pp. 272–288, Marcel Dekker, Inc., New York.

Dubravsky, N. B., Hunter, N., Mason, K., and Withers, H. R., 1978, Dibutryl cyclic adenosine monophophate: Effect on radiosensitivity of tumors and normal tissues in mice, *Radiology* **126**:799.

Dunning, W. F., and Curtis, M. R., 1958, Effect of a high protein and high nucleic acid diet on occurrence of 2-acetylaminofluorene-induced cancer in rats, *Proc. Soc. Exp. Biol. Med.* **99**:88.

Dwosh, I. L., Roncari, D. A. K., Marliss, E., and Fox, I. H., 1977, Hypouricemia in disease: A study of different mechanisms, *J. Lab. Clin. Med.* **90**:153.

Earp, H. S., 1980, The role of insulin, glucagon, and cAMP in the regulation of hepatocyte guanylate cyclase activity, *J. Biol. Chem.* **255**:8979.

Egg, D., Altmann, H., Gunther, R., Klein, W., and Kocsis, F., 1978, The influence of some prostaglandins on DNA synthesis and DNA excision repair in mouse spleen cells in vitro, *Prostaglandins* **15**:437.

Emerit, I., and Cerutti, P., 1981, Clastogenic activity from Bloom syndrome fibroblast cultures, *Proc. Natl. Acad. Sci. USA* **78**:1868.

Emerit, I., and Cerutti, P. A., 1982, Tumor promoter phorbol 12-myristate 13-acetate induces a clastogenic factor in human lymphocytes, *Proc. Natl. Acad. Sci. USA* **79**:7509.

Emmett, B., and Hochachka, P. W., 1981, Scaling of oxidative and glycolytic enzymes in mammals, *Resp. Physiol.* **45**:273.

Enstrom, J. E., and Pauling, L., 1982, Mortality among health-conscious elderly Californians, *Proc. Natl. Acad. Sci. USA* **79**:6023.

Evans, G. W., and Wiederanders, R. E., 1967, Blood copper variation among species, *Am. J. Physiol.* **213**:1183.

Everitt, A. V., 1980, The neuroendocrine system and aging, *Gerontology* **26**:108.

Everitt, A. V., and Burgess, J. A., 1976, *Hypothalamus, Pituitary and Aging,* C. C. Thomas, Co., Springfield, Illinois.

Everitt, A. V., Syedsman, N. J., and Jones, F., 1980, The effects of hypophysectomy and continous food restriction begun at ages 70 and 400 days, on collagen aging, proteinuria, incidence of pathology and longevity in the male rat, *Mech. Ageing Dev.* **12**:161.

Fahmy, M. J., and Fahmy, O. G., 1980a, Intervening DNA insertions and the alteration of gene expression by carcinogens, *Cancer Res.* **40**:3374.

Fahmy, M. J., and Fahmy, O. G., 1980b, Altered control of gene activity in the soma by carcinogens, *Cancer Res.* **40**:3374.

Fanelli, G. M., and Beyer, K. H., 1974, Uric acid in nonhuman primates with special reference to its renal transport, *Ann. Rev. Pharmacol.* **14:** 355.

Farber, E., and Cameron, R., 1980, The sequential analysis of cancer development, *Adv. Cancer Res.* **31:**125.

Feldman, M. O., 1976, Dendritic changes in aging rat brain: Pyrimidal cell dendrite length and ultrastructure, in: *The Aging Brain and Senile Dementia* (K. Nandy and I. Sherwin, eds.), pp. 23–37, Plenum Press, New York.

Finch, C. E., 1972, Enzyme activities, gene function and ageing in mammals (review), *Exp. Gerontol.* **7:**53.

Flavell, R., 1982, Sequence amplification, deletion and rearrangement: Major sources of variation during species divergence, in: *Genome Evolution* (G. A. Dover and R. B. Flavell, eds.), pp. 301–323, Academic Press, New York.

Flohé, L., 1982, Glutathione peroxidase brought into focus, in: *Free Radicals in Biology*, Vol. V (W. A. Pryor, ed.), pp. 223–254, Academic Press, New York.

Florine, D. L., Ono, T., Cutler, R. G., and Getz, M. J., 1980, Regulation of endogenous murine leukemia virus-related nuclear and cytoplasmic RNA complexity in C57BL/6J mice of increasing age, *Cancer Res.* **40:**519.

Flower, S. S., 1931, Contribution to our knowledge of the duration of life in vertebrate animals, *Proc. Zool. Soc.* **10:**145.

Folkers, K., 1974, Relationships between coenzyme Q and vitamin E, *Amer. J. Clin. Nutr.* **27:**1026.

Foote, C. S., Chang, Y. C., and Denny, R. W., 1970, Chemistry of singlet oxygen X. carotenoid quenching parallels biological protection, *J. Am. Chem. Soc.* **92:**5216.

Forman, H. J., and Fisher, A. B., 1981, Antioxidant defenses, in: *Oxygen and Living Processes* (D. L. Gilbert, ed.), pp. 235–255, Springer-Verlag, New York.

Francis, A. A., Lee, W. H., and Regan, J. D., 1981, The relationship of DNA excision repair of ultraviolet-induced lesions to the maximum life span of mammals, *Mech. Aging Dev.* **16:**181.

Freeman, B. A., and Crapo, J. D., 1982, Biology of disease. Free radicals and tissue injury, *Lab. Invest.* **47:**412.

Frieden, E., 1980, Caeruloplasmin: A multi-functional metalloprotein of vertebrate plasma, in: *Biological Roles of Copper, Ciba Found. Symp. 79*, pp. 93–124. Excerpta Medica, Amsterdam.

Fucci, L., Oliver, C. N., Coon, M. J., and Stadtman, E. R., 1983, Inactivation of key metabolic enzymes by mixed-function oxidation reactions: Possible implication in protein turnover and ageing, *Proc. Natl. Acad. Sci. USA* **80:**1521.

Fuentes, O. R., 1978, Enhancement of rat liver catalase activity dietary cholesterol, *Life Sci.* **23,** 2039.

Fukuzawa, K., Chida, H., Tolumura, A., and Tsukatani, H., 1981, Antioxidant effect of a-tocopherol incorporation into lecithin liposomes on ascorbic acid-Fe2+-induced lipid peroxidation, *Arch. Biochem. Biophys.* **206:**173.

Gajdusek, D., 1972, Slow virus infection and activation of latent infections in aging, *Adv. Gerontol. Res.* **4:**201–218.

Gardner, T. S., 1963, The possible roles of oral yeast ribonucleic acid (Y-RNA) in geriatrics and gerontology, *Gerontologia* **7:**109.

Garrett-Laster, M., Oaks, L., Russell, R. M., and Oaks, E., 1981, A lowering effect of a pharmacological dose of vitamin E on serum vitamin A in normal adults, *Nutr. Res.* **1:**559.

Gemsa, D., Leser, H.-G., Seitz, M., Deimann, W., and Bärlin, E., 1982, Membrane perturbation and stimulation of arachidonic acid metabolism, *Molec. Immunol.* **19:**1287.

Gerli, G. C., Beretta, L., Bianchi, M., Agostoni, A., Gualandri, V., and Orsini, G. B., 1982,

Erythrocyte superoxide dismutase, catalase, and glutathione peroxidase in glucose-6-phosphate dehydrogenase deficiency, *Scand. J. Haematol.* **29:**135.
Getz, M. J., and Florine, D. L., 1981, Age-related changes in the expression of endogenous mouse C-type RNA tumor virus genes, in: *Biological Mechanisms in Aging* (Schimke, R. T., ed.), pp. 226–244, National Inst. on Aging, NIH Pub. No. 81-2194.
Gillespie, D., Donehower, L., and Strayer, D., 1982, Evolution of primate DNA organization, in: *Genome Evolution* (G. A. Dover and R. B. Flavell, eds.), pp. 113–133, Academic Press, New York.
Goetzl, E. J., 1980, Vitamin E modulates the lipoxygenation of arachidonic acid in leukocytes, *Nature* **288:**183.
Goodrick, C. L., 1980, Effects of long-term voluntary wheel exercise on male and female Wistar rats. I. Longevity, body weight and metabolic rate, *Gerontology* **26:**22.
Goodwin, J. S., Husby, G., and Williams, R. C., 1980, Prostaglandin E and cancer growth, *Cancer Immunol. Immunother.* **8:**3.
Goodwin, T. W., 1954, *Carotenoids. Their Comparative Biochemistry,* Chem Pub. Co., New York.
Goodwin, T. W., 1962, Carotenoids: Structure, distribution, and function, in: *Comparative Biochemistry,* Vol. IV (M. Florkes and H. S. Mason, eds.), pp. 643–675, Academic Press, New York.
Gould, S. J., 1971, Geometric similarity in allometric growth; A contribution to the problem of scaling in the evolution of size, *Am. Naturalist* **105:**113.
Gould, S. J., 1975, Allometry in primates, with emphasis on scaling and the evolution of the brain, *Contrib. Primatol.* **5:**244.
Gould, S. J., 1976a, Human babies as embryos, *Natural History* **85:**22.
Gould, S. J., 1976b, The child as man's real father, *Natural History* **84:**18.
Gould, S. J., 1977, *Ontogeny and Phylogeny,* The Belknap Press, Cambridge, Massachusetts.
Gaaff, G., Stephenson, J. H., Glass, D. B., Haddox, M. K., and Goldberg, N. D., 1978, Activation of soluble splenic cell guanylate cyclase by prostaglandin endoperoxides and fatty acid hydroperoxides, *J. Biol. Chem.* **253:**7676.
Gross, S., 1979, Antioxidant relationship between selenium-dependent glutathione peroxidase and tocopherol, *Am. J. Ped. Hematol./Oncol.* **1:**61.
Gutteridge, J. M. C., Richmond, R., and Halliwell, B., 1980, Oxygen free-radicals and lipid peroxidation: inhibition by the protein ceruloplasmin, *FEBS Lett.* **112:**269–272.
Gutteridge, J. M. C., Rowley, D. A., and Halliwell, B., 1981, Superoxide-dependent formation of hydroxyl radicals in the presence of iron salts, *Biochem. J.* **199:**263.
Haldane, J. B. S., 1932, The time of action of genes, and its bearing on some evolutionary problems, *Am. Naturalist* **66:**5.
Haldane, J. B. S., 1947, Evolution: past and future, *Atl. Monthly* **179:**45.
Hamilton, J. B., and Mestler, G. E., 1969, Mortality and survival: Comparison of eunuchs with intact men and women in a mentally retarded population, *J. Gerontol.* **24:**395.
Hamilton, W. D., 1966, The moulding of senescence by natural selection, *J. Theor. Biol.* **12:**12.
Hane, S., Robertson, O. H., Wexler, B. C., and Krupp, M. A., 1966, Adrenocortical respond to stress and ACTH in Pacific salmon and steelhead trout at successive stages in the sexual cycle, *Endocrinology* **78:**791.
Harman, D., 1956, Aging: A theory based on free radical and radiation chemistry, *J. Gerontol.* **11:**298.
Harman, D., 1962, Role of free radicals in mutation, cancer, aging, and the maintenance of life, *Radiat. Res.* **16:**753.
Harman, D., 1969, Prolongation of life: Role of free radical reactions in aging, *J. Am. Geriatr. Soc.* **17:**721.

Harman, D., 1981, The aging process, *Proc. Natl. Acad. Sci. USA* **78:**7128.

Harman, D., 1982, The free-radical theory of aging, in: *Free Radicals in Biology* (W. A. Pryor, ed.), pp. 255–275, Academic Press, New York.

Harris, H., 1979, Some thoughts about genetics, differentiation, and malignancy, *Somatic Cell Genet.* **5:**923.

Hart, R. W., and Setlow, R. B., 1974, Correlation between deoxyribonucleic acid excision repair and lifespan in a number of mammalian species *Proc. Natl. Acad. Sci. USA* **71:**2169.

Hartman, P. E., 1979, Metaplasias, *Genetics* **91:**s45.

Hartman, P. E., 1982, Nitrates and nitrites: Ingestion, pharmacodynamics and toxicology, in: *Chemical Mutagens, Principles and Methods for Their Detection,* Vol. 7, (F. J. de Serres and A. Hollander, eds.), pp. 211–294, Plenum Press, New York.

Hayaishi, O., and Shimizu, T., 1982, Metabolic and functional significance of prostaglandins in lipid peroxide research, in: *Lipid Peroxides in Biology and Medicine* (K. Yagi, ed.), pp. 41–53, Academic Press, New York.

Hazelton, G. A., and Lang, C. A., 1980, Glutathione contents of tissues in the aging mouse, *Biochem. J.* **188:**25.

Hegner, D., 1980, Age-dependence of molecular and functional changes in biological membrane properties, *Mech. Ageing Dev.* **14:**101.

Heron, D. S., Shinitzky, M., Hershkowitz, M., and Samuel, D., 1980, Lipid fluidity markedly modulates the binding of serotonin to mouse brain membranes, *Proc. Natl. Acad. Sci. USA* **77:**7463.

Hill, D. L., and Grubbs, C. J., 1982, Retinoids as chemopreventive and anticancer agents in intact animals, *Anticancer Res.* **2:**111.

Hill, K. E., and Burk, R. F., 1982, Effect of selenium deficiency and vitamin E deficiency on glutathione metabolism in isolated rat hepatocytes, *J. Biol. Chem.* **257:**10668.

Hirsch, G. P., 1979, Spontaneous mutation balance reproductive selective advantage and genetically determine longevity, *Mech. Ageing Dev.* **9:**355.

Hochschild, R., 1971, Effect of membrane stabilizing drugs on mortality in Drosophila melanogaster, *Exp. Geront.* **6:**133.

Hochschild, R., 1973, Effects of various drugs on longevity in female C57BL/6J mice, *Gerontologia* **19:**271.

Holley, R. W., 1980, Control of animal cell proliferation, *J. Supramolec. Structure* **13:**191.

Holliday, M. A., Potter, D., Jarrah, A., and Bearg, S., 1967, The relation of metabolic rate to body weight and organ size, *Ped. Res.* **1:**185.

Hope, W. C., Dalton, C., Machlin, L. F., Filipski, R. J., and Vane, F. M., 1975, Influence of dietary vitamin E on prostaglandin biosynthesis rat blood, *Prostaglandins* **10:**557.

Hornsby, P. J., 1980, Regulation of cytochrome P-450-supported llb-hydroxylation of deoxycortisol by steroid, oxygen, and antioxidants in adrenocortical cell cultures, *J. Biol. Chem.* **255:**4020.

Hornsby, P. J., and Gill, G. N., 1981a, Regulation of responsiveness of cultured adrenal cells to adrenocorticotropin and prostaglandin E: cell density, cell division, and inhibitors of protein synthesis, *Endocrinology* **108:**183.

Hornsby, P. J., and Gill, G. N., 1981b, Regulation of glutamine and pyruvate oxidation in cultured adrenocortical cells by cortisol, antioxidants, and oxygen: Effects on cell proliferation, *J. Cell Physiol.* **109:**111.

Hornsby, P. J., Simonian, M. H., and Gill, 1979, Aging of adrenocortical cells in culture, *Int. Rev. Cytol.* **10:**131.

Horwitt, M. K., 1976, Vitamin E: A reexamination, *Am. J. Clin. Nutr.* **29:**569.

Hsu, J. M., 1981, Lead toxicity as related to glutathione metabolism, *J. Nutr.* **111:**26.

Huberman, E., Heckman, C., and Langenbach, R., 1979, Stimulation of differentiated functions in human melanoma cells by tumor-promoting agents and dimethyl sulfoxide, *Cancer Res.* **39:**2618.

Heubner, R., and Todaro, G., 1969, Oncogenes of RNA tumor viruses as determinants of cancer, *Proc. Natl. Acad. Sci. USA* **64:**1087.

Hughes, B. A., Roth, G. S., and Pitha, J., 1980, Age-related decrease in repair of oxidative damage to surface sulfhydryl groups on rat adipocytes, *J. Cell Physiol.* **103:**349.

Iida, H., Imai, A., Nozawa, Y., and Kimura, T., 1982, Lipid peroxidation in bovine adrenocortical mitochondria: Arachidonic acid as substrate, *Biochem. Med.* **28:**365.

Iritani, N., and Ideda, Y., 1982, Activation of catalase and other enzymes by corn oil intake, *J. Nutr.* **112:**2235.

Iser, O., ed., 1971, *Carotenoids,* Birkhauser Verlag, Basel.

Jacobs, D., 1976, Hyperuricemia as a risk factor in coronary heart disease, in: *Purine Metabolism in Man. II Physiology, Pharmacology and Clinical Aspects* (M. M. Muller, E. Kaiser, and J. E. Seegmiller, eds.), pp. 231–237, Plenum Press, New York.

Jakoby, W. B., 1978, The glutathione S-transferases: A group of multifunctional detoxification proteins, *Adv. Enzymol. Related Areas Molec. Biol.* **46:**383.

Jakoby, W. B., ed., 1980, *Enzymatic Basis of Detoxification,* Academic Press, New York.

Jakoby, W. B., Bend, J. R., and Caldwell, J., eds., 1982, *Metabolic Basis of Detoxification,* Academic Press, New York.

Jensen, G. E., and Clausen, J., 1981, Glutathione peroxidase activity in vitamin E and essential fatty acid-deficient rats, *Ann. Nutr. Metab.* **25:**27.

Jerison, H. J., 1973, *Evolution of the Brain and Intelligence,* Academic Press, New York.

Johnson, H. A., and Pavelec, M., 1972, Thermal noise in cells. A cause of spontaneous loss of cell function, *Am. J. Pathol.* **69:**119.

Johnson, H. D., Kintner, L. D., and Kibler, H. H., 1963, Effects of 48 F (8.9C) and 83 F (28. 4 C) on longevity and pathology of male rats, *J. Gerontol.* **18:**29.

Jones, M. L., 1962, Mammals in captivity–Primate longevity, *Lab. Primate Newslett.* **1:**3.

Jones, M. L., 1968, Longevity of primates in captivity, *Intern. Zoo Yearbook* **8:**183.

Jukes, T. H., and King, J. L., 1975, Evolutionary loss of ascorbic acid synthesizing ability, *J. Human Evol.* **4:**85.

Kalamegham, R., Naidu, N. A., and Krishnaswamy, K., 1981, Metabolism of xenobiotics in undernourished rats—Regulation by dietary energy and protein levels, *Nutr. Rept. Intern.* **24:**754.

Kark, J. D., Smith, A. H., Switzer, B. R., and Hames, C. G., 1981a, Retinol, carotene, and the cancer/cholesterol association, *Lancet* **1:**1371.

Kark *et al.,* 1981b, Research news, *Science* **211:**1410.

Karmali, R. A., 1980, Review: Prostaglandins and cancer, *Prostaglandins Med.* **5:**11.

Karrer, P., and Jucker, E., 1950, *Carotenoids,* Elsevier Pub. Co., Amsterdam.

Kedziora, J., Lukaszewicz, R., Koter, M., Bartosz, G., Pawlowska, B. and Aitkin, D., 1982, Red blood cell glutathione peroxidase in simple trisomy 21 and translocation 21/22, *Experientia* **38:**543.

Keehner, J. M. 1979, The relationship of serum uric acid to intelligence, achievement, and need for achievement in U.S. adolescents, twelve through seventeen years of age, *Dissert. Abstr.* 1283-A.

Keith, A., ed., 1948, *A New Theory of Human Evolution,* Watts and Co., London.

Kelleher, J., and Losowsky, M. S., 1978, Vitamin E in the elderly, in: *Tocopherol, Oxygen and Biomembranes* (C. deDuve and O. Hayaishi, eds.), pp. 311–327, Elsevier/North Holland, Amsterdam.

King, M. C., and Wilson, A. C., 1976, Evolution at two levels in humans and chimpanzees, *Science* **188:**107.

Kirk, J. E., 1962, Variations in tissue content of vitamins and hormones, *Vitam. Horm.* **20**:82.

Kleiber, M., 1975, *The Fire of Life*, Kreiger, Huntington, New York.

Klevay, L. M., 1980, Hyperuricemia in rats due to copper deficiency, *Nutr. Repts. Intern.* **22**:617.

Knox, W. E., 1976, *Enzyme Patterns in Fetal, Adult and Neoplastic Rat Tissues*, Karger, Basel.

Kohn, R. R., 1978, *Principles of Mammalian Aging*, Prentice-Hall, Inc., Englewood Cliff, New Jersey.

Kohn, R. R., 1982a, Cause of death in very old people, *J. Am. Med. Assoc.* **247**:2793.

Kohn, R. R., 1982b, Evidence against cellular aging theories, in: *Testing the Theories of Aging* (R. C. Adelman and G. S. Roth, eds.), pp. 221–231, CRC Press, Boca Raton, Florida.

Krinsky, N. I., 1971, IX. Function, in: *Carotenoids* (O. Isler, ed.), pp. 669–716, Birkhauser Verlag, Basel.

Krinsky, N. I., 1982, Photobiology of carotenoid protection, in: *The Science of Photomedicine* (J. D. Regan and J. A. Parrish, eds), pp. 397–405, Plenum Press, New York.

Krinsky, N. I., and Deneke, S. M., 1982, Interaction of oxygen and oxy-radicals with carotenoids, *J. Natl. Cancer Inst.* **69**:205.

Kuehl, F., and Egan, R. W., 1980, Prostaglandins, arachidonic acid, and inflammation, *Science* **210**:978.

Kunkel, H. O., Spalding, J. F., de Franciscis, G., and Futrell, M. F., 1956, Cytochrome oxidase activity and body weight in rats and in three species of large animals, *Am. J. Physiol.* **86**:203.

Kurian, M., and Iyer, G. Y. N., 1976, Lipid peroxidation in erythrocytes from different animal species, *Indian J. Biochem. Biophys.* **13**:67.

Lack, D., 1954, *The Natural Regulation of Animal Numbers*, Clarendon, Oxford.

Lamb, M. J., 1977, *Biology of Aging*, J. Wiley & Sons, New York.

Landfield, P. W., 1978, An endocrine hypothesis of brain aging and studies on brain-endocrine correlations and monosynaptic neurophysiology during aging, *Adv. Exp. Med. Biol.* **113**:179–199.

Landfield, P. W., 1980, Correlative studies of brain neurophysiology and behavior during aging, in: *The Psychobiology of Aging: Problems and Perspectives* (Stein, D. G., ed.), pp. 227–251, Elsevier North Holland, Amsterdam.

Landfield, P. W. Rose, G., Sandles, L., Wohlstadter, T. C., and Lynch, G., 1977, Patterns of astroglial hypertrophy and neuronal degeneration in the hippocampus of aged, memory-deficient rats, *J. Gerontol.* **32**:3.

Landfield, P. W., Waymire, J. C., and Lynch, G., 1978, Hippocampal aging and adrenocorticoids: Quantitative correlations, *Science* **202**:1098.

Landfield, P. W., Sundberg, D. K., Smith, M. S., Eldridge, J. C., and Morris, M., 1980, Mammalian aging: Theoretical implications of changes in brain and endocrine systems during mid- and late-life in rats, *Peptides* **1**:185.

Landfield, P. W., Baskin, R. K., and Pitler, T. A., 1981, Brain aging correlates: Retardation by hormonal-pharmacological treatments, *Science* **214**:581.

Lawrence, R. A., and Burk, R. F., 1978, Species, tissue and subcellular distribution of non Se-dependent glutathione peroxidase activity *J. Nutr.* **108**:211.

Lee, Y. H., Layman, D. K., Bell, R. R., and Norton, H. W., 1981, Response of glutathione peroxidase and catalase to excess dietary iron in rats, *J. Nutr. 111:2195.*

Lehman, H. C., 1953, *Age and Achievement*, Princeton Univ. Press, Princeton, New Jersey.

Lehman, H. C., 1962, More about age and achievement, *The Gerontologist*, **2**:141.

Lehnert, S., 1975a, Modification of postirradiation survival of mammalian cells by intracellular cyclic AMP, *Radiat. Res.* **62**:107.

Lehnert, S., 1975b, Relation of intracellular cyclic AMP to the shape of mammalian cell survival curves, in: *Cell Survival After Low Doses of Radiation: Theoretical and Clinical Implication* (T. Alper, ed.), pp. 226–236, J. Wiley & Sons, New York.
Lehnert, S., 1979a, Radioprotection of mouse intestine by inhibitors of cyclic AMP phosphodiesterase, *Int. J. Rad. Oncol. Biol. Phys.* **5**:825.
Lehnert, S., 1979b, Modification of radiation response of CHO cells by methol-isobutyl xanthine 1. Reduction of D., *Radiat. Res.* **78**:1.
Leibovitz, B. E., and Siegel, B. V., 1980, Aspects of free radical reactions in biological systems: Aging, *J. Gerontol. 35:45.*
Leitner, Z. A., Moore, T., and Sharman, I. M., 1960, Vitamin A and vitamin E in human blood. 1. Levels of vitamin A and carotenoids in British men and women, 1948–1957, *Br. J. Nutr.* **14**:157.
Lewin, S., 1976, *Vitamin C: Its Molecular Biology and Medical Potential,* Academic Press, New York.
Lindstedt, S. L., and Calder, W. A., 1981, Body size, physiological time, and longevity of homeothermic animals, *Q. Rev. Biol.* **56**:1.
Lorenz, K., 1971, *Studies in Human and Animal Behavior,* Vol. 2, Harvard Univ. Press, Cambridge, Massachusetts.
Lotan, R., 1980, Effects of vitamin A and its analogs (retinoids) on normal and neoplastic cells, *Biochim. Biophys. Acta* **605**:33.
Lovejoy, C. W., 1981, The origin of man, *Science* **211**:341.
Ludwig, F. C., 1980, Editorial, What to expect from gerontological research? *Science* **209**.
Macdougall, L. G., 1972, Red cell metabolism in iron deficiency anemia, III. The relationship between glutathione peroxidase, catalase, serum vitamin E, and susceptibility of iron-deficient red cells to oxidative hemolysis, *J. Pediatrics* **80**:775.
Machlin, L. J., 1978, Vitamin E and prostaglandins (PG), in: *Tocopherol, Oxygen and Biomembranes* (C. deDuve and O. Hayaishi, eds.), pp. 1799–189, Elvelsier, New York.
Machlin, L. J., 1980, Epilogue, in: *Vitamin. A Comprehensive Treatise* (L. J. Machlin, ed.), pp. 637–645, Marcel Dekker, Inc. New York.
De Marchena, O., Guarnieri, M., and McKhann, G., 1974, Glutathione peroxidase levels in brain, *J. Neurochem.* **22**:773.
Maréchal, R., Lion, Y., and Duchesne, J., 1973, Radicaux libres organiques et longevite maximale chez les mammiferes et les oiseaux, *C.R. Acad. Sci. Paris, Ser. D* **277**:1085.
Martin, C. W., Fjermestad, J., Smith-Barbado, P., and Reddy, B. S., 1980, Dietary modification of mixed function oxidases, *Nutr. Rep. Intern.* **22**:395–408.
Masugi, F., and Nakamura, T., 1976, Effect of vitamin E deficiency on the level of superoxide dismutase, glutathione peroxidase, catalase and lipid peroxide in rat liver, *Int. J. Vit. Nutr. Res.* **46**:186.
Matkovics, B., Gyurkovits, K., László, A., and Szabo;ó, L., 1982a, Altered peroxide metabolism in erythrocytes from children with cystic fibrosis, *Clin. Chim. Acta* **125**:59.
Matkovics, B., Varga, Sz. I., Lzabo, L., and Witas, H., 1982b, The effect of diabetes on the activities of the peroxide metabolism enzymes, *Horm. Metabol. Res.* **14**:77.
Mattenheimer, H., 1971, *Mattenheimer's Clinical Enzymology. Principles and Applications,* Ann Arbor Science Pub. Inc., Ann Arbor.
Maynard Smith, J., 1959, A theory of aging, *Nature* **184**:956.
Maynard Smith, J., 1962, Review lectures on senescence I. The causes of ageing, *Proc. R. Soc. London Ser. B* **157**:115.
Maynard Smith, J., 1966, Theories of aging, in: *Topics in the Biology of Aging* (P. L. Krohn. ed.), pp. 1–27, Interscience, New York.

McCay, P. B. and Gibson, D. D., 1982, Control of lipid peroxidation by a heat-labile cytosolic factor, in: *Lipid Peroxides in Biology and Medicine* (K. Yagi, ed.), Academic Press, New York, pp. 179–197.

McClure, H. M., 1973, Tumors in nonhuman primates: Observations during a six-months period in the Yerkes Primate Center Colony, *Am. J. Phys. Anthrop.* **38:**425.

McClure, H. M., 1975, Neoplasia in Rhesus monkeys, in: *The Rhesus Monkey II. Management, Reproduction and Pathology* (G. H. Bourne, ed.), pp. 369, Academic Press, New York.

Medawar, P. B., 1955, The definition and measurement of senescence, in: *Ciba Foundation Colloq. on Ageing,* Vol. I. *General Aspects* (G. W. E. Wolstenholme and M. P. Cameron, eds.), pp. 4–15, Churchill, London.

Medawar, P. B., 1957, *The Uniqueness of the Individual,* Methuen and Co., Inc., London.

Medvedev, Zh. A., 1966, Error theories of aging, in: *Perspectives in Experimental Gerontology* (N. W. Shock, ed.), pp. 336–350, C. C. Thomas, Pub., Springfield, Illinois.

Meister, A., 1982, Roles and functions of glutathione, *Biochem. Soc. Trans.* **10:**78.

Mikkelsen, W. M., Dodge, H. J., and Valkenburg, H., 1965, The distribution of serum uric acid values in a population unselected as to gout or hyperuricemia: Tecumseh, Michigan 1959–1960. *Am. J. Med.* **39:**242.

Miller, J. K., Bolla, R., and Denckla, W. D., 1980, Age-associated changes in initiation of ribonucleic acid synthesis in isolated rat liver nuclei, *Biochem. J.* **188:**55.

Mitnick, P. D., and Beck, L. H., 1979, Hypouricemia and malignant neoplasms, *Arch. Int. Med.* **139:**1186.

Mitruka, B. M., and Rawnsley, H. M., 1981, *Clinical, Biochemical and Hematological Reference Values in Normal Experimental Animals and Normal Humans,* Masson Pub., New York.

Mittal, C. K., and Murad, F., 1977, Properties and oxidative regulation of guanylate cyclase, *J. Cyclic Nucleotide Res.* **3:**381.

Monod, J., 1971, *Chance and Necessity,* Alfred A. Knopf, New York.

Montagu, A., 1981, *Growing Young,* McGraw-Hill, New York.

Moore, T., 1957, *Vitamin A,* Elsevier Pub. Co., Amsterdam.

Mueller, E. F., Kasl, S. V., Brooks, G. W., and Cobb, S., 1970, Psychological correlates of serum urate levels, *Psychol. Bull.* **73:**238.

Murad, F., Mittal, C. K., Arnold, W. P., Katsuki, S., and Kimura, H., 1978, Guanylate cyclase: Activation by azide, nitro compounds, nitric oxide and hydroxyl radical and inhibition by hemoglobin and myoglobin, *Adv. Cyclic Nucleotide Res.* **9:**145–158.

Murad, F., Arnold, W. P., Mittal, C. K., and Braughhler, J. M., 1979, Properties and regulation of guanylate cyclase and some proposed functions for cyclic GMP, *Adv. Cyclic Nucleotide Res.* **11:**175–204.

Murphy, T. P., Wright, K. E., and Pudelkiewicz, W. J., 1981, An apparent rachitogenic effect of excessive vitamin E intake in the chick, *Poultry Sci.* **60:**1873.

Myers, D. D., 1978, Review of disease patterns and lifespan in aging mice: Genetic and enviormental interaction, in: *Birth Defects on Aging* (D. Bergsma, D. E. Harrison, and N. E. Paul, eds.) pp. 41–53, Allan R. Liss, New York.

Nagatsuka, S., and Nakazawa, T., 1982, Effects of membrane-stabilizing agents, cholesterol and cepharanthin, on radiation-induced lipid peroxidation and permeability in liposomes, *Biochim. Biophys. Acta* **691:**171.

Naito, H. K., 1979, Role of vitamin C in health and disease, in: *Nutritional Elements and Clinical Biochemistry* (M. A. Brewster and H. K. Naito, eds.), pp. 69–115, Plenum Press, New York.

Nakakimura, H., Kakimoto, M., Wada, S., and Mizuno, K., 1980, Studies on lipid peroxidation in biological systems I. Effects of various factors on lipid peroxide level in blood, *Chem. Pharm. Bull.* **28:**2101.

Nandy, K., 1976, Immune reactions in aging brain and senile dementia, in: *The Aging Brain and Senile Dementia* (K. Nandy and I. Sherwin, eds.), pp. 181–196, Plenum Press, New York.

Napier, J. R., and Napier, P. H., 1967, *A Handbook of Living Primates,* Academic Press, New York.

Newberne, P. M., and Rogers, A. E., 1981, Vitamin A, retinoids, and cancer, in: *Nutrition and Cancer: Etiology and Treatment* (G. R. Newell and N. M. Ellison, eds.), pp. 217–232, Raven Press, New York.

Newberne, P. M., Bresnahan, M. R., and Kuka, N., 1969, Effects of two synthetic antioxidants, vitamin E and ascorbic on the choline-deficient rat, *J. Nutr.* **97:**219.

Novogrodsky, A., Ravid, A., Rubin, A. L., and Stenzel, K. H., 1982, Hydroxyl radical scavengers inhibit lymphocyte mitogenesis, *Proc. Natl. Acad. Sci. USA* **79:**1171.

O'Dell, T. B., Wright, H. N., and Bieter, R. N., 1953, Chemotherapeutic activity of nucleic acids and high protein diets against the infection caused by the MM virus in mice, *J. Pharm. Exp. Therap.* **107:**232.

Odumosu, A., 1972, The role of iron in the metabolism of tissue ascorbic acid, *Br. J. Pharmacol.* **45:**136P–137P.

Oertel, G. W., Menzel, P., and Bauke, D., 1970, Effects of dehydroepiandrosterone and its conjugates upon the activity of glucose-6-phosphate dehydrogenase, *Clin. Chim. Acta* **27:** 197.

Ogino, N., Yamamoto, S., Hayaishi, O., and Tokuyama, T., 1979, Isolation of an activator for prostaglandin hydroperoxidase from bovine vascicular gland cytosol and its identification as uric acid, *Biochim. Biophys. Res. Commun.* **87:**184.

Ohkawa, H., Ohishi, N., and Yagi, K., 1978, Reaction of linoleic acid hydroperoxide with thiobarbituric acid, *J. Lipid Res.* **19:**1053.

Ohno, S., 1980, Gene duplication, junk DNA, intervening sequences and the universal signal for their removal, *Rev. Brasiliera Genet.* 3–99.

Ohno, T., and Myoga, K., 1981, The possible toxicity of vitamin C in the guinea pigs, *Nutr. Rep. Int.* **24:**291.

Ohuchi, K., and Levine, L., 1980, a-tocopherol inhibits 12-0-tetradecanoyl-phorbol-13 acetate-stimulated deacytelation of cellular lipids, prostaglandin production, and changes in cell morphology of Madin-Darby canine kidney cells, *Biochim. Biophys. Acta* **619:**11.

Olson, J. A., and Hayaishi, O., 1965, The enzymatic cleavage of b-carotene into vitamin A by soluble enzymes of rat liver and intestine, *Proc. Natl. Acad. Sci. USA* **54:**1364.

Omaye, S. T., and Turnbull, J. D., 1980, Effect of ascorbic acid on heme metabolism in hepatic microsomes, *Life Sci.* **27:**441.

Ono, T., and Cutler, R. G., 1978, Age-dependent relaxation of gene repression: Increase of endogenous murine leukemia virus-related and globin-related RNA in brain and liver of mice, *Proc. Natl. Acad. Sci. USA* **75:**4431.

Orgel, L. E., 1970 ,The maintenance of the accuracy of protein synthesis and its relevance to ageing: a correction, *Proc. Natl. Acad. Sci. USA* **67:**1476.

Orgel, L. E., 1973, Ageing of clones of mammalian cells, *Nature* **243:**441.

Orgel, L. E., and Crick, F. H. C., 1980, Selfish DNA: The ultimate parasite, *Nature* **284:**604.

Pacifici, G. M., Boobis, A. R., Brodie, M. J., McManus, M. E., and Davies, D. S., 1981a, Tissue and species differences in enzymes of epoxide metabolism, *Xenobiotica* **11:**73.

Pacifici, M., Fellini, S. A., Holtzer, H., and DeLuca, S., 1981b, Changes in the sulfated proteoglycans synthesized by "ageing" chondrocytes. I. Dispersed cultured chondrocytes and in vivo cartilages, *J. Biol. Chem.* **256:**1029.

Pall, M. L., 1981, Gene-amplification model of carcinogenesis, *Proc. Natl. Acad. Sci. USA* **78:**2465.

Palmer, S., 1977, Influence of vitamin A nutriture on the immune response: Findings in children with Down's syndrome, *Int. J. Vit. Nutr. Res.* **48:**188.

Park, K. -S., and Asaka, A., 1980, Plasma and urine uric acid levels: Heritability estimates and correlation with IQ, *Jap. J. Hum. Genet.* **25:**193.

Parker, R. J., Berkowitz, B. A., Lee, C. -H., and Denckla, W. D., 1978, Vascular relaxation, aging and thyroid hormones, *Mech. Ageing Dev.* **8:**397.

Pashko, L. L., and Schwartz, A. G., 1982, Inverse correlation between species' life span and species' cytochrome P-488 content of cultured fibroblasts, *J. Gerontol.* **37:**38.

Pashko, L. L., Schwartz, A. G., Abou-Gharbia, M., and Swern, D., 1981, Inhibition of DNA synthesis in mouse epidermis and breast epithelium by dehydroepiandrosterone and related steroids, *Carcinogenesis* **2:**717.

Pearl, R., 1928, *The Rate of Living,* Knopf, New York.

Pekoe, G., Vandyke, K., Peden, D., Mengoli, H., and English, D., 1982, Antioxidation theory of non-steroidal antiinflammatory drugs based upon the inhibition of luminol-enhanced chemiluminescence from the myeloperoxidase reaction, *Agents Actions* **12:**371–376, 1982.

Peto, R., Doll, R., Buckley, J. D., and Sporn, M. B., 1981, Can dietary β-carotene materially reduce cancer rates? *Nature* **290:**201.

Petrović, V. M., Gudz, T., and Saičić, Z., 1981, Selective effect of noradrenaline on superoxide dismutase activity in the brown adipose tissue and liver of the rat, *Experientia* **37:**14.

Petrović, V. M., Spasić, M., Saičić, Z., Milić, B., and Radojičič, R., 1982, Increase in superoxide dismutase activity induced by thyroid hormones in the brains of neonate and adult rats, *Experientia* **38:**1335.

Placer, Z., and Slabochová, Z., 1961, Inhibition of the lipoxidase system by serum proteins, *Biochim. Biophys. Acta* **48:**426.

Plonka, A., and Metodiewa, D., 1980, ESR evidence of superoxide radical dismutation by human ceruloplasmin, *Biochem. Biophys. Res. Commun.* **95:**978.

Ponder, B. A. J., 1980, Genetics and cancer, *Biochim. Biophys. Acta* **605:**369.

Portmann, A., 1945, Die ontogenese des menschen: Als probleme der evolutionsforschung, *Verh. Schweiz. Naturf. Ges.* **125:**44.

Prins, H. K., and Loos, J. A., 1969, Glutathione, in: *Biochemical Methods in Red Cell Genetics* (J. Yunis, ed.), pp. 115–137, Academic Press, New York.

Pryor, W. A., 1978, The formation of free radicals and the consequences of their reactions in vivo, *Photochem. Photobiol.* **28:**787.

Ramsdell, C. M., and Kelley, W. N., 1973, The clinical significance of hypouricemia, *Ann. Intern. Med.* **78:**239.

Reddy, C. C., Tu, C. -P. D., Burgess, J. R., Ho, C. -Y., Scholtz, R. W., and and Massaro, E. J., 1981, Evidence for the occurrence of senium-independent glutathione peroxidase activity in rat liver microsomes, *Biochem. Biophys. Res. Commun.* **101:**970.

Reddy, C. C., Thomas, C. E., Scholz, R. W., and Massaro, E. J., 1982, Effects of inadequate vitamin E and/or selenium nutrition on enzymes associated with xenobiotic metabolism, *Biochem. Biophys. Res. Commun.* **107:**75.

Reddy, V. N., Giblin, F. J., and Matsuda, H., 1980, Defense system of eye lens against oxidative damage, in: *Red Blood Cell and Lens Metabolism* (Srivastava, S. K., ed.), pp. 139–154, Elsevier/North Holland, Amsterdam.

Reiss, U., and Gershon, D., 1979, Methionine sulfoxide reductase: A novel protective enzyme in liver and its potentially significant role in aging, in: *Liver and Aging* (K. Kitani, ed.), pp. 35–64, Elsevier/North Holland, Amsterdam.

Reznick, A. Z., Steinhagen-Thiessen, E., and Gershon, D., 1982, The effect of exercise on enzyme activities in cardiac muscles of mice of various ages, *Biochem. Med.* **28**:347.

Richardson, A., 1981, A comprehensive review of the scientific literature on the effect of aging on protein synthesis, in: *Biological Mechanisms in Aging* (R. T. Schimke, ed.), pp. 339–358, National Institute on Aging, NIH Pub. No. 81-2194.

Robinson, A. B., and Richheimr, S. L., 1975, Instability and function: Ascorbic acid and glutaminyl and asparaginyl residues, *Ann. N.Y. Acad. Sci.* **258**:314.

Roch-Ramel, F., and Peters, G., 1978, Urinary excretion of uric acid in nonhuman mammalian species, in: *Handbook Exp. Pharmacol.,* Vol. 51: *Uric Acid* (W. N. Kelley and I. M. Weiner, eds.), pp. 211–255, Springer-Verlag, New York.

Roch-Ramel, F., Diez-Chomety, F., de Rougemont, D., Tellier, M. Widmer, J., and Peters, G., 1976, Renal excretion of uric acid in rat: A micropuncture and perfusion study, *Am. J. Physiol.* **230**:768.

Rook, A., 1965, Endocrine influences on hair growth, *Br. Med. J. 1:609.*

Rosen, R., 1978a, Feedforwards and global system failures: A general mechanism for senescence, *J. Theor. Biol.* **74**:579.

Rosen, R., 1978b, Cells and senescence, *Int. Rev. Cytol.* **54**:161.

Rosen, R., 1981, Dynamic aspects of senescence, in: *Biological Mechanisms in Aging* (R. T. Schimke, ed.), pp. 108–136, National Inst. on Aging, NIH Pub. No. 81-2194.

Roth, G. S., 1979, Hormone receptor changes during adulthood and senescence. Significance for aging research, *Fed. Proc.* **38**:1910.

Rounds, D. E., 1961, RNA as a protective agent against irradiation of cell cultures, *Ann. N.Y. Acad. Sci.* **95**:994.

Rubinstein, H. M., Dietz, A. A., and Srinavasan, R., 1969, Relation of vitamin E and serum lipids, *Clin. Chim. Acta* **23**:1.

Rubner, M., 1908, Probleme des Wachstums und der Lebensdauer, in: *Gesellschaft für Innere Medizine und Kinderheilkunde,* Vol. 7, pp. 58–72, Mitteilungen, Beiblatt, Wien.

Russanov, E. M., and Kassabova, T. A., 1982, Enzymes of oxygen metabolism and lipid peroxidation in erythrocytes from copper-deficient rats, *Int. J. Biochem.* **14**:321.

Samaan, N. A., 1979, Ectopic hormone producing tumors, in: *Influences of Hormones in Tumor Development,* Vol. 1 (J. A. Kellen, and R. Holf, eds.), pp. 95–113, CRC Press, Boca Raton, Florida.

Sacher, G. A., 1959, Relation of lifespan to brain and body weight in mammals, in: *Ciba Found. Colloq. on Ageing,* Vol. 5 (G. E. W. Wolstenholme and C. M. O'Connor, eds.), pp. 115–133, Churchill Press, London.

Sacher, G. A. 1960a, The dimensionality of the life span, in: *The Biology of Aging* (B. L. Strehler, ed.), pp. 251–252, A.I.B.S., Washington, D.C.

Sacher, G. A., 1960b, The effect of physiologic fluctuations on mortality and aging, in: *The Biology of Aging* (B. L. Strehler, ed.), pp. 246–257, A.I.B.S. Washington, D.C.

Sacher, G. A., 1962, The role of physiological fluctuations in the aging process and the relation of longevity to the size of the central nervous sytem, in: *Aging and Levels of Biological Organization* (A. M. Brues and G. A. Sacher, eds.), pp. 266–305, Univ. Chicago Press, Chicago.

Sacher, G. A., 1965, On longevity regarded as an organized behavior: The role of brain structure, in: *Contributions to the Psychobiology of Aging* (R. Kastenbaum, ed.), pp. 99–110, Springer, New York.

Sacher, G. A., 1966, Abnutzungstheorie, in: *Perspectives in Experimental Gerontology* (N. W. Shock, ed.), pp. 326–335, C. C. Thomas, Pub., Springfield, Illinois.
Sacher, G. A., 1968, Molecular versus systemic theories on the genesis of ageing, *Exp. Gerontol.* **3**:265.
Sacher, G. A., 1970, Allometric and factorial analysis of brain structure in insectivores and primates, in: *The Primate Brain* (C. R. Noback and W. Montagna, eds.), pp. 245–287, Appleton-Century-Crofts, New York.
Sagone, A. L., Greenwald, J., Kraut, E. H., Bianchine, J., and Singh, D., 1983, Glucose: A role as a free radical scavenger in biological systems, *J. Lab. and Clin. Med.* **101**:97.
Sass. M. D., Caruso, C. J., and O'Connell, D. J., 1965, Decreased glutathione in aging red cells, *Clin. Chim. Acta* **11**:334.
Sato, P., and Udenfriend, S., 1978, Studies on ascorbic acid related to the genetic basis of scurvey, *Vitam. Horm.* **36**:33.
Sato, Y., Hotta, N., Sakamoto, N., Matsuoka, S., Ohishi, N., and Yagi, K., 1979, Lipid peroxide level in plasma of diabetic patients, *Biochem. Med.* **21**:104.
Schapira, F., 1973, Isozymes and cancer, *Adv. Cancer Res.* **18**:77.
Scheibel, M. E., Lindsay, R. D., Tomiyasu, U., and Scheibel, A. B., 1975, Progressive dendritic changes in aging human cortex, *Exp. Neurol.* **47**:392.
Schultz, A. H., 1936, Characters common to higher primates and characters specific for man, *Q. Rev. Biol.* **2**:259.
Schultz, A. H., 1949, Ontogenetic specializations of man, *Arch. Julius Klaus Stiftung* **24**:197.
Schultz, A. H., 1950, The physical distinctions of man, *Proc. Am. Philos. Soc.* **94**:428.
Schultz, A. H., 1960, Age changes in primates and their modifications in man, in: *Human Growth* (J. Tanner, ed.), pp. 1–20, Pergamon Press, London.
Schultz, A. H., 1966, Changing views on the nature and interrelations of the higher primates, *Emory Univ. Primate Res. Centr. Newslett.* **3**:15.
Schultz, A. H., 1968, The Recent hominoid primates, in: *Perspectives on Human Evolution,* Vol. 1 (S. L. Washburn and P. C. Jay, eds.), pp. 122–195, Holt, Rinehart and Winston, New York.
Schwartz, A. G., 1979, Inhibition of spontaneous breast cancer formation in female C3H(A/a) mice by long-term treatment with dehydroepiandrosterone, *Cancer Res.* **39**:1129.
Schwartz, A. G., and Perantoni, A., 1975, Protective effect of dehydroepiandrosterone against aflatoxin Bl- and 7,12-dimethylbenz(a)anthracene-induced cytotoxicity and transformation in cultured cells, *Cancer Res.* **35**:2482.
Scott, M., Bolla, R., and Denckla, W. D., 1979, Age-related changes in immune function of rats and the effect of long-term hypophysectomy, *Mech. Ageing Dev.* **11**:127.
Seal, U. S., 1964, Vertebrate distribution of serum ceruloplasmin and sialic acid and the effects of pregnancy, *Comp. Biochem. Physiol.* **13**:143.
Seegmiller, J. E.,, 1979, Disorders of purine and pyrimidine metabolism in: *Contemporary Metabolism,* Vol. 1 (E. Freinkel, ed.), pp. 1–85. Plenum Press, New York.
Seregi, A., Serfőző, P., and Mergl, Z., 1983, Evidence for the localization of hydrogen peroxide-stimulated cycloxygenase activity in rat brain mitochondria: A possible coupling with monoamine oxidase, *J. Neurochem.* **40**:407.
Sharma, S. K., and Krisna Murti, C. R., 1976, Ascorbic acid—A naturally occurring mediator of lipid peroxide formation in rat brain, *J. Neurochem.* **27**:299.
Shmookler Reis, R. J., Lumpkin, C. K., McGill, J. R., Riabowol, K. T., and Goldstein, S., 1983, Extrachromosomal circular copies of an 'inter-Alu' unstable sequence in human DNA are amplified during in vitro and in vivo ageing, *Nature* **301**:394.
Shock, N. W., 1952, Ageing of homeostatic mechanisms, in: *Cowdry's Problems of Ageing* (A. I. Lansing, ed.), pp. 415–446, Williams & Wilkins, Baltimore, Maryland.

Shock, N. W., 1961, Physiological aspects of aging in man, *Ann. Rev. Physiol.* **23**:97.
Shock, N. W., 1970, Physiologic aspects of aging, *J. Am. Diet. Assoc.* **56**:491.
Sinet, P. M., 1982, Metabolism of oxygen derivatives in Down's syndrome, in: *Alzheimer's Disease, Down's Syndrome and Aging,* Ann. N.Y. Acad. Sci., Vol. 396 (F. M. Sinex and C. R. Merril, eds.), pp. 83–94.
Sklan, D., Rabinowitch, H. D., Donoghue, S., 1981, Superoxide dismutase: Effect of vitamins A and E, *Nutr. Rept. Int.* **24**:551.
Snipes, C. A., Forest, M. G., and Migeion, C. J., 1969, Plasma androgen concentrations in several species of Old and New World monkeys, *Endocrinology* **85**:941.
Solyom, L., Enesco, H. E., and Beaulieu, C., 1968, The effect of RNA, uric acid and caffein on conditioning and activity in rats, *J. Psychiatr. Res.* **6**:175.
Somero, G. N., and Childress, J. J., 1980, A violation of the metabolism-size scaling paradigm: Activities of glycolytic enzymes in muscle increase in larger-size fish, *Physiol. Zool.* **53**:322.
Sparagana, M., and Phillips, G., 1972, Dehydroepiandrosterone 3b-hydroxy-5-androsten-17-one) metabolism in gout, *Steroids* **19**:477.
Sparberg, M., 1981, Blood chemistry and hematology, in: *Clinical Data for Gorillas, Orange-Utans, and Chimpanzees at the Lincoln Park Zoological Gardens: Preliminary Report* (E. R. Maschgan, ed.), pp. 45–54, Dept. Biochem. Michael Reese Med. Center, Chicago, Illinois.
Spector, I., 1974, Animal longevity and protein turnover rate, *Nature* **249**:66.
Spector, W. S., ed., 1956, *Handbook of Biological Data* Division of Biology and Agriculture, National Academy of Sciences, National Research Council.
Sporn, M. B., and Newton, D. L., Chemoprevention of cancer with retinoids, *Fed. Proc.* **38**:2528, 1979.
Stahl, W. R., 1962, Similarity and dimensional methods in biology, *Science* **137**:205.
Stanley, S. M., 1979, *Macroevolution,* W. H. Freeman and Co., San Francisco.
Stanley, S. M., 1981, *The New Evolutionary Timetable,* Basic Books, New York.
Stetten, D., 1958, Gout and metabolism, *Sci. Am.* **198(6)**:73.
Stich, H. F., Wei, L., Whiting, R. F., 1979, Enhancement of the chromosome-damaging action of ascorbate by transition metals, *Cancer Res.* **39**:4145.
Stohs, S. J., Hassing, J. M., Al-Turk W. A., and Masoud, A. N., 1980, Glutathione levels in hepatic and extrahepatic tissues of mice as a function of age, *Age* **3**:11.
Stohs, S. J., Al-Turk, W. A., and Angle, C. R., 1982, Glutathione S-transferase and glutathione reductase activities in hepatic and extrahepatic tissues of female mice as a function of age, *Biochem. Pharmacol.* **31**:2113.
Stone, W. L., and Dratz, E. A., 1980, Increased glutathione-S-transferase activity in antioxidant-deficient rats, *Biochim. Biophys. Acta* **631**:503.
Straus, D. S., 1981, Somatic mutation, cellular differentiation, and cancer causation, *J. Natl. Cancer Inst.* **67**:233.
Strehler, B. L., 1978, *Time, Cells and Aging,* Academic Press, New York.
Suematsu, T., Kamada, T., Abe, H., Kikuchi, S., and Yagi, D., 1977, Serum lipoperoxide level in patients suffering from liver diseases, *Clin. Chim. Acta* **79**:267.
Sugimura, T., Matsushima, T., Kawachi, T., Kogure, K., Tanaka, T., Miyake, S., Hozumi, M., Sato, S., and Sato, H., 1972, Disdifferentiation and decarcinogenesis, *Gann Monogr. Cancer Res.* **3**:31.
Summer, K. -H., and Greim, H., 1981, Hepatic glutathione S-transferases: Activities and cellular localization in rat, Rhesus monkey, chimpanzee and man, *Biochem. Pharm.* **30**:1719.
Sun, A. Y., and Sun, G. Y., 1982, Dietary antioxidants and aging on membrane functions, in: *Nutritional Approaches to Aging* (G. B. Moment, ed.), pp. 135–156, CRC Press, Boca Raton, Florida.

Swell, L., Field, J. R., and Treadwell, C. R., 1960, Correlation of arachidonic acid of serum cholesterol esters in different species with susceptibility to atherosclerosis, *Proc. Soc. Exp. Biol.* **104:**325.

Taketa, L., Watanabe, A., and Kosaka, K., 1976, Undifferentiated gene expression in liver injuries, in: *Onco-Developmental Gene Expression* (W. H. Fishman and S. Sell, eds.), pp. 219–226, Academic Press, New York.

Tappel, Al. L, 1980, Measurement of and protection from in vivo lipid peroxidation, in: *Free Radicals in Biology,* Vol. IV (W. A. Pryor, ed.), pp. 1–47, Academic Press, New York.

Tappel, M. E., Chaudieri, J., and Tappel, A. L., 1982, Glutathione peroxidase activities of animal tissues, *Comp. Biochem. Physiol.* **73B:**945.

Terry, R. D., 1980, Some biological aspects of the aging brain, *Mech. Ageing Dev.* **14:**191.

Thompson, S. Y., Ganguly, J., and Kon, S. K., 1949, The conversion of β-carotene to vitamin A in the intestine, *Br. J. Nutr.* **3:**50.

Todaro, G. J., 1980, Interspecies transmission of mammalian retroviruses, in: *Molecular Biology of RNA Tumor Viruses* (J. R. Stephenson, ed.), pp. 47–76, Academic Press, New York.

Tolmasoff, J. M., Ono, T., and Cutler, R. G., 1980, Superoxide dismutase: Correlation with lifespan and specific metabolic rate in primate species, *Proc. Natl. Acad. Sci. USA* **77:**2777.

Tomlinson, B. E., 1977, Morphological changes and dementia in old age, in: *Aging and Dementia* (W. L. Smith and M. Kinsbourne, eds.), pp. 25–56, Spectrum Pub., Inc., Jamaica, New York.

Totter, J. R., 1980, Spontaneous cancer and its possible relationship to oxygen metabolism, *Proc. Natl. Acad. Sci. USA* **77:**1763.

Tovey, 1980, M. G., Viral latency and its importance in human disease, *Pathol. Biol.* **28:**631.

Townsley, J. D., and Pepe, G. J., 1981, Serum dehydroepiandrosterone and dehydroepiandrosterone sulphate in baboon (Papio papio) pregnancy, *Acta Endocrinol.* **85:**415.

Treton, J. A., and Courtois, Y., 1982, Correlations between DNA excision repair and mammalian lifespan in lens epithelial cells, *Cell Biol. Int. RCp.* **6:**253.

Trosko, J. E., and Chang, C. -C., 1981, The role of radiation and chemicals in the induction of mutations and epigenetic changes during carcinogenesis, *Adv. Rad. Biol.* **9:**1.

Urbach, C., Hickman, K., and Harris, P. L., 1952, Effect of individual vitamins A, C, E, and carotene administered at high levels on their concentration in blood, *Exp. Med.* **10:**7.

Valentovic, M. A., Gairola, C., and Lubawy, W. C., 1982, Lung, aorta, and 14C-arachidonic acid in vitamin E deficient rats, *Prostaglandins* **24:**215.

Veatch, R., 1978, *Lifespan,* Basic Books, New York.

Vesely, D. L., and Hill, D. E., 1980, Estrogens and progesterone increase fetal and maternal guanylate cyclase activity, *Endocrinology* **107:**2104.

Vidlakova, M., Erazimova, J., Horky, J., and Placer, Z., 1972, Relationship of serum antioxidative activity to tocopherol and serum inhibitor of lipid peroxidation, *Clin. Chim. Acta* **36:**61.

Walford, R. L., 1974, Immunological theory of aging: Current status, *Fed. Proc.* **33:**2020.

Walker, E. P., 1975, *Mammals of the World,* Johns Hopkins Univ. Press, Baltimore, Maryland.

Wallace, D. C., 1967, The inevitability of growing old, *J. Chronic. Dis.* **20:**475.

Wallace, D. C., 1975, A theory of the cause of aging, *Med. J. Australia* **1:**829.

Washburn, S. L., 1981, Longevity in primates, in: *Aging: Biology and Behavior* (J. L. McGaugh and S. B. Kiesler, eds.), pp. 11–29, Academic Press, New York.

Waterlow, J. C., and Jackson, A. A., 1981, Nutrition and protein turnover in man, *Br. Med. Bull.* **37:**5.

Watkin, D. M., 1982, The physiology of aging, *Am. J. Clin. Nutr.* **36:**750.

Weinberg, R. A., 1980, Origins and roles of endogenous retroviruses, *Cell* **22:**643.

Weinstein, B., Irreverre, F., and Watkin, D. M., 1965, Lung carcinoma, hypouricemia and aminoaciduria, *Am. J. Med.* **39:**520.

Weisburger, J. H., Cohen, L. A., and Wynder, E. L., 1977, On the ethiology and metabolic epidemiology of the main human cancers, in: *Origins of Human Cancer*. Book A. *Incidence of Cancer in Humans* (H. H. Hiatt, J. D. Watson, and J. A. Winsten, eds.), pp. 567–602, CSH Pub., New York.

Wexler, B. C., 1976, Comparative aspects of hyperadrenocorticism and aging, in: *Hypothalamus, Pituitary and Aging* (A. V. Everitt and J. A. Burgess, eds.), C. C. Thomas, Springfield, Illinois, pp. 333–361.

Willett, J. D., Rahim, I., Geist, M., and Zuckerman, B. M., 1980, Cyclic nucleotide exudation by nematodes and the effect on nematode growth, development and longevity, *Age* **3**:82.

Williams, G. C., 1957, Pleiotropy, natural selection, and the evolution of senescence, *Evolution* **11**:398.

Wilson, A. C., 1976, Gene regulation in evolution, in: *Molecular Evolution* (F. J. Ayala, ed.), pp. 225–234, Sinauer Assoc. Inc., Sunderland, Massachusetts.

Wilson, A. C., Sarich, V. M., and Maxson, L. R., 1974, The importance of gene rearrangement in evolution: Evidence from studies on rates of chromosomal, protein and anatomical evolution, *Proc. Natl. Acad. Sci. USA* **71**:3028.

Wilson, A. C., Carlson, S. S., and White, T. J., 1977, Biochemical evolution, *Ann. Rev. Biochem.* **46**:573.

Wilson, E. O., 1975, *Sociobiology*, Belknap Press, Cambridge, Massachusetts.

Wisniewski, H. M., and Soifer, D., 1979, Neurofibrillary pathology: Current status and research perspectives, *Mech. Ageing Dev.* **9**:119.

Wodinsky, J., 1977, Hormonal inhibition of feeding and death in octopus: Control by optic gland secretion, *Science* **198**:948.

Wyngaarden, J. B., and Kelley, W. N., 1979, Disorders of purine and pyrimidine metabolism, in: *Contemporary Metabolism*, Vol. 1 (N. Freinkel, ed.), pp. 1–130, Plenum Press, New York.

Yagi, K., ed., 1982, *Lipid Peroxides in Biology and Medicine*, Academic Press, New York.

Younes, M., and Siegers, C. -P., 1981, Mechanistic aspects of enhanced lipid peroxidation following glutathione depletion in vivo, *Chem. Biol. Interact.* **34**:257.

Zigler, J. S., and Goosey, J., 1981, Aging of protein molecules: Lens crystallins as a model system, *Trends Biochem. Sci.* **6**:133.

Zs.-Nagy, I., 1979, The role of membrane structure and function in cellular aging: A review, *Mech. Ageing Dev.* **9**:237.

Zuckerkandl, E., 1976a, Gene control in eukaryotes and the C-value paradox. "Excess" DNA as an impediment to transcription of coding sequences, *J. Molec. Evol.* **9**:73.

Zuckerkandl, E., 1976b, Programs of gene action and progressive evolution, in: *Molecular Anthropology* (M. Goodman and R. E. Tashian, eds.), pp. 387–447, Plenum Press, New York.

2

Receptors and Aging

GERALD D. HESS and GEORGE S. ROTH

1. INTRODUCTION

Recent years have seen major advances in neuroendocrinology, neurobiology, and membrane biochemistry. Hormone and neurotransmitter receptors were poorly understood and many were unknown until a few years ago. During the past decade there has been a growing realization that neurotransmitters convey nerve impulses and that hormone receptors ultimately convey "messages" to cellular control centers (Pradham, 1980; Cuatrecases, 1974). While some researchers have attempted to document the age-related deterioration of physiological control systems in a general way, this study considers the role of hormone and neurotransmitter receptors in the age-related dysfunctions of once "finely tuned" control systems. Current studies of age changes in receptors come from such diverse fields as pharmacology, neuroendocrinology, and experimental psychology. These and related subdisciplines are becoming increasingly aware of the importance of such studies viewed from a gerontological perspective.

It is our opinion that knowledge of age-related receptor changes is necessary if a more complete understanding of the impact of aging on both the endocrine and nervous systems is to be achieved. Since these are the two major control systems of the body, age-related changes in their functional integrity have widespread implications for other systems in the body.

During the next decade, ongoing refinement of research models and techniques by endocrinologists and neurobiologists promises to yield increasingly more detailed understanding of the normal functional characteristics of these

---

GERALD D. HESS • Department of Natural Sciences, Messiah College, Grantham, Pennsylvania 17027. **GEORGE S. ROTH** • National Institute on Aging, National Institutes of Health, Gerontology Research Center, Baltimore City Hospitals, Baltimore, Maryland 21224.

systems. This is not to suggest that the cause(s) of aging are necessarily to be found in the endocrine and/or nervous systems. Although some researchers continue to search for "the cause" of aging, many others believe that aging results from gradual and simultaneous deterioration in a number of the body's functional components (Finch, 1976; Shock, 1977a,b). It is reasonable to believe that receptor changes in the endocrine and nervous systems are involved here.

It is our purpose to summarize the trends emerging from studies of age-related receptor changes that have been published in recent years. This is not intended to be an exhaustive treatise on this topic, but we do wish it to be a balanced presentation of what is known about receptors and aging. Because the literature on receptors and aging generated in recent years is voluminous, findings will be considered in terms of specific hormone receptor groups. Three main categories of receptors will be addressed: (1) surface-binding neurotransmitter receptors, (2) surface-binding hormone receptors, and (3) cytoplasmic hormone receptors.

Since receptors are required for transmitters and hormones to exert their effects on cells, there should be some correlation between certain receptor changes and biological responses. Changes in response could also involve alterations at a level beyond the receptor. The number of such "postreceptor" changes with age reported in the literature to date is limited, but an attempt will be made to bring related changes of this type into focus at the end of each section.

It is now a well-documented fact that neurotransmitters and hormones exert their effects either by way of receptors on the surface or in the cytoplasm of target cells (Roth and Adelman, 1975; Roth, 1979b; Cuatrecasas, 1974). Likewise, there exists a sizable body of evidence that tissue responsiveness to hormones and neurotransmitters (Dilman, 1971; Finch, 1976; Everett, 1973; Pradham, 1980) declines with age. This being true, it is important to consider changes in receptors as possible causes for these declines. These declines could involve alterations in receptor number (concentration), binding-affinity, functional integrity, etc., or as mentioned above, changes beyond the receptor. In many cases, (hormone) receptor binding activates second (and third) messenger substances which travel to precise locations within cells and bring about ultimate cellular responses (Pastan and Willingham, 1981; Roth and Adelman, 1975).

Neurotransmitters convey chemical information from neuron to neuron. Currently, several substances (e.g., acteyleholine, norepinephrine, dopamine, etc.) are known to be neurotransmitters on the basis of substantial experimental evidence. A number of other substances are now implicated as neurotransmitters, and experimental verification of their role will be forthcoming in the future. Vesicles (prepackaged sacs) of these transmitter substances are released from neuronal end bulbs in response to nerve impulses and attach to receptors on the surface of adjoining neurons. In some cases, receptor binding results in altered membrane permeability, which, depending on the transmitter in question, can either generate or inhibit nerve impulses in the affected neuron. For other transmitters, binding initiates a sequence of events leading to some type of increased cellular activity (as for example contraction of smooth or cardiac muscle). Still other substances (e.g., opioids) are thought to act as neuromodulators rather than as neurotransmitters. Acting postsynaptically, they are thought to enhance or diminish responses previously induced in the respective neurons (Nicoll *et al.,* 1977). For both transmitter and modulator substances, actions are exerted via binding of protein or proteinlike substances to receptors located on cell surfaces.

By definition, hormones are synthesized and released at sites distant from where they exert their actions. These actions, which are exerted via binding to receptors, may lead to changes in enzymatic and general metabolic activity within cells and/or effects upon membrane permeability. Many of the hormones are protein or proteinlike molecules that bind to surface receptors (Cuatrecasas, 1974; Kaplan, 1981). The actual cellular response results from information carried by second messengers such a cAMP, Ca^{2+}, etc., to specific enzymes or related components in the case of hormones that exert metabolic effects. For surface-binding hormones, response mechanisms beyond the receptor could also be responsible for age-related changes in tissue responsiveness.

Steroid hormones such as the androgens, estrogens, and glucocorticoids, readily traverse cell membranes. Receptors for these hormones are located within the cell cytoplasm rather than on cell surfaces. Hormone–receptor complexes of this type travel to the cell nucleus where they enhance the rate of transcription (Latham and Finch, 1976). This in turn produces increased enzyme availability and subsequent increases in the related metabolic activities of the cell.

While there are differences for specific hormones and transmitters, techniques for quantitating receptors are characterized by many similarities. These will be presented here. It should be noted that these techniques measure functional receptors (i.e., those that will specifically bind to the specific ligand under study).

Properly washed homogenates of known (protein) concentration are incubated in a mixture containing a radioactive isotope of the transmitter or hormone under study. Usually, at least five specific and progressively higher concentrations of radioligand are added to successive tubes of homogenate. These are incubated in duplicate or triplicate to enhance precision. Since the ligand also binds nonspecifically (i.e., to sites other than receptors) to homogenate, a comparable sequence of tubes containing an established quantity of nonisotopic ligand as well as the successively higher concentrations of radioligand are incubated at the same time. Prior addition of the unlabeled ligand blocks receptors, thereby allowing only nonspecific binding of the radioligand to occur in these tubes. In this way corrections for nonspecific binding can be made at each concentration of radioligand subsequent to counting in a liquid scintillation spectrometer. Unbound radioligand is removed by a special washing technique prior to counting.

The resulting data are analyzed by Scatchard analysis. A plot of bound-to free- hormone ratio versus concentration of bound radiologand defines a straight line whose slope represents the binding affinity between ligand and receptor and whose X-intercept defines the receptor concentration for the tissue under study. (For additional details concerning techniques for the binding studies as well as the subsequent analysis of the data, see Roth, 1974.)

2. INTERPRETING RECEPTOR DATA

Before discussing the effect of age on specific receptor groups, several important factors relating to data interpretation must be considered. A survey of the tables presented here suggests that cases of reported decreases in receptor number with age outnumber cases with no age-related changes. Hopefully this represents an accurate accounting of the situation. A bias could be introduced by the tendency to study and report only those receptor systems in which a decrease in number has occurred.

Receptor levels are known to be influenced by endogenous concentrations of their neurotransmitters or hormones. In addition, other hormones can also alter receptor concentration. For example, adrenergic receptor levels are altered not only by endogenous catecholamine levels but also by hormones such as thyroxine, progesterone, or cortisone (Hoffman and Lefkowitz, 1980).

For beta-adrenergic and dopaminergic receptors, reduced levels of the endogenous neurotransmitter sometimes result in increased receptor concentrations (Hoffman and Lefkowtiz, 1980) and vice-versa. This response likely constitutes a self-induced protective mechanism for the cells involved. It is illustrated best in so-called sensitization and desensitatization studies that have been carried out for adrenergic and dopaminergic receptors (Hoffman and Lefkowitz, 1980; Joseph et al., 1981). For certain hormones, there seems to be a more direct relationship between endogenous hormone levels and receptor concentration. That is, declining hormone levels result in lowered concentrations and vice-versa (Boesel and Shain, 1980).

Another caution relates to the manner in which animals are grouped into age categories. Studies in rats and mice, for example, cannot be considered to be studies of senescence unless animals at least 18 months of age (depending on the mean lifespan of the strain) are utilized for the aged (senescent) group.

In these species, changes observed between birth and 3 months of age are considered to be developmental rather than gerontological. At 12 months of age, characteristics of mature adult animals generally exist, although for some parameters age-related changes may already be evident. For other systems, physiological changes are not observed unless animals approaching 24 months of age are compared to year-old animals.

Ideally, studies of aging should be longitudinal (i.e., observations from the same animal thoughout its entire lifespan). Since most techniques involved in receptor studies require sacrifice of the animal in order to obtain tissue, longitudinal studies are frequently not practical. Researchers using rats and mice for studies of aging have somewhat arbitrarily established three age groups: (a) young (3–6 months), (b) adult (approximately 12 months), and (c) senescent (approximately 24 months). Depending on sex and/or strain, the maximum lifespan of rats generally ranges from 24 to 36 months (p. 187, Comfort, 1979) with an average of 30 months (Hess, unpublished observation).

There is some variation from laboratory to laboratory in the way age groups of rats and mice are designated. For other species, there are fewer studies and therefore even less standardization of age categories. In humans, many receptor studies are based on tissues acquired at autopsy. The possible impact of postmortem changes must be included in the interpretation of studies based on human tissue.

A final comment relates to the fact that receptor studies generally consider both receptor number (concentration) and binding affinity. In the literature reviewed for this article, relatively few changes in receptor binding affinity were reported. Therefore only changes in receptor concentration will be considered here. In the future, it is possible that changes in receptor binding affinity will also need to be included.

Many variables must be taken into account when studying the relationship between receptors and aging. Although an exhaustive discussion of these factors is not appropriate in this review, we have raised these issues so that they may be taken into account as readers interpret the studies presented here. In the sections that follow, various groups of neurotransmitter and hormone receptors will be considered.

3. SURFACE-BINDING NEUROTRANSMITTER RECEPTORS

3.1. Adrenergic Receptors

The influence of increasing age on beta-adrenergic receptor concentrations in the central nervous system has been studied extensively in recent years. The physiological role of these receptors is discussed at length in reviews by

Pradham (1980) and Hoffman and Lefkowitz (1982), which should be consulted by those desiring detailed coverage of this material.

Studies by Misra *et al.* (1980), Enna and Strong (1981), and by Greenburg and Weiss (1978) summarized in Table I have shown that beta-adrenergic receptor concentrations in the rat cerebral cortex decline with age. Piantanelli *et al.* (1981) found changes comparable to these in the cerebral cortex of aging mice. Pittman *et al.* (1980) have shown that beta-adrenergic receptor levels also decrease with age in the cerebellar cortex of rats while Enna and Strong (1981) and Maggi *et al.* (1979) have found the same to occur in both mice and humans. Other regions of the central nervous system in the rat in which beta-adrenergic receptors decrease with age are the brain stem (Enna and Strong, 1981), the corpus striatum and pineal gland (Greenburg and Weiss, 1978), and the micro vessels of the brain (Trabucchi *et al.,* 1981). While receptor losses could be attributed to age-related decreases in the number of cells, it is likely that the number of receptors per cell also decreases with age.

As seen in Table I, not all studies of neural tissue have revealed declining beta-adrenergic receptor concentrations with age. Maggi *et al.* (1979) found no change in the cerebral cortex of aging rats while Pittman *et al.* (1980) report an age-related increase in beta-receptor (beta-one subtype) concentrations of the rat cerebellum.

Blood cells provide a ready system for studies of receptors. While such studies sometimes consider red cells at various stages of their 120 day lifespan, those reported here are grouped according to the age of blood cell donors rather than the "age" of blood cells as such. Byland *et al.* (1977) reported that beta-receptors are lost from rat erythrocytes with increasing age while Shocken and Roth (1977) observed similar changes in beta-adrenergic receptor concentrations of human lymphocytes. In contrast to these findings are reports that beta-adrenergic receptor concentrations did not change with increasing age in rat erythrocytes when corrected for decreased cell size (Bilizekian and Gammon, 1978) and human lymphocytes (Abrass and Scarpace, 1981; Landmann *et al.,* 1981). It should be noted that in the early study, Shocken and Roth (1977) examined a binding site with a K_D (i.e., binding affinity) of about 20 nM, while in the more recent studies a site with high binding affinity (K_D of about 2nM) was involved. Hence the difference in results can be attributed to the fact that Shocken and Roth (1977) present data from a different binding site than do the others.

Beta-adrenergic receptor concentrations have been quantitated in a number of other tissues as well. Piantanelli *et al.* (1980) found that they decreased in submandibular gland tissue of aging rats, but in our laboratory, Ito *et al.* (1981) found no age-related change in rat parotid glands. Beta-adrenergic receptors have also been studied as a function of age in adipocytes. Giudicelli

Table I. Receptor Concentration: Senescence versus Mature Adulthood

Tissue	Age effect	Species	Reference
	Beta-adrenergic receptors		
Cerebral cortex	Decrease	Rat	Misra *et al.* (1980)
Cerebral cortex	Decrease	Rat	Enna and Strong (1981)
Cerebral cortex	Decrease	Mouse	Piantanelli *et al.* (1981)
Cerebellum (β_2)	Decrease	Rat	Pittman *et al.* (1980)
Cerebellum	Decrease	Rat	Greenberg and Weiss (1981)
Cerebellum	Decrease	Rat	Maggi *et al.* (1979)
Cerebellum	Decrease	Mouse	Enna and Strong (1981)
Cerebellum	Decrease	Human	Enna and Strong (1981)
Cerebellum	Decrease	Human	Maggi *et al.* (1979)
Pineal gland	Decrease	Rat	Greenberg and Weiss (1981)
Corpus striatum	Decrease	Rat	Greenberg and Weiss (1981)
Brain cortex microvessels	Decrease	Rat	Trabucchi *et al.* (1981)
Brain stem	Decrease	Rat	Enna and Strong (1981)
Erythrocytes	Decrease	Rat	Bylund *et al.* (1977)
Lymphocytes	Decrease	Human	Schocken and Roth (1977)
Submandibular gland	Decrease	Mouse	Piantanelli *et al.* (1980)
Adipocytes	Decrease	Rat	Giudicelli and Pequery (1978)
Cerebral cortex	No change	Rat	Maggi *et al.* (1979)
Cerebral cortex	No change	Rat	Bilizekian and Gammon, (1978)
Heart	No change	Rat	Zitni and Roth (1981)
Heart	No change	Rat	Guarnieri *et al.* (1980)
Heart	No change	Rat	Scarpace and Abrass (1981)
Erythrocytes	No change	Rat	Bilizekian and Gammon (1978)
Lymphocytes	No change	Human	Abrass and Scarpace (1981)
Lymphocytes	No change	Human	Landman *et al.* (1981)
Adipocytes	No change	Rat	Dax *et al.* (1981)
Parotid gland	No change	Rat	Ito *et al.* (1981)
Cerebellum (β_1 subtype)	Increase	Rat	Pittman *et al.* (1980)
	Alpha-adrenergic receptors		
Cerebral cortex	Decrease	Rat	Misra *et al.* (1980)
Heart	Decrease	Rat	Partilla *et al.* (1982)
Parotid cells (α_1 subtype)	Increase	Rat	Ito *et al.* (1982)
	Dopaminergic receptors		
Corpus striatum	Decrease	Rat	Joseph *et al.* (1978)
Corpus striatum	Decrease	Rat	Joseph *et al.* (1981)
Corpus striatum	Decrease	Rat	Levin *et al.* (1981)
Corpus striatum	Decrease	Rat	Memo *et al.* (1980)
Corpus striatum	Decrease	Rat	Algeri *et al.* (1981)
Corpus striatum	Decrease	Rat	Hruska *et al.* (1981)
Corpus striatum	Decrease	Rat	Marquis *et al.* (1981)
Corpus striatum	Decrease	Mouse	Marquis *et al.* (1981)
Corpus striatum	Decrease	Mouse	Severson and Finch (1980)
Corpus striatum	Decrease	Human	Thal *et al.* (1980)

(*continued*)

Table I. (continued)

Tissue	Age effect	Species	Reference
Corpus striatum	Decrease	Human	Severson *et al.* (1982)
Hypothalamus	Decrease	Rabbit	Thal *et al.* (1980)
Frontal cortex	Decrease	Rabbit	Thal *et al.* (1980)
Anterior limbic cortex	Decrease	Rabbit	Thal *et al.* (1980)
Corpus striatum	Increase	Rat	Marquis *et al.* (1981)
Corpus striatum	Increase	Mouse	Marquis *et al.* (1981)
Retina	Increase	Rat	Riccardi *et al.* (1981)
	Opioid receptors		
Frontal poles	Decrease	Rat	Hess *et al.* (1981)
Frontal poles	Decrease	Rat	Messing *et al.* (1981)
Corpus striatum	Decrease	Rat	Hess *et al.* (1981)
Corpus striatum	Decrease	Rat	Messing *et al.* (1981)
Hippocampus	Decrease	Rat	Hess *et al.* (1981)
Hippocampus	Decrease	Rat	Messing *et al.* (1981)
Anterior cortex	Decrease	Rat	Messing *et al.* (1981)
Hypothalamus	Decrease	Rat	Messing *et al.* (1980)
Midbrain	Decrease	Rat	Messing *et al.* (1980)
Thalamus	Decrease	Rat	McDougal *et al.* (1980)
Cerebral cortex	No change	Rat	Hess *et al.* (1981)
Amygdala	No change	Rat	Hess *et al.* (1981)
Amygdala	No change	Rat	Messing *et al.* (1981)
Thalamus	No change	Rat	Messing *et al.* (1981)
Midbrain	No change	Rat	Messing *et al.* (1981)
Retina	Increase	Rat	Riccardi *et al.* (1981)
	GABA receptors		
Hypothalamus	Decrease	Rat	Calderini and Toffanno (1983)
Hypothalamus	Decrease	Rat	Calderini *et al.* (1981a,b)
Corpus striatum	Decrease	Rat	Calderini and Tofanno, in press
Corpus striatum	Decrease	Rat	Calderini *et al.* (1981a,b)
Substantia nigra	Decrease	Rat	Calderini and Tofanno (1983)
Substantia nigra	Decrease	Rat	Calderini *et al.* (1983a,b)
Spinal cord	Decrease	Rat	Calderini and Tofanno (1983)
Spinal cord	Decrease	Rat	Calderini *et al.* (1981a,b)
Brain	No change	Rat	Maggie *et al.* (1979)
Cortex	No change	Rat	Calderini and Tofanni (1983)
Cortex	No change	Rat	Calderini *et al.* (1981a,b)
Cerebellum	No change	Rat	Calderini and Toffani, in press
Hippocampus	No change	Rat	Calderini and Toffano (1983)
Hippocampus	No change	Rat	Calderini *et al.* (1981a,b)
Pons	No change	Rat	Calderini and Toffano (1983)
Pons	No change	Rat	Calderini *et al.* (1981a,b)

Table I. (continued)

Tissue	Age effect	Species	Reference
	Serotonin receptors		
Cerebral cortex	Decrease	Human	Shih and Young (1978)
Forebrain	Decrease	Rat	Hershkowitz *et al.*, in press
	Benzodiazapin receptors		
Kidney	Decrease	Rat	Pedigo *et al.* (1981)
Cerebral cortex	No change	Rat	Pedigo *et al.* (1981)
Cerebral cortex	No change	Rat	Memo *et al.* (1981)
Brain	No change	Mouse	Heusner and Bosman (1981)
Corpus striatum	No change	Rat	Memo *et al.* (1981)
Cerebellum	No change	Rat	Pedigo *et al.* (1981)
Cerebellum	No change	Rat	Memo *et al.* (1981)
Hypothalamus	No change	Rat	Memo *et al.* (1981)
Hippocampus	Increase	Rat	Memo *et al.* (1981)
	Glutamate receptors		
Hippocampus	Increase	Rat	Baudry *et al.* (1981)
	Cholinergic receptors		
Cerebral cortex	Decrease	Rat	James and Kannungo (1976)
Cerebral cortex	Decrease	Rat	Strong *et al.* (1980)
Cerebral cortex	Decrease	Rat	Lippa *et al.* (1981)
Cerebral cortex	Decrease	Human	Perry (1980)
Cerebral cortex	Decrease	Human	White *et al.* (1977)
Dorsal hippocampus	Decrease	Rat	Strong *et al.* (1980)
Dorsal hippocampus	Decrease	Rat	Lippa *et al.* (1981)
Hippocampus	Decrease	Rat	Lippa *et al.* (1980)
Hippocampus	Decrease	Rat	Nordberg and Winblad (1981)
Hippocampus	Decrease	Human	Nordberg and Winblad (1981)
Hippocampus	Decrease	Human	Nordberg and Wahlström (1981)
Basal ganglia	Decrease	Rat	Morin and Westerlain (1980)
Corpus striatum	Decrease	Rat	Strong *et al.* (1980)
Corpus striatum	Decrease	Rat	Lippa *et al.* (1981)
Cerebellar cortex	Decrease	Rat	James and Kanungo (1976)
Cerebellar cortex	Decrease	Rat	Strong *et al.* (1980)
Cerebellar cortex	Decrease	Rat	Lippa *et al.* (1981)
Adenohypophysis	Decrease	Rat	Avissar *et al.* (1981)
Cerebral cortex	No change	Rat	Avissar *et al.* (1981)
Cerebral cortex	No change	Human	Davies and Verth (1978)
Hippocampus	No change	Human	Davies and Verth (1978)
Hippocampus	No change	Mouse	Strong *et al.* (1980)
Hippocampus	No change	Rat	Strong *et al.* (1980)
Hippocampus	No change	Rat	Lippa *et al.* (1981)
Basal ganglia	No change	Human	Davies and Verth (1978)

(continued)

Table I. (continued)

Tissue	Age effect	Species	Reference
Preoptic area	No change	Rat	Avissar *et al.* (1981)
Medial hypothalamus	No change	Rat	Avissar *et al* (1981)
Posterior hypothalamus	No change	Rat	Avissar *et al.* (1981)
Medulla	No change	Rat	Avissar *et al.* (1981)
	Insulin receptors		
Liver	Decrease	Rat	Pagano *et al.* (1981)
Liver	Decrease	Rat	Freeman *et al.* (1973)
Isolated hepatocytes	Decrease	Rat	Pagano *et al.* (1981)
Isolated hepatocytes	Decrease	Rat	Freeman *et al.* (1973)
Adipocytes	Decrease	Rat	Olefsky and Reaven (1975)
Adipocytes	Decrease	Human	Pagano *et al.* (1981)
Erythrocytes	Decrease	Human	Dons *et al.* (1981)
Erythrocytes	Decrease	Human	Kosmakos *et al.* (1980)
Monocytes	Decrease	Human	Muggeo *et al.* (1981)
Skin fibroblasts	Decrease	Human	Ito (1979)
Liver	No change	Mouse	Sorrentino and Florini (1976)
Heart	No change	Mouse	Sorrentino and Florini (1976)
Monocytes	No change	Human	Rowe (personal communication)
Monocytes	No change	Human	Helderman (personal communication)
Monocytes	No change	Human	Fink *et al.* (1982)
T-lymphocytes	No change	Human	Helderman (1980)
Skin fibroblasts	No change	Human	Hollenberg and Schneider (1979)
Skin fibroblasts	Increase	Human	Rosenbloom *et al.* (1976)
	Glucagon receptors		
Adipocyte	Decrease	Rat	Livingston *et al.* (1974)
	Growth hormone receptors		
Liver	No change	Mouse	Sorrentino and Florini (1976)
	Prolactin receptors		
Ventral prostate	Decrease	Rat	Barkey *et al.* (1977)
Ventral prostate	Decrease	Rat	Boesel and Shain (1980)
Ventral prostate	Decrease	Rat	Kledzik *et al.* (personal communication)
Seminal vesicles	Decrease	Rat	Barkey *et al.* (1977)
Ovary	No change	Rat	Saiga-Narumi and Kawashima (1978)
	Gonadotropin receptors		
Leydig cells	Decrease	Rat	Pirke *et al.* (1978)
Leydig cells	Decrease	Rat	Tsitouras *et al.* (1979)
Leydig cells	Decrease	Rat	Geisthövel *et al.* (1981)
Testes	Decrease	Rat	Vassileva-Popova (1974)
Testes	No change	Rat	Steger *et al.* (1979)
Ovary	No change	Rat	Saiga-Narumi and Kawashima (1978)

Table I. (continued)

Tissue	Age effect	Species	Reference
Ovary	No change	Rat	Steger *et al.* (1978)
Interstitial cells	Increase	Rat	Clasen *et al.* (1981)
	Glucocorticoid receptors		
Cerebral cortex	Decrease	Rat	Roth (1974)
Isolated neurons	Decrease	Rat	Roth (1976)
Hippocampus	Decrease	Rat	Carmickle *et al.) (1979)*
Hippocampus	Decrease	Rat	Defiore and Turner (1981)
Hypothalamus	Decrease	Rat	Carmickle *et al.* (1979)
Heart	Decrease	Rat	Mayer *et al.* (1981)
Skeletal muscle	Decrease	Rat	Roth (1974)
Skeletal muscle	Decrease	Rat	Mayer *et al.* (1981)
Liver	Decrease	Rat	Bolla (1980)
Liver	Decrease	Rat	Petrovic and Markovic (1975)
Liver	Decrease	Rat	Parchman *et al.* (1978)
Liver	Decrease	Rat	Singer *et al.* (1973)
Adipose tissue	Decrease	Rat	Roth (1974)
Adipocytes	Decrease	Rat	Roth and Livingston (1976)
Thymus	Decrease	Rat	Petrovic and Markovic (1975)
Splenic lymphocytes	Decrease	Rat	Roth (1975)
Lung fibroblasts (WI-38)	Decrease	Human	Rosner and Cristofalo (1981)
Lung fibroblasts (WI-38)	Decrease	Human	Forciea and Cristofalo (1981)
Lung fibroblasts (WI-38)	Decrease	Human	Kalimi and Seifter (1979)
Lung fibroblasts (WI-38)	Decrease	Human	Kondo *et al.* (1978)
Cerebral cortex	No change	Rat	Carmickle *et al.* (1979)
Cerebral cortex	No change	Rat	Defiore and Turner (1981)
Cerebral cortex	No change	Mouse	Nelson *et al.* (1976b)
Hippocampus	No change	Mouse	Nelson *et al.* (1976b)
Liver	No change	Rat	Roth (1974)
Liver	No change	Mouse	Latham and Finch (1976)
Adipocytes	No change	Rat	Kalimi and Banerji (1981)
	Androgen receptors		
Cerebral cortex	Decrease	Rat	Chouknyiska and Vassileva-Popova (1977)
Hypothalamus	Decrease	Rat	Haji *et al.* (1981)
Hypothalamus	Decrease	Rat	Chouknyiska and Vassileva-Popova (1977)
Pituitary	Decrease	Rat	Haji *et al.* (1981)
Pituitary	Decrease	Rat	Chouknyiska and Vassileva-Popova (1977)
Liver	Decrease	Rat	Roy *et al.* (1974)
Testes	Decrease	Rat	Chouknyiska and Vassileva-Popova (1977)
Lateral prostate	Decrease	Rat	Shain and Boesel (1977)
Lateral prostate	Decrease	Rat	Shain and Axelrod (1973)
Lateral prostate	Decrease	Rat	Robinette and Mawhinney (1977)
Lateral prostate	Decrease	Rat	Haji *et al.* (1981)
Ventral prostate	Decrease	Rat	Shain and Boesel (1977)

(*continued*)

Table I. (continued)

Tissue	Age effect	Species	Reference
Ventral prostate	Decrease	Rat	Shain and Axelrod (1973)
Ventral prostate	Decrease	Rat	Robinette and Mawhinney (1977)
Ventral prostate	Decrease	Rat	Haji *et al.* (1981)
Penis	Decrease	Rat	Rajfer *et al.* (1980)
Anterior prostate	No change	Rat	Robinette and Mawhinney (1977)
Dorsal prostate	No change	Rat	Robinette and Mawhinney (1977)
Prostate	No change	Dog	Shain and Boesel (1978)
Seminal vesicles	No change	Rat	Robinette and Mawhinney (1977)
Seminal vesicles	Increase	Rat	Robinette and Mawhinney (1978)
	Estrogen receptors		
Cerebral hemispheres	Decrease	Rat	Carmickle *et al.* (1979)
Cerebral cortex	Decrease	Rat	Defiore and Turner (1981)
Amygdala	Decrease	Rat	Carmickle *et al.* (1979)
Hypothalamus	Decrease	Rat	Carmickle *et al.* (1979)
Hypothalamus	Decrease	Rat	Haji *et al.* (1981)
Pituitary	Decrease	Rat	Haji *et al.* (1981)
Uterus	Decrease	Rat	Haji *et al.* (1981)
Uterus	Decrease	Rat	Gesell and Roth (1981)
Uterus	Decrease	Rat	Saiduddin and Zassenhaus (1979)
Uterus	Decrease	Rat	Holinka *et al.* (1975)
Uterus	Decrease	Mouse	Nelson *et al.* (1976a)
Myometrium	No change	Hamster	Blaha and Leavitt (1976)
Testes	No change	Rat	Lin *et al.* (1981)
Seminal vesicles	Increase	Rat	Robinette and Mawhinney (1978)
	Progesterone receptors		
Myometrium	No change	Hamster	Blaha and Leavitt (1976)
Uterus	No change	Rat	Saiduddin and Zassenhaus (1979)
	Thyroid hormone receptors		
Cerebral hemispheres	Decrease	Rat	Timiras and Bignani (1976)
Brain	No change	Rat	Cutler (1981)
Liver	No change	Rat	Cutler (1981)

and Pequery (1978) found that the concentration of beta-adrenergic receptors in rat adipocytes decreased with age but Dax *et al.* (1981), in a similar study with rats, found no age-related change in receptor concentrations. The differences in beta-adrenergic receptor patterns observed here could be accounted for by the differences in dietary conditions and/or animal maintenance conditions.

Cardiac muscle is another tissue in which binding to the beta receptor exerts a stimulatory effect. Increases in both force and rate of cardiac muscle contraction result. A number of authors have reported that ventricular beta receptor concentration remains constant despite increasing age in the rat (Zitnik and Roth, 1981; Guarnieri *et al.*, 1980; Scarpace and Abrass, 1981). This is interesting in light of the observation that various aspects of cardiovascular function decline with age in several species (for a review of this subject see Shock, 1977a). Apparently these declines are not related to loss of cardiac reception of catecholamines. In fact there is some evidence that losses in cardiac functional integrity are related to calcium availability instead (Guarnieri *et al.*, 1980).

Although cardiac responses are an exception, pharmacologists generally consider beta-adrenergic receptor to exert an inhibitory effect. By contrast, alpha-adrenergic receptors are thought to exert excitatory effects in the periphery, although the designation excitatory or inhibitory is more difficult to apply to the central nervous system. As seen in Table I, studies by Misra *et al.* (198)) have shown that alpha-adrenergic receptor concentration in the cerebral cortex of the rat declines with age. The same trend has been observed for α_1 receptors in cardiac ventricular muscle of aging rats by Partilla *et al.* (1982). In contrast to these observations is the finding by Ito *et al.* (1982) that α_1 adrenergic receptor concentrations in rat parotid gland cells actually increase between 3 and 12 months of age.

3.2. Dopaminergic Receptors

Dopamine is another well-documented neurotransmitter substance. Several classes of dopamine receptors have been postuluated based upon receptor location and mechanism of action. The D1 class of dopaminergic receptors is thought to be postsynaptic and associated with adenylate cyclase (Kebabian and Calne, 1979). Another category of dopaminergic receptors is the D2 class, which is believed to be located presynaptically in part (Hazum *et al.*, 1980) and to be devoid of any association with adenylate cyclase activity. Ongoing studies based on the use of various agonists and antagonists are being used to further characterize the pharmacological specificity of these dopaminergic receptor subclasses (Martin, 1981; Tyers, 1980).

There is widespread evidence that dopamine receptor concentrations decrease with age in the corpus striatum of humans (Severson *et al.*, in press), rabbits (Thal *et al.*, 1980), mice (Marquis *et al.*, 1981; Severson and Finch, 1980), and rats (Joseph *et al.*, 1978; Joseph *et al.*, 1981; Levin *et al.*, 1981, Memo *et al.*, 1980; Algeri *et al.*, 1981; Hruska *et al.*, 1981, Marquis *et al.*, 1981) as shown in Table I. A preferential loss of D1 receptors seems to occur although D2 receptors might also be decreasing in number. A more detailed discussion of age-related changes in different dopaminergic receptor subclasses may be found elsewhere (Roth, 1982).

Marquis *et al.* (1981) have reported that dopaminergic receptor concentrations in the corpus striatum of both mice and rats increase with age. Differences here could be due to measurement of lower affinity, higher capacity binding sites since concentrations appeared relatively high in this report. Riccardi *et al.* (1981) have also observed an age-related increase in dopaminergic receptor concentrations in the retina of rats. Because manipulation of dopamine production in localized brain regions can alter dopamine receptor levels, such techniques provide another means for study of dopaminergic receptors. Chronic treatment with a dopamine antagonist, such as 6-hydroxydopamine, causes an increase in dopamine receptors (called supersensitization) of both young and old rats (Joseph *et al.,* 1981). In both mature and senescent rats, receptor concentrations increased about 40% following 6-hydroxydopamine treatment. While there was no evidence of an age difference in the ability of these Wistar rats to develop receptor supersensitivity, absolute receptor levels were reduced by 40% in the senescent rats.

Levin *et al.* (1981) have observed that dopaminergic receptor concentrations (as measured by ^{3}H-ADTN specific binding) of 24-month rats that have undergone dietary restriction are quite similar to those of 3–6-month-old rats fed *ad libitum.* Thal *et al.* (1980) report that dopaminergic receptor concentrations also decline with age in the hypothalamus, frontal cortex, and anterior limbic cortex of the rabbit. Since the hypothalamus represents an important regulatory center for a number of homeostatic processes, including the endocrine system, it is tempting to speculate that dopamine receptor losses here represent a crucial factor in the widely reported decrement in physiological function associated with age.

3.3. Opioid Receptors

The recent discovery of the opiate receptor has sparked new interest in neurobiology among both physiologists and pharmacologists. Extensive study of the differential pharmacological responses of the opiate receptor has led to postulation of several opiate receptor subtypes (Martin, 1981). Of the few studies of opiate receptor action during aging that have been undertaken, most have focused on sensitivity to pain. Studies in our laboratory have shown that the threshold to both thermal and electrical pain stimulation in rats increases with age (Hess *et al.,* 1981). The ability of naloxone to at least partially eliminate these age differences suggests that the endogenous opioid system may undergo changes during aging.

The effect of aging on opiate receptor concentration has been studied by several laboratories (Table I). Messing *et al.* (1980, 1981) reported decreased opiate receptor concentrations in frontal poles, anterior cortex, and striatum of senescent male Fisher 344 rats and thalamus and midbrain of senescent female rats based on ^{3}H-dihydromorphine specific binding. They found no significant difference between senescent and mature adult male rats for the amygdala, thalamus, and midbrain of male rats. Interestingly, the binding affinity of frontal poles was about twice as high in the senescent rats as in mature adults. While the studies from our laboratory were based on ^{3}H-etorphine specific binding (Hess *et al.*, 1981), they are in basic agreement with the findings of Messing *et al.* (1980, 1981). Opiate receptor concentrations were decreased in the frontal poles, striatum, and hippocampus of aged male Wistar rats while no significant difference was observed in the amygdala or anterior cortex. Binding affinities for opiate receptors in these brain regions showed no age-related difference. McDougal *et al.* (1980a) reported a progressive decrease in ^{3}H-naloxone binding in the hypothalamus of male Fisher 344 rats. No significant age differences were observed in other brain regions. Riccardi *et al.* (1981) found that opiate receptor concentrations in the retina of the rat increased with age. While these studies indicate a trend toward decreasing opiate receptor levels with age, there is an apparent discrepancy between receptor levels and pain sensitivity. In another study of opiate action during aging, McDougal *et al.* (1980b) found that older rats exhibit a smaller hypothermic response than young rats to high morphine doses, although at low morphine doses older rats had a greater hyperthermic response. Both sensitivity to pain and thermoregulation are complex responses. Additional studies are needed to elucidate the role of opiate receptors in their mediation of these responses and consequent impact of age changes in these receptors.

3.4. Serotonin, GABA, and Other Receptors

In recent years a variety of neurotransmitter substances have been identified in neural tissue. While only a few have been clearly shown to fill this role, there is evidence that a number of others also serve in this capacity. It is quite likely that the list of neurotransmitters will continue to grow as the discipline of neurobiology grows. It is generally held that neurotransmitters bind to receptors on the membrane surface although in some cases their physiological role is not thoroughly understood.

Serotonin and GABA (gamma-aminobutyric acid) both serve as neurotransmitters (Table I). In one of the few studies to consider serotonin, Shih and Young (1978) discovered an age-related decrease in serotonin receptor concentration of human cerebral cortex. Hershkowitz *et al.* (in press) described a similar patten in the rat forebrain. The physiological importance of serotonin receptors requires further elucidation.

GABA is a substance whose whole role as a neurotransmitter has been known or at least implied for a number of years. It is generally thought to serve as an inhibitory transmitter substance and to be released by the endings of inhibitory neurons. Calderini *et al.* (1981a,b), and Calderini and Toffano (1983) found that GABA receptor concentrations in the rat decrease with age in the hypothalamus, corpus striatum, substantia nigra, and spinal cord. They observed no change with age, however, in the cortex, cerebellum, hippocampus, and pons. Studies by Maggi *et al.* (1979) did not focus on specific anatomical locations but found that GABA receptors in the rat brain remain unchanged with increasing age.

Receptors for several other substances, whose role as neurotransmitters has yet to be established, have also been studied as a function of age. Glutamate is one such candidate whose receptor concentration was determined by Baudry *et al.* (1981). In these studies, glutamate receptor levels were found to increase with age in the rat hippocampus (Table I).

Benzodiazapine compounds have been known and used for their central nervous system effects (treatment of anxiety) for many years (Table I). Memo *et al.* (1981) discovered that benzodiazapine receptor concentrations increased with age in the rat hippocampus. Studies by both Memo *et al.* (1981) and Pedigo *et al.* (1981) failed to reveal any age-related differences in benzodiazapine receptor concentrations of rat cerebral cortex or cerebellum. Memo *et al.* (1981) also reported no change in benzodiazapine receptor levels of the rat hypothalamus and corpus striatum. They did, however, find that benzodiazapine receptor levels decreased with age in the rat kidney. In the mouse, Heusner and Bosman (1981) found no age change in benzodiazapine receptor levels in the brain.

3.5. Cholinergic Receptors

The role of acteylcholine as a neurotransmitter is well established. Cholinergic receptors are found in both the central and peripheral nervous systems. While a number of investigators have studied the effects of increasing age on cholinergic receptor levels, unanimity does not characterize their findings (Table I). James and Kanungo (1976), Strong *et al.* (1980), and Lippa *et al.* (1981) have all reported that acetycholine receptor concentrations decline with increasing age in the rat cerebral cortex. Perry (1980) and White *et al.* (1977) found that acetycholine receptor concentrations also decline with age in the cerebral cortex of humans. In contrast to these findings, studies in the rat by Avissar *et al.* (1981) and in the human by Davies and Verth (1978) failed to show any age-related change in cerebral cortex acetylcholine receptor levels. Cholinergic receptors in regions of the brain associated with the limbic system have also been considered. According to data from Lippa *et al.* (1980) as well as from Nordberg and Winbland (1981), cholinergic receptor levels decline with age in the rat hippocampus. Additional studies by Strong *et al.* (1980) and Lippa *et al.* (1981) indicate that acetylcholine receptor concentrations also decrease in the dorsal hippocampus of aging rats. Decreased acetycholine receptor levels have also been observed in hippocampal tissue of human origin (Nordberg and Winblad, 1981; Nordberg and Wahlström, 1981). However, the reports from Strong *et al.* (1980) and Lippa *et al.* (1981) indicate no change with age in cholinergic receptors of the rat hippocampus. Similarly, no age-related change was observed in cholinergic receptor levels of the hippocampus in either humans (Davies and Verth, 1978) or mice (Strong *et al.*, 1980). Cholinergic receptor levels in other brain regions associated with the limbic system were also investigated for (Strong *et al.*, 1980; Lippa *et al.*, 1971) and basal ganglia (Morin and Westerlain, 1980) of the rat, no age effect was evident in basal ganglia of human origin (Davies and Verth, 1978).

Data have also been reported for the hypothalamus and pituitary of the rat. Avissar *et al.* (1981) found that levels of cholinergic receptors did not change with age in the preoptic area, medial hypothalamus, or posterior hypothalamus. Cholinergic receptor levels in the adenohypophysis (anterior pituitary) did, however, decline with age. It is significant that several groups (James and Kanungo, 1977; Strong *et al.*, 1980; and Lippa *et al.*, 1981) have reported decreased cholinergic receptors in the cerebellar cortex of aged rats. Although this is speculation, such findings could explain the deterioration of motor control for the limbs that is so often seen in aged individuals. Interestingly, Avissar *et al.* (1981) did not observe any change in cholinergic receptor levels of the rat medulla with increasing age.

3.6. Receptor-Related Responses

Age-related changes in neurotransmitter receptor concentrations are of interest because these changes ultimately affect the nervous system's ability to transmit messages. Evidence of changes in neurophysiological response as the result of decreases in receptor concentration are presented in Table II. These generally involve decreased responses in the face of decreased receptor concentrations.

Bylund *et al.* (1977) have demonstrated a 50% decrease in both beta-adrenergic receptor concentrations and adenylate cyclase activity of rat erythrocyte membranes when comparing cells from 1–5-month rats to those from 5-month or 15-month animals. Similar findings have been reported in the cerebellar cortex of the rat (Schmidt and Thornberry, 1978; Walker and Walker, 1973). Other systems for which decreases in both beta-adrenergic receptor number and physiological response have been observed include stimulation of lipolysis

Table II. Receptor-Related Responses

Tissue	Reduced response	Species	Reference
	Beta adrenergic responses		
Cerebral cortex	Histone phosphorylation	Rat	Das and Kanungo (1980)
Cerebellar cortex	Stimulation of adenylate cyclase	Rat	Schmidt and Thornberry (1978)
Cerebellar cortex	Stimulation of adenylate cyclase	Rat	Walker and Walker (1973)
Cerebellar cortex	Stimulation of Purkinje cell firing	Rat	Morwaha *et al.* (1980)
Erythrocyte	Stimulation of adenyl cyclase	Rat	Bylund *et al.* (1977)
Adipocyte	Stimulation of lipolysis	Rat	Giudicell and Pequery (1978)
Submandibular gland	Stimulation of DNA synthesis	Rat	Piantanelli *et al.* (1980)
	Alpha-adrenergic responses		
Heart	Stimulation of heart rate	Rat	Kelliher and Stoner (1978)
Heart	Stimulation of blood pressure	Rat	Kelliher and Stoner (1978)
	Dopaminergic responses		
Corpus striatum	Stimulation of adenylate cyclase	Rat	Walker and Walker (1973)
Corpus striatum	Stimulation of adenylate cyclase	Rat	Puri and Volicer (1977)
Corpus striatum	Stimulation of adenylate cyclase	Rat	Thal *et al.* (1980)
Corpus striatum	Induction of rotational behavior	Rat	Cubells and Joseph (1981)
Corpus striatum	Stimulation of sterotypic behavior	Mouse	Randall *et al.* (1981)
	Cholinergic responses		
Dorsal hippocampus	Stimulation of pyramidal cell firing	Rat	Lippa *et al.* (1981)
	Gonadotropin responses		
Leydig cells	Steroidogenesis	Rat	Pirke *et al.* (1978)
Leydig cells	Steroidogenesis	Rat	Tsitouras *et al.* (1979)
Leydig cells	Steroidogenesis	Rat	Geisthövel *et al.* (1981)
	Glucagon responses		
Adipocytes	Stimulation of lipolysis	Rat	Livingston *et al.* (1974)
	Insulin responses		
Skin fibroblasts	Stimulation of glucose oxidation	Human	Ito (1979)

Table II. (continued)

Tissue	Reduced response	Species	Reference
Skin fibroblasts	Stimulation of glucose oxidation	Rat	Villee *et al.* (1979)
Monocytes	Glucose tolerance	Human	Muggeo *et al.* (1981)
Hepatocytes	Glucose tolerance	Rat	Pagano *et al.* (1982)
Adipocytes	Stimulation of glucose oxidation	Rat	DiGirolamo and Mendlinger (1972)
Adipocytes	Stimulation of mRNA and protein synthesis	Rat	Vydelingum *et al.* (1981)
Adipocytes	Glucose tolerance	Human	Pagano *et al.* (1981)
	Glucocorticoid responses		
Liver	Stimulation of RNA synthesis	Rat	Miller and Bolla (1981)
Liver	Stimulation of RNA synthesis	Rat	Martin *et al.* (1980)
Liver	Stimulation of RNA synthesis	Rat	Murota and Koshihara (1977)
Adipocytes	Inhibition of glucose oxidation	Rat	Roth and Livingston (1976)
Lung fibroblasts (WI-38)	Extension of *in vitro* division potential	Human	Cristofalo (1972)
Splenic leukocytes	Inhibition of uridine uptake	Rat	Roth (1975)
	Androgen responses		
Liver	Induction of $\alpha_{2}\mu$ globulin	Rat	Roy *et al.* (1974)
Ventral prostate	Maintenance of cell content	Rat	Shain and Boesel (1977)
Ventral prostate	Induction of L-ornithine decarboxylase	Rat	Shain and Moss (1981)
	Estrogen responses		
Uterus	Induction of phosphofructokinase	Rat	Singhal *et al.* (1969)
Uterus	Induction of glucose-6-phosphatase	Rat	Singhal *et al.* (1969)
Uterus	Decidualization response	Rat	Saiduddin and Zassenhaus (1979)
Uterus	Weight maintenance	Rat	Saiduddin and Zassenhaus (1979)
Cerebral cortex	Induction of acetycholinesterase	Rat	Moudgil and Kanungo (1973)
Cerebral cortex	Histone acetylation	Rat	Thakur *et al.* (1978)
Hypothalamus/ pituitary	Stimulation of prolactin release	Rat	Huang *et al.* (1980)

in rat adipocytes (Giudicelli and Pequery, 1978) and stimulation of Purkinje cell firing in the rat cerebellar cortex (Morwaha *et al.*, 1980). Das and Kanungo (1980) found decreased histone phosphorylation with age in the rat cerebral cortex, while Piantanelli *et al.* (1980) observed decreased stimulation of DNA synthesis in mouse submandibular gland. Kelliher and Stoner (1978) showed that stimulation of heart rate and blood pressure were reduced in aged rats following administration of the alpha-adrenergic agonists norepinephrine or phenylephrine. Dopamine-induced adenylate cyclase activity in the corpus striatum of rat (Puri and Volicer, 1977) and rabbit (Thal *et al.*, 1980) have also been observed to decrease with age (Table II). Dopamine-induced rotational behavior (Cubells and Joseph, 1981) in the rat and dopmaine-induced stereotypic behavior in the mouse (Randall *et al.*, 1981) decrease in concert with declining dopamine receptor levels in the corpus striatum of these animals. Lippa *et al.* (1981) report that pyramidal cell firing in the dorsal hippocampus of old rats is reduced along with the cholinergic receptor concentration.

4. SURFACE-BINDING HORMONE RECEPTORS

4.1. Overview

The endocrine system has long been felt to play a significant role in the aging process. Over the years there have been both proponents and opponents of the theory that changes in endocrine function are responsible for aging. While this theory is no longer accepted (Gregerman and Bierman, 1981), the point can be made that changes in endocrine function (due to aging or other causes) are capable of altering various aspects of body function. The endocrine system, along with the nervous system, plays a key role in regulating body functions. If the aging process is to be understood, it is imperative that the functional characteristics of these systems in senescent animals be carefully described. Although molecular biologists have made substantial contributions to our understanding of age-related changes in living organisms, there needs to be a renewed effort to study physiological changes in aging animals from the perspective of the functioning organism; the recent advances in molecular biology notwithstanding, a meaningful understanding of the aging process demands an integration of molecular concepts into a holistic picture.

Endocrinologists are by definition concerned about the dynamics of blood hormone levels that characterize both normal and abnormal functioning of the body. Modern techniques such as radioimmunoassay and competitive-binding assays have made possible the accumulation of a large data base for many different hormones. While blood hormone levels are influenced by both internal and external factors, the ultimate question is this: How does the aging process affect hormone effectiveness at the tissue level? One way to answer this question is made possible by the techniques for estimating the concentration and binding affinity of hormone receptors. The current knowledge about hormone receptor concentrations as influenced by aging will be summarized in the following paragraphs and their corresponding tables. While this is useful information, there are other steps between hormone binding and cellular response for essentially all hormones. Many surface-binding hormones exert their effect by way of cyclic AMP or other second messengers (Aurbach, 1982). While some discussion of postreceptor changes in hormone response will be given here, much additional work is needed on the relationship between second messenger functioning and hormone responsiveness during aging. Thus, information about receptors for the different hormones is of value in understanding the biology of aging. The sizable body of knowledge related to age-related changes in receptor concentrations of surface-binding hormones will now be considered. The reader needs to keep in mind potential influences such as changes in endogenous hormone levels, declining numbers of target cells, and/or the presence of biologically inactive receptors to be accounted for by future studies.

4.2. Insulin

One major function of the endocrine system is to regulate the distribution of glucose and its alternative energy sources to the various body tissues in the face of changing demand and availability. The fundamental role of insulin in this regulatory process is well known. Insulin receptors are widely distributed among different tissues and may even be present on cells of the nervous system. In general, insulin receptor concentrations either decrease or remain unchanged on the cells of aging animals (Table I). However, Rosenbloom *et al.* (1976) report that insulin receptor concentrations in human skin fibroblasts actually increase with aging in contrast to the above generalization. Hollenberg and Schneider (1979) also studies insulin receptor concentration in skin fibroblasts derived from adult humans but found no change with donor age. Ito *et al.* (1979) found an age-related decline in the insulin receptor concentration of exponentially growing human skin fibroblasts.

Just as insulin might be considered the primary hormone involved in regulating the varied energy sources of the body, the liver might be considered the fundamental organ of metablism. Research by Pagano *et al.* (1981), utilizing rat liver tissue, demonstrated an age-related decline in insulin receptor concentration.

Another study by Freeman *et al.* (1973) that used isolated hepatocytes of rats rather than whole liver tissue, confirms the decline in insulin receptor concentration with age. In contrast to these studies on rats, Sorrentino and Florini

(1976) found no age-related decrement in the insulin receptor concentration of mouse liver. Another important site for insulin action is adipose tissue. Since adipose tissue is a metabolically active tissue, changes in insulin receptor concentrations bear the potential for altering the body's metabolic patterns. Insulin receptor concentrations in the adipose tissue of rats as reported by Olefsky (1975) and in human adipose tissue as reported by Pagano *et al.* (1981) showed an age-related decline. Interestingly, the concentrations of receptors for the insulin antagonist glucagon found in rat adipocytes, has also been reported to decrease with age (Livingston *et al.*, 1974). Balanced operation of metabolic activities in the mammalian body depends on precise functioning of the endocrine system. If receptor integrity changes with increasing age, the potential for changes in precision of control also exists. Insulin can be considered one of the pivotal hormones in the entire metabolic control process. Understanding the impact of aging here goes beyond measurements of insulin receptor concentration. Insulin receptor binding is known to inhibit the activity of cAMP but also to stimulate cGMP formation (Cuatrecasas, 1974). Therefore, potential changes in "second messenger" must also be considered when dealing with age-related changes in tissue responsiveness to insulin.

In addition to liver and adipocytes, insulin receptor concentration have been determined for several kinds of blood cells. Muggeo *et al.* (1981) quantitated the insulin receptor concentration of monocytes obtained from humans in three age categories (16–34 years, 35–50 years, and 51–75 years) and found that insulin receptor concentration decreased as donor age increased. Dous *et al.* (1981) obtained erythrocytes from adult humans (21–36 years of age) and used density gradient determinations to separate the cells from each donor into age groups. [^{125}I]insulin binding to red cells was found to decrease exponentially as a function of red cell age. Decreased binding in these studies is attributed to decreased receptor numbers in older cells. Similar findings have been reported by Kosmakos *et al.* (1980). Sorrentino and Florini, (1976), who observed no difference in insulin receptor binding capacity of liver obtained from mice of different ages, also found no age-related changes in insulin binding capacity of cardiac muscle. Similarly, growth hormone binding capacity was found to exhibit no age-related changes in these studies (Table I).

4.3. Gonadotropins and Prolactin

The endocrine system regulates reproductive function in addition to metabolism. The development and maintenance of anatomical structures associated with reproduction occurs in direct response to the male and female steroid sex hormones. Control of steroid sex hormone secretion rests with the gonadotropins (FSH and LH) secreted by the anterior pituitary gland. These hormones along with prolactin, another anterior pituitary hormone, are small protein molecules that bind to receptors on the cell surface.

Prolactin is required for lactation to occur in female mammals. In rodents prolactin is also required for postovulatory luteinization. The role of prolactin in the male reproductive system, if any, is currently unknown. Nonetheless, several laboratories have demonstrated (Barkey *et al.*, 1977; Boesel and Shain, 1980; Kledzik *et al.*, personal communication) the existence of prolactin receptors in ventral prostate and seminal vesicle tissue of male rats and have further shown that prolactin receptor concentration decreases with increasing age (Table I). Saiga-Narumi and Kawashima (1978) have measured ovarian prolactin receptors in female rats but found no age-related change in their concentration. Since changes in prolactin secretion are reported in aging rats (see Meites and Sonntag, 1981) it is interesting to note the constancy of prolactin receptors with increasing age observed in these studies.

The gonadotropins, FSH and LH secreted by the anterior pituitary, control reproductive function in both sexes. Reports from several laboratories (Steger *et al.*, 1978; Saiga-Narumi and Kawashima, 1978) indicate that gonadotropin receptor numbers in rat ovaries do not change with increasing age. It is unclear whether female reproductive failure in senescence results from changes at the level of the hypothalamus or the pituitary (Finch, 1976). Vassileva-Popova (1974) found that gonadotropin receptors of rat testes decreased with age. In other studies of age-related changes in male rat gonads, decreased gonadotropin receptor concentrations have been observed in the androgen secreting Leydig (interstitial) cells by Pirke *et al.* (1978), Tsitouras *et al.* (1979), and Geisthövel *et al.* (1981). By way of contrast, Clausen *et al.* (1981) found increased gonadotropin receptor concentrations when they compared interstitial cells derived from old male rats to those from younger rats. Steger *et al.* (1979) found no age-related change in gonadotropin (LH) binding to testicular tissue. The reason for this lack of agreement between results of different laboratories is not apparent at this time.

4.4. Receptor-Related Responses (Table II)

Numerous studies of altered responsiveness to surface binding hormones in aging animals have been reported (see Table II). While corresponding receptor studies have been performed, one should not conclude that decreased responsiveness is always caused by receptor losses. Although this may in some cases be true, most of the experiments to date have not been designed to establish this point.

The ability of gonadotropins to stimulate steroidogenesis in Leydig cells derived from testicular tissue of senescent rats has been shown to decline by

Pirke *et al.* (1978), Tsitouras *et al.* (1979), and by Geisthövel *et al.* (1981). Livingston *et al.* (1974) demonstrated that glucagon's ability to stimulate lipolysis in adipoytes also declines with increasing age. Because of is importance to homeostasis, insulin is one of the more extensively studied hormones. The effect of aging on insulin response has also been studied. Ito (1979) and Villee *et al.* (1979) found that the ability of insulin to stimulate glucose oxidation in rat skin fibroblasts was reduced with increasing age.

In other studies (Pagano *et al.*, 1981, 1982) an age-related decrease in glucose tolerance has been observed in human adipocytes and rat hepatocytes. Muggeo *et al.* (1981) have reported the same phenomenon for monocytes of human origin. In other studies of insulin responsiveness as a function of age, DiGirolamo and Mendlinger (1972) discovered that insulin stimulation of glucose oxidation in adipocytes derived from aged rats was decreased. Another report from the laboratory of Vydelingum *et al.* (1981) reveals an age-related decline in the ability of insulin to stimulate mRNA and protein synthesis in rat adipocytes.

5. INTRACELLULAR HORMONE RECEPTORS

5.1. Overview

Steroid hormones secreted by the gonads and adrenal cortex are lipid soluble and therefore readily cross the cell membrane. Included in this group are the female sex hormones (estrogens and progesterone), the male sex hormones (androgens), as well as the glucocorticoids and mineralocorticoids produced by the adrenal cortex. Receptors for these hormones are located within the cell cytoplasm rather than on the cell surface (Latham and Finch, 1976). The glucocorticoids exert important effects on metabolic activities of various tissues. Among their more widely recognized metabolic effects are the ability to inhibit cellular glucose uptake and to promote release of fatty acids from adipose tissue as well as their utilization by tissues. In high concentrations, glucocorticoids promote catabolism of tissue proteins and subsequent use of the resulting amino acids for gluconeogenesis by the liver. Glucocorticoids are also involved in the body's stress response (Latham and Finch, 1976).

5.2. Glucocorticoids

Many of the metabolic effects exerted by glucocorticoids occur in the liver. For example, glucocorticoids stimulate the uptake of amino acids and their conversion to glucose (gluconeogenesis). This response is particularly important during stress and/or starvation. Bolla (1980), Petrovic and Markovic (1975), as well as Parchman *et al.* (1978) reported a dramatic age-related decline in liver of glucocorticoid receptors (Table I). Latham and Finch (1976) isolated a glucocorticoid binding protein from liver cytosol of C57B1/6J mice that showed no significant age differences in concentrations, however. This finding is consistent with our finding (Roth, 1974) that glucocorticoid receptors in rat liver did not change with age.

Adipose tissue represents another important site of metabolic activity. Composed primarily of adipocytes, it serves as a storage site for fatty acids, and its metabolic activity is regulated by a number of hormones including the glucocorticoids. In one set of studies from our laboratory, both adipose tissue (Roth, 1974) and isolated adipocytes (Roth and Livingston, 1976) from rats were found to have an age-related decline in glucocorticoid receptor concentration. In contrast to our findings, Kalimi and Banerji (1981) recently studied glucorcorticoid receptor concentration in rat adipocytes and detected no age-related change. Responsiveness was not examined in this study. The possible effect of glucocorticoids on the nervous system is supported by their presence in neural tissues. Work from our laboratory (Roth, 1974; Roth, 1976) indicates that glucocorticoid receptor concentrations decrease with age both in tissues derived from the cerebral cortex and in isolated neurons of rats. Carmickle *et al.* (1979) have reported a decrease in hypothalamic glucocorticoid receptor concentration. Earlier studies (Hess and Riegle, 1972) indicated a loss of sensitivity in the glucocorticoid feedback control mechanism but did not pinpoint the site involved. Carmickle *et al.* (1979) also found that glucocorticoid receptor concentrations declined in the hippocampus with increasing age, but observed no change in the cerebral cortex of rats. Defiore and Turner (1981) also found that glucocorticoid binding is decreased in the hippocampus but not the cerebral cortex of aged rats. In the mouse, studies of cerebral cortex, hypothalamus, and hippocampus by Nelson *et al.* (1976) failed to reveal an age-related change in glucocorticoid receptor concentration. The recurring disagreement between rat data and mouse data suggests that there may be significant species differences in the aging process of these widely used laboratory animals. Other factors such as laboratory differences in experimental technique or dietary–maintenance regimens could also be involved in these differences.

Skeletal muscle represents a major proportion of body tissue mass. Roth (1974) and Mayer *et al.* (1981) report that glucocorticoid receptors decrease with age in rat skeletal muscle. While the physiological significance of this change in the aging body remains to be demonstrated, it does suggest that a major portion of body tissue mass becomes less responsive to glucocorticoid hormones with increasing age. The physiological importance of the heart is obvious, and the tendency for declining cardiac function with age is both generally known and documented in the literature (Kohn, 1977). Mayer *et al.* (1981) showed that glucocorticoid receptor concentration also declines in cardiac muscle of aging rats. Studies of the glucocorticoids role in normal cardiac function are needed to interpret the significance of this finding.

Connective tissue is also widely distributed throughout the mammalian body. Reports from Cristofalo's laboratory (Rosner and Cristofalo, 1981; Forciea and Cristofalo, 1981) indicate that WI-38 fibroblasts of human origin lose glucocorticoid receptors as they age *in vitro*. Petrovic and Markovic (1975) observed an age-related decline in glucocorticoid receptor concentrations of the thymus. While pharmacological levels of glucocorticoid are known to suppress tissue associated with the immune response, the influence of physiological glucocorticoid levels on thymic function is uncertain. In a related study, Roth (1975) found that glucocorticoid receptors are lowered in splenic (leukocytes) lymphocytes of senescent rats. In the past, it was common procedure to adrenalectomize rats a few days before using their tissues for glucocorticoid receptor assays in order to minimize competition with endogenous corticoids. More recently, many investigators have decided to examine both cytoplasmic and nuclear receptors in intact animals since there is some evidence that adrenalectomy may change glucocorticoid receptor concentrations independent of endogenous glucocorticoid levels (Boer and Oddos, 1979).

5.3. Androgens and Estrogens

The age-related deterioration of reproductive function in female laboratory animals is well documented (Finch, 1976; Meites and Sonntag, 1981). Much of the work performed to date deals with changes in gonadotropin levels and/or the hypothalamic control of their release from the pituitary. There have been more studies on the age-related changes of female reproductive endocrinology than there have been of male reproductive endocrinology. While there is some decline in reproductive function of the aging human male, reproductive capability is frequently retained into old age (Talbert, 1977). Chouknyiska and Vassileva-Popova (1977) report a decrease in androgen receptor levels in both the cerebral cortex and the testes of aging rats (Table I). Although neither confirmed nor contradicted by other studies, one can speculate that both effects are physiologically important in that androgens are required for normal spermatogenesis and that androgens are certain to exert behavioral influences. A decline in the androgen receptor concentration to both ventral and lateral prostate tissue has been reported by several laboratories (Shain and Boesel, 1977; Shain *et al.*, 1973; Robinette and Mawhinney, 1977; Haji *et al.*, 1981). However, Robinette and Mawhinney (1977) observed no change in androgen receptor concentrations of anterior prostate or dorsal prostate of aging rats. They also report no change in androgen receptor levels in the seminal vesicles of aging rats, but in a more recent study (Robinette and Mawhinney, 1978) they observed increased androgen receptor concentrations in the seminal vesicles of aged rats. Thus androgen receptor age changes appear to be tissue specific.

Studies relating endocrinology and aging have focused primarily on rats and mice. For both species, lifespan is a maximum of several years and many data characterizing the lifespan are available (Comfort, 1979). A broadly based understanding of the aging process necessitates data from additional species, although both data characterizing the lifespan as well as experimental animals seem to be less readily available for other species. Studies by Shain and Boesel (1978) have indicated that androgen receptor concentration in the prostate of dogs does not change with age. In the rat penis, Rajfer *et al.* (1980) have identified an age-related decrease in androgen receptor concentration. In another study, Roy *et al.* (1974) measured androgen receptor in rat liver tissue and found that their concentration decreases with age. The largest group of female sex hormones are collectively known as estrogens. They exert effects on the uterine smooth muscle and lining, the secondary sex organs, and a range of other tissues including the nervous system. Data from a number of studies indicate that estrogen receptor concentrations in rat uterus decline with age (Haji *et al.*, 1981; Gesell and Roth, 1981; Saiduddin and Zassenhaus, 1979; Holinka *et al.*, 1975). The same pattern has been observed by Nelson *et al.* (1976) in aging mice. Gesell and Roth (1981) have shown that uterine estrogen receptors decrease by about 45% between maturity (6–12 months) and senescence (22–24 months). Immunochemical titration of receptors with specific antiserum showed that the ratio of immunoreactive to functional receptors is not changed with age. In the hamster, Blaha and Leavitt (1976) report that estrogen receptor concentrations do not change in the myometrium (uterine smooth muscle) with increasing age.

Of particular interest to neuroendocrinologists is the finding that estrogen receptor concentrations in the hypothalamus (Carmickle *et al.*, 1979; Haji *et al.*, 1981) and pituitary (Haji *et al.*, 1981) decrease with age in the rat. This finding correlates wtih RIA studies (see Finch, 1976) that reveal age-related losses in the ability to secrete reproductive hormones. Likewise, the cerebral hemispheres and cortex of the rat brain also lose estrogen receptors with increasing age according to data from Carmickle *et al.* (1979) and Defiore and Turner (1981). Carmickle and colleagues also found that estrogen receptors decrease with age in the amygdala of rats. Since this region is part of the limbic system, one might speculate that the amygdala modulates estrogen-influenced behavioral patterns in the rat. Although the physiological meaning of these findings in male rats is uncertain, Lin *et al.* (1981) report no age-related change in estrogen receptor concentration in the testes. On the other hand, Robinette and Mawhinney (1978) report that estrogen receptor concentrations increased with age in the seminal vesicles.

5.4. Progesterone and Thyroid Hormone

Although progresterone is an important female sex hormone, its effects seem to be less widely studied than those of estrogen. Saiduddin and Zassenhaus (1979) report no age-related change in progesterone receptors of the rat uterus (Table I). A similar observation was made by Blaha and Leavitt (1976) for uterine progesterone receptors in the golden hamster.

Of all the hormones, none has more far-reaching effects than thyroid hormone. Involved in regulation of growth and development as well as metabolism, virtually all tissues possess thyroid hormone receptors. Timiras and Bignani (1976) report an age-related decrease in thyroid receptors of the rat cerebral hemispheres. However, Cutler (1981) reports the lack of age-related changes in thyroid hormone receptors of either rat brain or rat liver.

5.5. Receptor-Related Responses

Receptor-binding studies have comprised much of the work to date on age-related changes of steroid hormone action, but there are some data related to changes in physiological response. These changes are consistent with reports of decreased receptor concentration, although in some cases studies of receptor changes and response changes have been conducted by separate laboratories.

Steroid hormones are known to act by increasing protein synthesis via transcription in the nucleus. Three groups (Murota and Koshihara, 1977; Martin *et al.*, 1980; Miller and Bolla, 1981) have shown that the stimulation of RNA synthesis declines with age in rat liver (Table II). Glucocorticoids, in particular, play an important role in regulating carbohydrate metabolism. Roth and Livingston (1976) demonstrated that the inhibition of glucose oxidation, normally characteristic of glucocorticoids, is diminished in adipocytes derived from senescent rats. Cristofalo (1972) reported that the extension of *in vitro* division potential normally afforded by glucocorticoids is reduced in human lung fibroblasts that have already undergone a large number of divisions *in vitro*.

Roy *et al.* (1974) reported that androgen receptor activity in the hepatic cytosol is closely related to the urinary output of α_2u-globulin, an androgen-dependent protein produced by the liver. They suggest that synthesis of this protein is induced by its own ligands, the androgens. Their studies indicate that induction of this protein is reduced in liver derived from senescent rats. The enzyme L-ornithine decarboxylase is required for polyamine production, which in turn is needed for proliferative cell growth. Shain and Moss (1981) have demonstrated a dramatic decline in the ability to induce this enzyme in the ventral prostate of senescent rats. Shain and Boesel (1977) had previously reported a decrease in the maintenance of cell content in the ventral prostate of senescent rats.

Steroid hormones are also known to increase metabolic activity in tissues via induction (synthesis) of metabolic pathway enzymes. Singhal *et al.* (1969) have shown that estrogen-dependent induction of both phosphofructokinase and glucose-6-phosphatase is reduced in uterine tissue of senescent rats (Table II). Additional studies of the estrogen influence in uterine tissue by Saiduddin and Zassenhaus (1979) revealed an age-related decline in the decidualization response and uterine weight maintenance, both of which are estrogen dependent.

Other studies have examined estrogen-dependent responses in the nervous system. Moudgil and Kanungo (1973) demonstrated that induction of acetylcholinesterase by estrogen in the rat cerebral cortex is lowered in aged animals, while Thakur *et al.* (1978) found that estrogen-dependent histone acetylation in cerebral cortex tissue declines with age. Huang *et al.* (1980) also observed that the ability of estrogen to stimulate prolactin release via the hypothalamic–pituitary axis is diminished in senescent rats.

REFERENCES

Abrass, I. B., and Scarpace, P. J., 1981, Human lymphocyte beta-adrenergic receptors are unaltered with age, *J. Gerontol.* **36:**298–301.

Algeri, S., Cimino, M., Stramentinoli, G., and Vantini, G., 1981, Age-related modification of dopaminergic and α-adrenergic receptor systems: Restoration of normal activity by modifying membrane fluidity with s-adenosylmethionine, *Abstracts of Symp. on Aging Brain and Ergot Alkaloids* **23**.

Aurbach, G. D., 1982, Polypeptide and amine hormone regulation of adenylate cyclase, *Ann. Rev. Physiol.* **44:**653–666.

Avissar, S., Egozi, Y., and Sokolovsky, M., 1981, Aging process decreases the density of muscarinic receptors in rat adenohypophysis, *FEBS Lett.* **133:**275–278.

Barkey, R. J., Shani, J., Amit, T., and Barzilai, D., 1977, Specific binding of prolactin to seminal vesicle, prostate and testicular homogenates of immature, mature and aged rats, *J. Endocrinol.* **74:**163–173.

Baudry, M., Arst, D. S., and Lynch, G., 1981, Increased [^{3}H] glutamate receptor binding in aged rats, *Brain Res.* **223:**195–198.

Bierman, E. L., Albers, J. J., and Chait, A., 1979, Effect of donor age on the binding and degradation of low density lipoproteins by cultured human arterial smooth muscle cells, *J. Gerontol.* **34:**483–488.

Bilizekian, J. P., and Gammon, D. E., 1978, The effect of age on beta-adrenergic receptors and adenylate cyclase activity in rat erythrocytes, *Life Sci.* **23**:253–260.

Blaha, G. C., and Leavitt, W. W., 1976, Uterine progesterone receptors in the aged golden hamster, *J. Gerontol.* **33**:810–814.

Boer, A., and Oddos, J., 1979, Cardiac receptors for glucocorticoids in the rat. Factors involved in [^{3}H] dexamethasone binding and nuclear translocation of [^{3}H] dexamethasone receptor complexes in heart and liver, *Biochim. Biophys. Acta* **596**:70–86.

Boesel, R. W., and Shain, S. A., 1980, Aging in the AXC rat: Androgen regulation of prostate prolactin receptors, *J. Androl.* **1**:269–276.

Bolla, R., 1980, Age-dependent changes in rat liver steroid hormone receptor proteins, *Mech. Ageing Dev.* **12**:119–122.

Bylund, D. B., Tellez-Inon, M. T., and Hollenberg, M. D., 1977, Age-related parallel decline in α-adrenergic receptors, adenylate cyclase and phosphodiesterase activity in rat erythrocyte membranes, *Life Sci.* **21**:403–410.

Calderini, G., and Toffano, G., 1983, Age-dependent changes in information processing, in: *Arteriosclerotic Brain Disease* (G. Crepaldi, R. Fellin, A. G. Olsson, and G. Toffano, eds.) pp. 13–19, Raven Press, New York.

Calderini, G., Bonetti, A. C., Aldino, A., Savoini, G., DiPerri, B., Biggio, G., Toffano, G., 1981a, Functional interaction between benzodiazepine and GABA recongition sties in aged rats, *Neurobiol. Aging* **2**:309–313.

Calderini, G., Aldenio, C., Crews, F., Gaiti, A., Scapognini, U., Algieri, S., Ponzio, F., and Toffano, G., 1981b, Aging and information processing, in: *Apomorphone and Other Dopaminominetics*, Vol. 2, *Clinical Pharmacology* (G. V. Corsini and G. L. Gessa, eds.), pp. 235–242, Raven Press, New York.

Carmickle, L. J., Kalimi, M., and Terry, R. D., 1979, Aging rat brain: Changes in steroid hormone receptors and morphometric characteristics, *Fed. Proc.* **39**:482.

Cautrecases, P., 1974, Membrane receptors, *Ann. Rev. Biochem.* **43**:169–232.

Chouknyiska, R., and Vassileva-Popova, J. G., 1977, Effect of age on the bonding of ^{3}H-testosterone with receptor protein from rat brain and testes, *C. R. Acad. Sci. Bulgaria* **30**:133–135.

Clausen, O. P. F., Purvis, K., and Hansson, V., 1981, Age-related changes in [^{125}I] hLH specific binding to rat interstitial cells, *Acta Endocrinol.* **90**:569–579.

Comfort, A., 1979, *The Biology of Senescence*, 3rd ed., Elsevier, New York.

Cristofalo, V. J., 1972, Animal cell cultures as a model system for the study on aging, *Adv. Gerontol. Res.* **4**:45–79.

Cubells, J. F., and Joseph, J. A., 1981, Neostriatal dopamine receptor loss and behavioral deficits in the senescent rat. *Life Sci.* **28**:1215–1218.

Cutler, R. C., 1981, Thyroid hormone receptors in the brain of rat as a function of age and hypophysectomy, in *Brain Neurotransmitters and Receptors in Aging and Age-Related Disorders* (S. J. Enna, T. Samorajski, and B. Berr, eds.), pp. 117–132, Raven Press, New York.

Das, R., and Kanungo, M. S., 1980, *In vitro* phosphorylation of chromosomal proteins of the brain of rats of various ages and its modulation by epinephrine, *Indian J. Biochem. Biophys.* **17**:217–221.

Davies, P., and Verth, A. H., 1978, Regional distribution of muscarine acetylcholine receptor in normal and Alzheimer's-type dementia brains, *Brain Res.* **138**:385–392.

Dax, E. M., Partilla, J. S., and Gregerman, R. I., 1981, Mechanism of the age-related decrease of epinephrine-stimulated lipolysis in isolated rat adipocytes: α-adrenergic receptor binding, adenylate cyclase activity, and cyclic cAMP accumulation, *J. Lipid Res.* **22**:934–943.

Defiore, C. H., and Turner, B. B., 1981, Glucocorticoid binding is decreased in hippocampus but not cortex of aged rats, *Abst. Soc. Neurosci.* **7**:947.

Dilman, V. M., 1971, Age associated elevation of hypothalamic threshold to feedback control, and its role in development, ageing and disease, *Lancet* **1**::1211.

DiGirolamo, M., and Mendlinger, S., 1972, Glucose metabolism and responsiveness to bovine insulin by adipose tissue from three mammalian species, *Diabetes* **21**:1151–1161.

Dons, R. F., Corash, L. M., and Gorden, P., 1981, The insulin receptor is an age-dependent integral component of the human erythrocyte membrane, *J. Biol. Chem.* **256**:2982–2987.

Enna, S. J., and Strong, R., 1981, Age-related alterations in central nervous system neurotransmitter receptor binding, in: *Brain Neurotransmitters and Receptors in Aging* (S. J. Enna, T. Samorajski, and B. Beer, eds.), pp. 133–142, Raven Press, New York.

Everett, A. V., 1973, The hypothalamic-pituitary control of aging and age-related pathology, *Exp. Gerontol.* **8**:265–277.

Finch, C. E., 1976, Physiological changes of aging in mammals, *Quart. Rev. Biol.* **51**:49–83.

Forciea, M. A., and Cristofalo, V. J., 1981, Glucocorticoid specific binding in WI-38 cells: Confirmation of an age-associated decline in receptors in a cell free preparation, *Gerontologist* **21**:179.

Freeman, C., Karoly, K., and Adelman, R. C., 1973, Impairments in the availability of insulin to liver *in vitro* and in binding of insulin to purified hepatic plasma membranes during aging. *Biochem. Biophys. Res. Commun.* **54**:1573–1580.

Geistövel, F., Brabant, G., Wickings, E. J., and Nieschiag, E., 1981, Changes in testicular HCG binding and Leydig cell function in rats throughout life, *Hormone Res.* **14**:47–55.

Gesell, M. S., and Roth, G. S., 1981, Decrease in rat uterine estrogen receptors during aging: Physio- and immunochemical properties, *Endocrinology* **109**:1502–1508.

Guidicelli, Y., and Pequery, R., 1978, α-adrenergic receptors and catecholamine sensitive adenylate cyclase in rat fat cell membranes: Influence of growth, cell size and aging, *Eur. J. Biochem.* **90**:413–419.

Greenberg, L. H., and Weiss, B., 1978, β-adrenergic receptors in aged rat brain: Reduced number and capacity of pineal glands to develop supersensitivity, *Science* **201**:61–63.

Gregerman, R. T., and Bierman, E. L., 1981, Aging and hormones, in: *Textbook of Endocrinology* (R. H. Williams, ed.), pp. 1192–1212, Saunders, Philadelphia.

Guarnieri, T., Filburn, C. R. Zitnik, G., Roth, G. S., and Lakatta, E. G., 1980, Contractile and biochemical correlates of β-adrenergic stimulation of the aged heart, *Am. J. Physiol.* **239**:H501–H508.

Haji, M., Kato, K., Nawata, H., and Ibabyashi, H., 1981, Age-related changes in the concentrations of cytosol receptors for sex steroid hormones in the hypothalamus and pituitary gland of the rat, *Brain Res.* **204**:373–386.

Hazum, E. L., Chang, K-J., and Cautrecasas, P., 1980, Cluster formation of opiate (enkephalin) receptors in neurotalastoma cells: Difference between agonists and antagonists and possible relationship to biological functions, *Proc. Natl. Acad. Sci. USA* **77**:3038–3041.

Hershkowitz, M., Heron, D, Samuel, D., and Shinitzky, M., The modulation of protein phosphorylation and receptor binding in synaptic membranes by changes in lipid fluidity: Implications for aging, *Progr. Brain Res.*, in press.

Hess, G. D., and Reigle, G. D., 1972, Effects of chronic ACTH stimulation on adrenocortical function in young and aged rats, *Am. J. Physiol.* **222**:1458–1461.

Hess, G. D., Joseph, J. A., and Roth G. S., Effect of age on sensitivity to pain and brain opiate receptors, *Neurobiol. Aging* **2**:1876–1878.

Heusner, J. E., and Bosman, H. B., 1981, Benzodiazepine receptor binding in aged mice, *Gerontologist* **21**:90.

Hoffman, B. B., and Lefkowitz, R. J., 1980, Radiologand binding studies of adrenergic receptors: New insights into molecular and physiological regulation, *Ann. Rev. Pharmacol. Toxicol.* **20**:581–608.

Hoffman, B. B., and Lefkowitz, R. J., 1982, Adrenergic receptors in the heart, *Ann. Rev Physiol.* **44:**475–484.

Holinka, C. F., Nelson, J. F., and Finch, C. E., 1975, Effect of estrogen treatment of estradiol binding capacity in uteri of aged rats. *Gerontologist* **15:**30.

Hollenberg, M. D., and Schneider, E. L., 1979, Receptors for insulin and epidermal growth factor—Urogastrone in adult human fibroblasts do not change with donor age, *Mech. Ageing Dev.* **11:**37–43.

Hruska, R. E., Weir, R., Pitman, K. T., and Silbergeld, E. K., 1981, Ergot derivatives are potent drugs at CNS aminergic receptors: Correlations to behavior and aging, *Abstracts Symp. on Aging Brain and Ergot Alkaloids* **41.**

Huang, H. H., Steger, R. W., Sonntag, W. E., and Meites, J., 1980, Loss of positive feedback by estrogen-progesterone on prolactin release in old castrated female rats, *Fed. Proc.* **39:**947.

Ito, H., 1979, Age-related changes in insulin action and binding to ^{125}I-insulin to the cultered human skin fibroblasts, *J. Jpn Diab. Soc.* **22:**517–526.

Ito, H., Baum, B. J., and Roth G. S., 1981a, β-adrenergic regulation of rat parotid gland exocrine protein secretion during aging, *Mech. Ageing Dev.* **15:**177–188.

Ito, H., Hoopes, M. T., Baum, B. J., and Roth G. S., 1981b, Age related alterations in α-adrenergic responsiveness of rat parotid cells, *Gerontologist* **21:**115.

Ito, H., Baum, B. J., Uchida, T., Hoopes, M. T., and Roth, G. S., 1982, Diminished alpha adrenergic responsiveness in rat parotid acinar cells with normal receptor characteristics, *J. Biol. Chem.* **246:**9532–9538.

James, T. C., and Kanungo, M. S., 1976, Alterations in atropine sites of the brain of rats as a function of age, *Biochem. Biophys. Res Commun.* **72:**170–175.

Joseph, J. A., Berger, E., Engel, B. T., and Roth, G. S., 1978, Age-related changes in the nigrostriatum: a behavioral and biochemical analysis *J. Gerontol.* **33:**643–649.

Joseph, J. A., Filburn, C. R., and Roth G. S., 1981, Development of dopamine receptor denervation supersensitivity in the neostriatun of the senescent rat, *Life Sci.* **29:**575–584.

Kalimi, M., and Banerji, A., 1981, Effect of age on the *in vitro* glucocorticoid binding activity of rat adipocytes, *Mech. Aging Dev.* **17:**19–25.

Kalimi, M., and Seifter, S., 1979, Glucocorticoid receptors in WI-38 fibroblasts: Characterization and changes with population doublings in culture, *Biochem. Biophys. Acta* **583:**352–361.

Kaplan, J., 1981, Polypeptide-binding membrane receptors: Analysis and classification, *Science* **212:**14–20.

Kebabian, J. W., and Calne, D. B., 1979, Multiple receptors for dopamine, *Nature* **277:**93–96.

Kelliher, G. J., and Stoner, S., 1978, Cardiovascular effects of norepinephrine (NE) and phenylephrine (PE) in Fischer 344 rats of different ages. *Gerontologist* **18:**88.

Kledzik, G. S., Marshall, S., Huang, H. H., Bruni, J. F., and Meites, J., Serum prolactin, testosterone and prolactin receptor activity in the ventral prostate of old and young male rats, personal communication.

Kohn, R. R., 1977, Heart and cardiovascular system, in: *Handbook of the Biology of Aging* (C. E. Finch and L. Hayflick, eds.), pp. 281–317, Van Nostrand Reinhold Company, New York.

Kondo, H., Kasuga, H., and Noumora, T., 1978, Specific binding of glucocorticoids to human diploid fibroblasts during *in vitro* aging, *Abstracts of the XII Intl. Congr. of Gerontology,* pp. 26–27.

Kosmakos, F. C., Nagulesparan, M., and Bennet, P. H., 1980, Insulin binding to erythrocytes: A negative correlation with red cell age, *J. Clin. Endocrinol. Metab.* **51:**46–50.

Landmann, R., Bittiger, H., and Buhler, F. R., 1981, High affinity beta-2adrenergic receptors in mononuclear leucocytes: Similar density in young and old normal subjects, *Life Sci.* **29**:1761–1771.

Latham, K. R., and Finch, C. E., 1976, Hepatic glucocorticoid binders in mature and senescent C57B1/6J mice. *Endocrinology* **98**:1480–1489.

Lee, H-C., Pay, M. A., and Gallop, P. M., 1982, Low-density liporotein receptor binding in aging human diploid fibroblasts in culture, *J. Biol. Chem.* **257**:8912–8918.

Levin, P., Janda, J. K., Joseph, J. A., Ingram, D. K., and Roth, G. S., 1981, Dietary restriction retards the age-associated loss of rat striatal dopaminergic receptors, *Science* **214**: 561–562.

Lin, T.,Chen, G. C. C., Murono, E. P., Osterman, J., and Nankin, H. R., 1981, The aging Leydig cell: VII. Cytoplasmic estradiol receptors, *Steroids* **38**:407–415.

Lippa, A. S., Pelman, R. W., Beer, B., Critchett, D. J., Dean, R. L., and Bartus, R. T., 1980, Brain cholinergic dysfunction and memory in age rats, *Neurobiol. Aging* **1**:13–19.

Lippa, A. S., Critchett, D. J., Ehlert, F., Yamamura, H. I., Enna, S. J., and Bartus, R. T., 1981, Age-related alteration in neurotransmitter receptors: An electrophysiological and biochemical analysis, *Neurobiol. Aging* **2**:3–8.

Livingston, J. N., Cautrecasas, P., and Lockwood, D. H., 1974, Studies of glucagon resistance in large rat adipocytes: ^{125}I-labeled glucagon binding and lipolytic capacity, *J. Lipid Res.* **15**:26–32.

Maggi, A., Schmidt, M. H., Ghetti, B., and Enna, S. J., 1979, Effect of aging on neurotransmitter receptor binding in rat and human brain, *Life Sci.* **24**:367–374.

Marquis, I., Lippa, A. S., and Pelham, R. W., 1981, Dopamine receptor alterations with aging in mouse and rat corpus striatum, *Biochem. Pharmacol.* **30**:1876–1878.

Martin, W. R., 1981, Mini-symposium II. Multpole opiod receptors, *Life Sci.* **28**:1547–1554.

Martin, R., Martin, H., and Rotzsch, W., 1980, Zur Molekularbiologie des Alterns 18: Mitteilung: Der Einfluss von Hormonen auf das Genon im Alter, *Z. Alternsforsch.* **35**:3–12.

Mayer, M., Amin, R., and Shafrir, E., 1981, Effect of age on myofibrillar protease activity and muscle binding of glucocorticoid hormones in the rat, *Mech. Ageing Dev.* **17**:1–10.

McDougal, J. N., Pedigo, N. W., Marques, P. R., Yamamura, H. J., and Burks, T. F., 1980a, Changes in ^{3}H-naloxone binding in the brains of senescent rats, *Abstracts of the Soc. for Neuroscience* **6**:78.

McDougal, J. N., Marques, P. R., and Burks, T. F., 1980b, Age-related changes in body temperature responses to morphine in rats, *Life Sci.* **27**:2679–2685.

Meites, J., and Sonntab, W. E., 1981, Hypothalamic hypophysiotropic hormones and neurotransmitter regulation: Current views, *Ann. Rev. Pharmacol. Toxicol.* **21**:295–322.

Memo, M., Lucchi, L, Spano, P. F., and Trabucchi, M., 1980, Aging process affects a single class of dopamine receptors, *Brain Res.* **202**:488–492.

Memo, M., Spano P. F., and Trabucchi, M., 1981, Brain benzodiazepine receptor changes during aging, *J. Pharm. Pharmacol.* **33**:64.

Messing, R. B., Vasquez, B. J., Spiehler, V. R., Martinez, J. L. Jensen, R. A., Rigter, H., and McGaugh, J. L., 1980, ^{3}H-dihdromorphine binding in brain regions of young and aged rats, *Life Sci.* **26**:921–927.

Messing, R. B., Vasquez, B. J., Sammaniego, B., Jensen, R. A., Martinez, J. L., and McGaugh, J. L., 1981, Alterations in dihydormorphine binding in cerebral hemispheres of aged male rats, *J. Neurochem.* **36**:784–790.

Miller, J. K., and Bolla, R., 1981, Influence of steroid hormone-receptor protein complexes on initiation of ribonucleic acid synthesis in liver nuclei isolated from rats of various ages, *Biochem. J.* **197**:373–375.

Misra, C. H., Shelat, H. S., and Smith, R. C., 1980, Effect of age on adrenergic and dopaminergic receptor binding in rat brain, *Life Sci.* **27**:521–526.

Morin, A. M., and Westerlain, C. G., 1980, Ageing and rat brain muscarinic receptors as measured by quinuclidinyl benzilate binding *Neurochem. Res.* **5**:301–308.

Morwaha, J., Hoffer, B., Wyatt, R., and Freedman, R., 1980, Age-related electrophysiological changes in rat cerebellum, *Abstracts of the Soc. for Neuroscience* **6**:280.

Moudgil, V. K., and Kanungo, M. S., 1973, Effect of age of the rat on induction of acetylcholinesterase of the brain by 17b-estradiol, *Biochem. Biophys. Acta* **329**:211–220.

Muggeo, M., Moghetti, P., Valerio, A., Faronato, P., Businaro, V., Sartore, P., and Crepoldi, G., 1981, Insulin sensitivity and insulin receptors on human monocytes in aging, *Abstracts of the XII Intl. Congr. of Gerontology,* p. 105.

Murota, S., and Koshihara, Y., 1977, Aging and receptor, *Protein, Nucleic Acid and Enzyme, Special edition of "Receptor Modulation"* pp. 190–197.

Nelson, J. F., Holinka, C. F., and Finch, C. E., 1976a, Age related changes in estradiol binding capacity of mouse uterine cytosol, *Abstracts of the 29th Annual Mtg. of the Geront. Soc.* **34**:86.

Nelson, J. F., Holinka, C. F., Latham, K. R., Allen, J. K., and Finch, C. E., 1976b, Coritcosterone binding in cytosols from brain regions of nature and senescent male C57B1/6J mice, *Brain Res.* **115**:345–351.

Nicoll, R. A., Siggens, G. R., Ling, N., Bloom, F. E., and Guilleman, R., 1977, Neuroual actions of endorphrins and enkephralins among brain regions: A comparative microinotophoretic study, *Proc. Natl. Acad. Sci. USA* **74**:2584–2588.

Nikitin, V. N., and Nesterenko, G. A., 1979, Tissue reception for steroid hormones in aging, *Fiziol. Zh.* **25**:306–312.

Nordberg, A., and Wahlstrom, G., 1981, Cholinergic receptors and aging, *Abstr. Eighth Intl. Congr. Pharmacol.* p. 827.

Nordberg, A., and Winblad, B., 1981, Cholinergic receptors in human hippocampus-regional distribution and variance with age, *Life Sci.* **29**:1937–1944.

Pagano, G., Cassander, M., Diana, A., Pisu, E., Bozzo, C., Ferroro, F., and Lenti, G., 1981, Insulin resistance in the aged: The role of periperal receptors, *Metabolism* **30**:46–49.

Pagano, G., Poli, G., Cassander, M., Chiarapotto, E., Pisu, E., Togliaferro, V., and Albano, E., 1982, Insulin resistance in old spontaneoulsy obese rats: Studies on insulin binding in rat isolated hepatocytes, *ICRS Med. Sci.-Blochem.* **10**:258.

Parchman, G. L., Cake, M. H., and Litwack, G. L., 1978, Functionality of the liver glucocorticoid receptor during the life cycle and development of a low-affinity membrane binding site, *Mech. Ageing Dev.* **7**:227–240.

Partilla, J. S., Hoopes, M. T., Ito, H., Dax, E. M., and Roth, G. S., 1982, Loss of ventricular α_1-adrenergic receptors during aging, *Life Sci.* **31**:2507–2512.

Pastan, I. H., and Willinghan, M. C., 1981, Journey to the center of the cell: Role of the receptosome, *Science* **214**:504–509.

Pedigo, N. W., Schoemaker, H., Morelli, M., McDougal, J. N., Malick, J. B., Burks, T. F., and Yamamura, H. I., 1981, Benzodiazepine receptor binding in young, mature, and senescent rat brain and kidney, *Neurobiol. Aging* **2**:83–88.

Perry, E. K., 1980, The cholinergic system in old age and Alzheimers disease, *Age Ageing* **9**:1–8.

Petrovic, J. S., and Markovic, R. Z., 1975, Changes in cortisol binding to soluble receptor proteins in rat liver and thymus during development and aging, *Dev. Biol.* **45**:176–182.

Piantanelli, L, Brogli, R., Bevilacqua, P., and Fabris, N., 1980, Beta-adrenorecptor changes in submandibular glands of old mice *Mech. Ageing Dev.* **14**:155–167.

Piantanelli, L., Gentile, S., and Viticchi, C., 1981, Beta-adrenoreceptor changes in brain cortex of aging mice, *Abstracts of the XII Intl. Cong. of Gerontology,* p. 138.

Piantanelli, L., Vitcchi, C., and Gentile, S., 1982, Age and thymus-dependent modulation of beta-adrenoreceptors in mouse brain cortex, *Abstracts "The Aging of the Brain"*, Montova, Italy, p. 47.

Pirke, L. M., Vogt, H., and Geiss, M., 1978, *In vitro* and *in vivo* studies on Leydig cell function in old rats, *Acta Endocrinol.* **89**:393–403.

Pittman, R. N., Minneman, K. P., and Molinoff, P. B., 1980, Alterations in β_1 -and β_2 -adrenergic receptor density in the cerebellum of aging rats, *J. Neurochem.* **35**:273–275.

Pradham, S. N., 1980, Central neurotransmitters and aging, *Life Sci.* **26**:1643–1656.

Puri, S. K., and Volicer, L., 1977, Effect of aging on cyclic AMP levels and adenylate cyclase and phosphodiesterase activities in the rat corpus striatum, *Mech. Ageing Dev.* **6**:53–58.

Rajfer, J., Namkung, P. C., and Petra, P. H., 1980, Identification, partial characterization and age-related changes of a cytoplasmic androgen receptor in the rat penis, *J. Steroid Biochem.* **13**:1489–1492.

Randall, P. K., Severson, J. A., and Finch, C. E., 1981, Aging and the regulation of striatal dopaminergic mechamisms in mice, *J. Pharmacol. Exp. Theraputics* **219**:695–700.

Riccardi, F., Covelli, V., Spano, P. F., Govoni, S., and Trabucchi, M., 1981, Rat dopaminergic function in the retina during aging, *Neurobiol. aging* **2**:229–231.

Robinette, C. L., and Mawhinney, H. G., 1977, Cytosol binding of dihydrotestosterone in young and senile rats, *Fed. Proc.* **36**:344.

Robinette, C. L., and Mawhinney, M. G., 1978, Effects of aging on the estrogen receptor in rat seminal vesicles, *Fed. Proc.* **37**:283.

Rosenbloom, A., Goldstein, S., and Yip, C., 1976, Insulin binding to cultured human fibroblasts increases with normal and precocious aging, *Science* **193**:412–414.

Rosner, B. A., and Cristofalo, V. J., 1981, Changes in specific dexamethasone binding during aging in WI-38 cells, *Endocrinology* **108**:1965–1971.

Roth, G. S., 1974, Age-related changes in specific glucocorticoid binding by steroid responsive tissues or rats, *Endocrinology* **94**:82–90.

Roth, G. S., 1975, Reduced glucocorticoid responsiveness and receptor concentration in splenic leukocytes of senescent rats, *Biochem. Biophys. Acta* **399**:145–146.

Roth, G. S., 1976, Reduced glucocorticoid binding site concentration in cortical neuronal perikarya from senescent rats, *Brain Res.* **107**:345–354.

Roth, G. S., 1979, Hormone receptor changes during adulthood and senescence: Significance for aging research, *Fed. Proc.* **38**:1910–1914.

Roth, G. S., 1982, Brain dopaminergic and opiate receptors and responsiveness during aging, in: *The Aging Brain and Ergot Alkaloids* (A. Agnoli, G., Crepaldi, P. E.Spano, and M. Trabucchi, eds.), pp. 53–60, Raven Press, New York.

Roth, G. S., and Adelman, R. C., 1975, Age related changes in hormone binding by target cells and tissues; possible role in altered adoptive responsiveness, *Exp. Gerontol.* **10**:1–11.

Roth, G. S., and Livingston, J. N., 1976, Reductions in glucocortiocid inhibition of glucose oxidation and presumptive glucocorticoid receptor content in rat adipocytes during aging, *Endocrinology* **99**:831–839.

Roy, A. K., Milin, B. S., and McMinn, D. M., 1974, Androgen receptors in rat liver: Hormonal and developmental regulation of the cytoplasmic receptor and its correlation with the androgen dependent synthesis of α2u globulin, *Biochem. Biophys. Acta* **345**:123–232.

Saiduddin, S., and Zassenhaus, H. P., 1979, Estrus cycles, decidual cell response and uterine estrogen and progesterone receptor in Fisher 344 virgin aging rats, *Proc. Soc. Exp. Biol. Med.* **161**:119–122.

Saiga-Narumi, T., and Kawashima, S., 1978, The binding of rat luteinizing hormone (rLH), rat follicle stimulating hormone (rFSH) and rat prolactin (rPRL) to dissociated cells of luteinized ovaries and intraplenic ovarian grafts, *Proc. Jpn Acad.* **54**:25–29.

Scarpace, P. I., and Abrass, I. B., 1981, Thyroid horme regulation of β-adrenergic receptor number in aged rats, *Endocrinology* **108:**1276–1278.

Schmidt, M. J., and Thornberry, J. F., 1978, Cyclic AMP and cyclic GMP accumulation *in vitro* in brain regions of young, old, and aged rats, *Brain Res.* **139:**169–177.

Schocken, D. D., and Roth G. S., 1977, Reduced beta adrenergic receptor concentrations in aging man, *Nature* **267:**856–858.

Severson, J. A., and Finch, C. E., 1980, Reduced dopaminergic binding during aging in the rodent striatum, *Brain Res.* **192:**147–162.

Severson, J. A., Marcusson, J., Winblad, B., and Finch, C. E., 1982, Age-correlated changes in dopaminergic binding sites in human basal ganglia, *J. Neurochem.* **39**(6)**:**1623–1631.

Shain, S. A., and Axelrod, L. R., 1973, Reduced high affinity 5α-dihydrotestosterone receptor capacity in the ventral prostate of the aging rat. *Steroids* **21:**801–812.

Shain, S. A., and Boesel, R. W., 1977, Aging-associated diminished rat prostate androgen receptor content concurrent with decreased androgen dependence, *Mech. Ageing Dev.* **6:**219–232.

Shain, S. A., and Boesel, R. W., 1978, Androgen receptor content of the normal and hyperplastic canine prostate, *J. Clin. Invest.* **61:**654–660.

Shain, S., and Moss, A. L., 1981, Aging in the AXC rat: Differential effects of chronic testosterone treatment on restoration of diminished prostate L-ornithine decarboxylase and s-adenosyl-L-methionine decarboxylase activities, *Endocrinology* **109:**1184–1191.

Shain, S. A., Boesel, R. W., and Axelrod, L. R., 1973, Aging in the rat prostate: Reduction in detectable ventral prostate androgen receptor content, *Arch. Biochem. Biophys.* **167:**247–263.

Shih, J. C., and Young, H., 1978, The alteration of serotonin binding sites in aged human brain, *Life Sci.* **23:**1441–1448.

Shock, N. W., 1977a, Biological theories of aging, in:*Handbook of the Psychology of Aging* (J. E. Birren and K. W. Schaie, eds.), pp. 103–115, Van Nostrand Reinhold Company, New York.

Shock, N. W., 1977b, Systems integration, in:*Handbook of the Biology of Aging* (C. E. Finch and L. Hayflic, eds.), pp. 639–665, Van Nostrand Reinhold Company, New York.

Singer, S., Ito, H., and Litwack, G., 1973, ^{3}H-cortisol binding by young and old human liver cytosol proteins *in vitro, Int. J. Biochem.* **4:**569–573.

Singhal, R. L., Valadares, H. R. E., and Ling, G. M., 1969, Estrogenic regulations of uterine carbohydrate metabolism during senescence, *Am. J. Physiol.* **217:**793–797.

Sorrentino, R. M., and Florini, J. R., 1976, Variations among individual mice in binding growth hormone and insulin to membranes from animals of different ages, *Exp. Aging Res.* **2:**191–205.

Steger, R. W., Peluso, J. J., Huang, H., Hafez, E. S. E., and Meites, J., 1978, Gonadotropin binding sites in the ovary of aged rats, *J. Reprod. Fert.* **48:**205–207.

Steger, R. W., Peluso, J. J., Bruni, J. F., Hafez, E. S. E., and Meites, J., 1979, Gonadotropin binding and testicular function in old rats, *Endokrinologie* **73:**1–5.

Strong, R., Hicks, P., Hsu, L., Bartus, R. T., and Enna, S. J., 1980, Age-related alterations in the rodent brain cholinergic system and behavior, *Neurobiol. Aging* **1:**59–63.

Talbert, G. B., 1977, Aging of the reproductive system, in: *Handbook of the Biology of Aging* (C. E. Finch and L. Hayflick, eds.), pp. 318–356, Van Nostrand Reinhold Company, New York.

Thakur, M. K., Das, R., and Kanungo, M. S., 1978, Modulation of acetylation of chromosomal proteins of the brain of rats of various ages by epinephrine and estradiol, *Biochem. Biophys. Res. Commun.* **81:**828–831.

Thal, L. J., Horowitz, S. G., Dvorkin, B., and Makman, M. H., 1980, Evidence for loss of brain ^{3}H-ADTN binding sites in rabbit brain with aging, *Brain Res.* **192:**185–194.

Timiras, P. S., and Bignani, A., 1976, Pathophysiology of the aging brain, in:*Special Review of Experimental Aging Research: Progress in Biology* (M. R. Elia, B. E. Eleftheriou, and P. K. Elias, eds.), pp. 351–389, Bar Harbor, Maine.

Trabucchi, M., Kobayashi, H., and Spano, P. F., 1981, β-adrenergic receptors in rat brain microvessels: Changes during aging, *Abstracts of the Society for Neuroscience* **7**:183.

Tsitouras, P. D., Kowatch, M. A., and Harman, S. M., 1979, Age-related alterations in isolated rat Leydig cell function: Gonadotropin receptors, adensoine 3′, 5′-monophosphate response, and testosterone secretion, *Endocrinology* **105**:1400–1405.

Tyers, M. B., 1980, A classification of apiate receptors that mediate antinociception in animals, *Br. J. Pharamacol.* **69**:503–512.

Vassileva-Popova, J., 1974, Developmental changes in gonadal binding of gonadotropins, *Proc. V Asia Oceania Congr. Endocrinol.* pp. 242–248.

Villee, D. B., Berger, R., and Wenninger, N., 1979, Insulin responsiveness in aging and peogeric fibroblasts, *Abstracts of the 61st Annual Mtg. of the Endocrine Society,* p. 109.

Vydelingum, N., Gambert, S. R., and Kissebah, A. H., 1981, Age related changes in rat adipocyte tissue ribsomal messenger RNA and protein synthesis: Effects of insulin, *Abstracts of the 63rd Annual Mtg. of the Endocrine Society,* p. 161.

Walker, J. B., and Walker, J. P., 1973, Properties of adenylate cyclase from senescent rat brain, *Brain Res.* **54**:391–399.

White, P., Hiley, C. R., Goodhardt, M. J., Carrasco, L. H., Keat, J. P. Williams, I. E. I., and Bowen, D. M., 1977, Neocortical cholenergic neurons in elderly people, *Lancet* **1**:668–670.

Zitnik, G., and Roth, G. S., 1981, Effects of thyroid hormones on cardiac hypertrophy and β-adrenergic receptors during aging, *Mech. Ageing Dev.* **15**:19–28.

3

Metabolism of the Brain
A Measure of Cellular Function in Aging

EDYTHE D. LONDON

1. INTRODUCTION

Because of the close relationships between cerebral functional activity and measures of cerebral blood flow and metabolism, investigators have used these measures to gain information about altered function in various experimental and pathologic states. This chapter reviews information regarding the associations between cerebral metabolism, blood flow, and function. Also reviewed are results from *in vitro* studies of oxidative metabolism and glucose metabolism in aging animals. Techniques for assessment of cerebral blood flow and oxidative metabolism *in vivo* are described, as are results obtained in animal and human studies of age differences in these parameters. The possible advantage of performing measurements on subjects in activated states, rather than the resting state, is discussed.

2. MEASURES OF CEREBRAL METABOLISM AND BLOOD FLOW AS INDICES OF BRAIN FUNCTIONAL ACTIVITY

Because the brain is one of the most metabolically active organs of the body, it requires enormous amounts of energy for processes including ion transport, storage, release, and reuptake of neurotransmitter, and synthesis of membranes and macromolecules (Kety, 1957; Sokoloff, 1960). Energy for these processes is supplied in the form of ATP, which is produced by the oxidation of

EDYTHE D. LONDON • National Institute on Drug Abuse, Addiction Research Center, and National Institute on Aging, Gerontology Research Center, Baltimore City Hospitals, Baltimore, Maryland 21224.

fuel molecules. The results of animal and human experiments have indicated that glucose is the main substrate for cerebral metabolism in adults, and that most of the glucose extracted by the brain is oxidized (Sokoloff, 1972; Siesjö, 1978). The oxidation of glucose is coupled to ATP formation and is stoichiometrically related to oxygen consumption under physiologic conditions. Measurements of the cerebral metabolic rate for glucose (CMR_{glc}), therefore, provide information on the cerebral metabolic rate for oxygen ($CMRO_2$) and on cerebral energy metabolism in general.

Presumably, alterations in cerebral function would affect energy demand and thereby influence oxidative metabolism. Therefore, it seems reasonable that measurements of $CMRO_2$ would reflect alterations in cerebral functional activity. In support of this view, an association of reduced $CMRO_2$ with impaired mental function has been described in many degenerative clinical conditions (Patterson *et al.*, 1950; Freyhan *et al.*, 1951; Schieve and Wilson, 1953; Garfunkel *et al.*, 1954; Lassen *et al.*, 1957). In some studies, the degree of reduction in $CMRO_2$ corresponded roughly with the degree of mental deterioration (Patterson *et al.*, 1950; Garfunkel *et al.*, 1954; Lassen *et al.*, 1957). However, this was not always true. Fazekas *et al.* (1952) noted that $CMRO_2$ in clinically normal elderly patients was significantly reduced from the mean in individuals below 50 years of age, despite completely adequate mental function.

The relations between glucose oxidation, ATP production and oxygen consumption imply that glucose utilization is related to cerebral functional activity. Using brain slices, McIlwain *et al.* (1951) established that glucose metabolism is tightly coupled to neuronal firing. This phenomenon was confirmed *in vivo* by various studies which have revealed a relation between cerebral functional activity and CMR_{glc} (Sokoloff, 1977). For example, glucose utilization was increased in optic centers of the rat brain in response to increased frequency of visual stimulation (Toga and Collins, 1981). Also, in a preliminary study of patients with senile dementia of the Alzheimer type, regional cerebral metabolic rates for glucose were significantly correlated with cognitive measures (Ferris *et al.*, 1980).

Measures of global cerebral blood flow (CBF) and regional cerebral blood flow (rCBF) also can serve as indices of cerebral functional activity. Roy and Sherrington (1890) first noted that CBF is regulated according to the requirements of cerebral metabolism. This observation has been confirmed in animals and in normal human subjects, as well as those with brain disease (Nilsson *et al.*, 1978). Regional CMR_{glc} ($rCMR_{glc}$) and regional CBF (rCBF) are closely related in conscious and anesthetized rats (Des Rosiers *et al.*, 1974). In man, an example of a similar relation has been obtained in the cerebral cortex, where rCBF is coupled to the regional cerebral metabolic rate for oxygen ($rCMRO_2$) during hard exercise (Raichle *et al.*, 1976).

CBF and rCBF are reduced in neuropathologic states such as senile dementia (Obrist et al., 1970; Hagberg and Ingvar, 1976; Ingvar and Lassen, 1979) and stroke (Schieve and Wilson, 1953). As with $CMRO_2$, the correlation of CBF with cognitive function is variable. Thus, Garfunkel *et al.* (1954) observed that CBF was minimal in children with very low developmental quotients, whereas Patterson *et al.* (1950) found a poor correlation between CBF and the degree of mental deterioration in demented patients.

3. IN VITRO MEASUREMENTS OF OXIDATIVE METABOLISM, GLUCOSE UTILIZATION, AND RELATED ENZYMES IN AGING

Sanadi (1977) has reviewed *in vitro* evidence for a reduction in cerebral oxidative metabolism with aging. Using a Warburg technique, Reiner (1947) demonstrated that oxygen consumption in whole brain homogenates from rats was maintained at the adult level until the age of the animals exceeded 2 years, when oxygen consumption dropped considerably. He also noted that glycolysis declined from a maximum which occurred between 2 and 4 months. Peng *et al.* (1977), who used similar techniques with homogenates of rat cerebral cortex, hypothalamus, hippocampus, and amygdala, observed age-related decrements in oxygen consumption by homogenates of all brain regions except the amygdala. The time course of the decline varied with the brain region. In contrast, using somewhat different methods with rat brain homogenates, Garbus (1955) found that oxygen consumption was equal before and after 24 months of age.

Studies of age effects on respiration and carbohydrate metabolism in brain slices have yielded different results depending on the species under investigation. Fox *et al.* (1975) incubated brain slices from young (average 8 months) and old (average 18 months) Syrian hamsters with ^{14}C-glucose, and found no age difference in oxygen uptake, $^{14}CO_2$ production, glucose utilization, or lactate and pyruvate formation. However, Parmacek *et al.* (1979) used the same techniques with cerebral cortex slices from young (average 6 months) and old (average 34 months) mice to show that in incubations at 37°C, slices from old mice had lower levels of oxygen uptake and $^{14}CO_2$ production with no differences in glucose utilization, lactate formation, or pyruvate formation. Although the reason for the decreased oxygen uptake and $^{14}CO_2$ production without associated changes in glucose utilization or in the production of metabolic intermediates remains obscure, all of the parameters showed age-associated decrements when incubations were carried out at 40°C. This suggests that age-related metabolic deficiencies which are not apparent under normal conditions can be manifested by "stressed" conditions.

Because of the well-documented role of pyruvate kinase as a rate-limiting and regulatory enzyme in cerebral carbohydrate metabolism (Takagaki,

1968), age-related changes in the activity of this enzyme could reflect alterations in the efficiency of the brain to utilize glucose. Schwark *et al.* (1971) found no significant difference in the activity of this enzyme in the brains of rats between the ages of 3 months and 1 year. Vitorica *et al.* (1981) also found no age-related decrease in pyruvate kinase activity throughout the rat brain; in fact, these authors noted a slight elevation at 16–21 months. In contrast, Chainy and Kanungo (1978) observed a marked decrease in the activity of pyruvate kinase in 14,000 $\times$ *g* supernatants from whole brain homogenates of 19.5- as compared with 9.5-month-old rats.

Activities of mitochondrial enzymes related to the function of the tricarboxylic acid cycle have been studied in brain homogenates and mitochondrial fractions from animals of various ages. In whole rat brain homogenates from 3-month- and 2-year-old rats, Patel (1977) found no differences in the activities of citrate synthase and pyruvate carboxylase. Similarly, Vitorica *et al.* (1981) saw no significant difference in the activities of citrate synthase and malate dehydrogenase among mitochondrial fractions from brains of 3- and 21-month-old rats. Conflicting data were presented by Ryder and Izquierdo (1979), who noted an age-dependent decrease in malate dehydrogenase from selected brain regions. A brain region-specific age effect also was observed for succinic dehydrogenase, which showed an age-dependent increase in mitochondrial fractions from striatum; however, Meier-Ruge *et al.* (1976) noted lower activities of succinic dehydrogenase in the granular layer of the cerebellum and cerebral cortex of 30-month-old rats compared with 5-month-old animals. Decrements in the activity of NAD-isocitrate dehydrogenase have been reported by Patel (1977), who noted lower activities in whole brain homogenates at 1 year vs. 3 months, and at 2 years vs. 1 year of age. Similarly, Vitorica *et al.* (1981) reported a gradual decline in NAD-isocitrate dehydrogenase after 3 months.

Although the results of the aforementioned studies are somewhat inconsistent, the general impression from *in vitro* assays is that cerebral oxidative metabolism and glucose utilization are reduced in aged rodents. There is evidence for age-induced deficiencies at various stages of carbohydrate metabolism, including anaerobic glycolysis as well as the tricarboxylic acid cycle. Findings have varied with differences in techniques as well as with the species or brain regions assayed.

4. METHODS FOR IN VIVO MEASUREMENT OF CBF, $CMRO_2$, AND CMR_{glc}

Details of the theory and techniques used in studies of CBF and metabolism have been reviewed previously (Lassen, 1959; Posner, 1973; Lassen, 1974; Marcus *et al.*, 1981). A pivotal advance in this area was the development of the inert gas method to measure CBF in unanesthetized human subjects

(Kety and Schmidt, 1945, 1948). This method, which yields an average value of blood flow per minute per unit weight of the brain as a whole, involves inhalation by the subject of low concentration of nitrous oxide (N_2O). The N_2O concentrations in arterial blood and cerebral venous blood, sampled from an internal jugular vein, are followed for a 10-min period. CBF is calculated from the resultant arterial and venous N_2O concentration curves by application of the Fick principle. This principle postulates that the quantity of any substance taken up in a given time by an organ from the blood which perfuses it is equal to the total amount of the substance carried to the organ by the arterial inflow less the amount removed by the venous drainage during the same period.

The Fick principle also is the basis of isotope clearance methods, which were developed by Lassen and Ingvar (1961). These methods involve the extracranial monitoring of an inert, diffusible radioisotope, such as ^{85}Kr or ^{133}Xe, injected into the internal carotid artery (Lassen *et al.,* 1963). The clearance curves obtained are subjected to a two-compartment analysis (Sveinsdottir, 1965) in which separate blood flows are derived for fast and slow tissue phases attributable to gray and white matter. The major limitation of these methods is the need to insert a catheter or needle into the carotid artery.

This problem has been obviated by the ^{133}Xe inhalation method introduced by Mallett and Veall (1963). Tracer amounts of ^{133}Xe are inhaled for 2 min and are monitored extracranially over the next 45–60 min. Obrist *et al.* (1967) used a three-compartment model to analyze the resultant clearance curves; thus, providing separate estimates of blood flow for gray and white matter, as well as a decay constant for a slow third component believed to arise from extracerebral tissue.

CBF also can be calculated by the intravenous radioiodinated serum albumin (RISA) technique, which can be used to estimate brain blood volume and mean transit time (Fujishima, 1967). ^{131}I-RISA is injected into the antecubital vein, and hemispheric RISA dilution curves are recorded by NaI scintillation detectors collimated separately for each hemisphere.

Given the value for CBF, one can estimate utilization or production by the brain of any substance which can be analyzed accurately in arterial and venous blood (Kety and Schmidt, 1948). The products of CBF and the arteriovenous differences for oxygen and glucose yield the global cerebral metabolic rates for oxygen ($CMRO_2$) and glucose (CMR_{glc}), respectively.

In conscious laboratory animals, the combination of autoradiographic techniques with the Fick principle has allowed estimates of rCBF. In 1955, Landau *et al.* described an autoradiographic technique for the determination of rCBF utilizing [^{131}I]trifluoroiodomethane, an inert gas. Dissolved in Tyrode's solution, the gas was administered intravenously over a 1-min period, during which concentrations of the gas in arterial blood were determined. At the end of the infusion, the animal was decapitated, and autoradiograms were prepared from 5-mm-thick brain sections which contained ^{133}I standards embedded in them. Measurements of rCBF could be made in all structures of the brain visible on autoradiographs. Subsequent techniques to study rCBF with the nonvolatile radiotracer [^{14}C]antipyrine and, more recently, [^{14}C]iodoantipyrine, have provided values very similar to those obtained with [^{131}I]trifluoroiodomethane (Reivich *et al.*, 1969; Sakurada *et al.*, 1978).

By 1977, a method was available for quantitation of regional cerebral metabolic rates for glucose (rCMR$_{glc}$) in animals (Sokoloff *et al.*, 1977). According to this technique, an intravenous pulse of [^{14}C]2-deoxy-D-glucose ([^{14}C]DG), serves as a tracer for the transport of glucose between the plasma and brain, and its phosphorylation by brain hexokinase. Because [^{14}C]DG-6-phosphate, which is formed from [^{14}C]DG, is a poor substrate for enzymes which occur in appreciable concentrations in the brain, it accumulates as it is formed and is trapped in cerebral tissue. The concentration of radioactivity in the tissue at the time the animal is killed can be measured by quantitative autoradiography, and is proportional to the rate of glucose utilization.

The operational equation for calculating rCMR$_{glc}$ is given in Fig. 1. The measured variables are as follows: (1) the history of the arterial plasma [^{14}C]DG concentration from zero time to the time of killing, (2) the arterial plasma glucose concentration over the same interval, and (3) the concentration

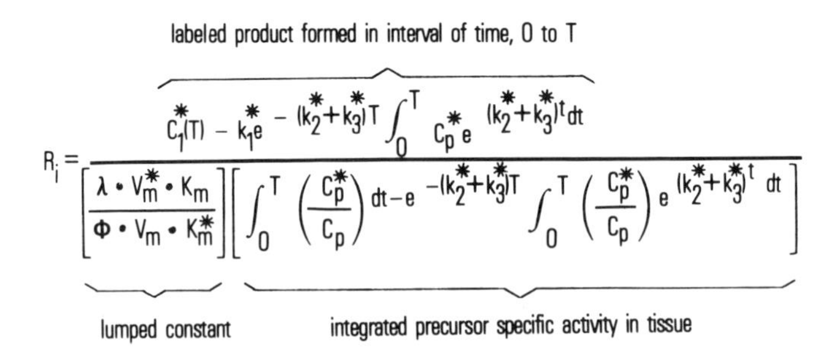

Figure 1. Operational equation of the [^{14}C]deoxyglucose ([^{14}C]DG) method. R$_i$ equals the rate of glucose utilization. C_i^* represents the total ^{14}C concentration in a single homogeneous tissue of the brain. C_p^* and C_p represent the concentrations of [^{14}C]DG and glucose in the arterial plasma, respectively. The constants k_1^*, k_2^*, and k_3^*, represent the rat constants for carrier-mediated transport for [^{14}C]DG from plasma to tissue and back from tissue to plasma, and for phosphorylation by hexokinase, respectively. T represents the time of termination of the experimental period; λ equals the ratio of the distribution space of [^{14}C]DG in the tissue to that of glucose; ϕ equals the fraction of glucose which once phosphorylated continues down the glycolytic pathway; and K_m^* and V_m^* and K_m and V_m represent the Michaelis–Menten kinetic constants of hexokinase for [^{14}C]DG and glucose, respectively (Sokoloff *et al.*, 1977).

of ^{14}C in brain regions at the time of killing. The rate constants for transport and phosphorylation of [^{14}C]DG (k_1^*, k_2^*, and k_3^*), and the lumped constant, which is an estimate of the relative predilection of the brain to take up and phosphorylate [^{14}C]DG as compared with glucose, are not calculated for each experimental situation. Although the rate constants may vary with different experimental situations, [^{14}C]DG is administered as a single intravenous bolus at zero time, and sufficient time (usually 45–50 min) is allowed for the clearance of [^{14}C]DG from the plasma so that the terms which contain the rate constants in the operational equation fall to levels too low to influence the final result (Sokoloff *et al.*, 1977). Therefore, standard values for the rate constants obtained in normal conscious rats (Sokoloff *et al.*, 1977) are used in calculating $rCMR_{glc}$ in various experimental paradigms involving rats and other species (Kennedy *et al.*, 1978; London *et al.*, 1983). The lumped constant is thought to be characteristic for each species, and standard published values for each species generally are used. Published values of the lumped constant are available for the albino rat (Sokoloff *et al.*, 1977), rhesus monkey (Kennedy *et al.*, 1978), cat (Sokoloff *et al.*, 1979b), and dog (Duffy *et al.*, 1979).

The deoxyglucose method has been applied to man by the use of [^{18}F]fluorodeoxyglucose [^{18}F]FDG (Phelps *et al.*, 1978; Reivich *et al.*, 1979), a labeled deoxyglucose analog which satisfies the following criteria: (1) is a suitable tracer labeled with a gamma-emitting radionuclide to enable detection of the activity through the skull; (2) the distribution of activity within the brain can be determined with three-dimensional resolution; (3) fulfills the criteria met by deoxyglucose; and (4) the radiation exposure is safe (Reivich *et al.*, 1979). [^{18}F]FDG is a good substrate for yeast hexokinase (Bessel *et al.*, 1972). Because it decays by positron emission, producing two 511-Kev photons upon annihilation, and has a short half-life (110 min), it meets the requirements for external detection and acceptable radiation dosimetry. External detection is performed by positron emission tomography (PET).

PET also has been used to study CBF and $CMRO_2$ by inhalation of $C^{15}O_2$ and $^{15}O_2$, respectively (Alpert *et al.*, 1979). The basic premise of these studies is that because ^{15}O has a short half-life, continuous inhalation of ^{15}O-labeled compounds rapidly produces a dynamic equilibrium in which the amount of tracer entering the brain is equal to that leaving due to washout and decay. With inhalation of $C^{15}O_2$, the label is transferred in the lungs to $H^{15}O_2$, and enters the brain as water of perfusion to be distributed in proportion to CBF. In contrast, label in the brain from inhaled $^{15}O_2$ represents primarily regional water of metabolism and circulating labeled water. Quantitation of absolute CBF or $CMRO_2$ requires determination of whole blood, plasma and red cell ^{15}O concentrations. $CMRO_2$ also has been studied by PET following injection of ^{15}O attached to hemoglobin (Ter-Pogossian *et al.*, 1969).

5. AGE EFFECTS ON CEREBRAL METABOLISM: ANIMAL STUDIES

Investigations of age differences in cerebral metabolic indices *in vivo* include studies of CBF, CMR_{glc}, $CMRO_2$, and $rCMR_{glc}$ in rodents and dogs, and rCBF in rats. The results indicate that age-related reductions in metabolic rate vary with the brain region and animal species investigated.

Gibson *et al.* (1978) studied whole brain glucose utilization with [U-^{14}C]glucose in BALB/C and C57BL mice at 3-, 10-, and 30 months of age. The concentration of lactate, an indirect measure of carbohydrate metabolism, was significantly higher in the 30-month-old mice; however, whole brain glucose utilization showed no differences. These findings in mice are at variance with subsequent results obtained in rats, where $rCMR_{glc}$, rather than global CMR_{glc}, was assessed.

Smith *et al.* (1980) used an autoradiographic [^{14}C]DG technique to evaluate age differences in $rCMR_{glc}$ in conscious resting Sprague-Dawley rats. In this study, $rCMR_{glc}$ was significantly lower at 14–16 months than at 4–6 months in areas of cerebral cortex; components of visual, auditory, and sensorimotor systems; the supragranular zone of the dentate gyrus; and some myelinated fiber tracts. In the same study, no significant differences were seen in $rCMR_{glc}$ of rats at 26–36 months of age compared with 14–16 month old rats.

London *et al.* (1981) studied $rCMR_{glc}$ in Fischer-344 rats of several ages by applying the [^{14}C]DG technique to hand-dissected brain regions (Table I). They reported that in some brain regions, rates of glucose utilization were

Table I. Regional Cerebral Metabolic Rates for Glucose in Fischer-344 Rats of Different Ages[a]

Brain region	Cerebral metabolic rates for glucose (μmol glucose/100 g tissue per min)			
	Age: 3 Months	12 Months	24 Months	34 Months
Frontal cortex	77 ± 4	59 ± 2[b]	61 ± 4	63 ± 5
Sensory-motor cortex	74 ± 4	60 ± 3[b]	59 ± 3	61 ± 5
Hypothalamus and thalamus	53 ± 4	41 ± 3	42 ± 4	39 ± 3
Striatum	73 ± 4	55 ± 3[b]	52 ± 4	58 ± 4
Hippocampus	56 ± 3	46 ± 3[b]	43 ± 3	46 ± 4
Inferior colliculus	94 ± 5	68 ± 3[b]	57 ± 4	56 ± 4
Superior colliculus	64 ± 3	50 ± 3[b]	47 ± 4	46 ± 4
Midbrain basis and tegmentum	56 ± 3	43 ± 2[b]	39 ± 3	40 ± 4
Medulla	46 ± 2	36 ± 3[b]	34 ± 3	32 ± 3
Pons	48 ± 3	33 ± 2[b]	33 ± 2	33 ± 2

[a]Data obtained from London *et al.* (1981). Each value is the mean ± S.E.M. for 7–10 rats.
[b]Significant difference from metabolic rate for glucose at previous age, $p \leq 0.05$.

Table II. Regional Cerebral Blood Flow in Fischer-344 Rats of Different Ages[a]

Brain region	Regional cerebral blood flow (ml/100 g per min)			
	Age: 3 Months	12 Months	24 Months	34 Months
Frontal cortex	127 ± 4	155 ± 10[b]	141 ± 8	149 ± 7
Parietal cortex	166 ± 8	188 ± 11	179 ± 15	213 ± 18
Striatum	123 ± 5	115 ± 8	97 ± 7	105 ± 5
Hypothalamus and thalamus	103 ± 3	117 ± 6	97 ± 6[b]	107 ± 5
Hippocampus	108 ± 4	118 ± 7	103 ± 6	117 ± 6
Inferior colliculus	118 ± 5	131 ± 8	97 ± 6[b]	115 ± 6
Superior colliculus	105 ± 4	109 ± 5	94 ± 7	103 ± 5
Midbrain basis and tegmentum	85 ± 3	91 ± 4	71 ± 5[b]	80 ± 4
Medulla	82 ± 2	84 ± 3	67 ± 4[b]	74 ± 3
Pons	86 ± 3	89 ± 4	70 + 4[b]	78 + 3

[a]Data obtained from Ohata *et al.* (1981). Each value is the mean ± S.E.M. for 21–22 rats.
[b]Significant difference from blood flow at previous age, $p \leq 0.05$.

lower at 12 months than at 3 months. Examples which showed apparent decrements of 25%–31% were the striatum, inferior colliculus and pons. These findings agreed generally with those of Smith *et al.* (1980). One difference was the observation by London *et al.* (1981) of a significantly lower $rCMR_{glc}$ in the frontal cortices of 12-month-old rats compared with younger animals. London *et al.* (1981) observed no age differences in $rCMR_{glc}$ when comparing 12-, 24-, and 34-month-old Fischer-344 rats. This agreed with the previous report of no significant age difference in $rCMR_{glc}$ in senescent Sprague-Dawley rats compared with middle-aged rats.

Because of the close association between $rCMR_{glc}$ and rCBF in various functional states (Des Rosiers, 1974; Sokoloff, 1977; Lassen *et al.*, 1978; Siesjö, 1978; Sokoloff, 1978), Ohata *et al.* (1981) used the [^{14}C]iodantipyrine method (Sakurada et al., 1978) to examine rCBF in conscious Fischer-344 rats of the same ages and exposed to the same experimental conditions used by London *et al.* (1981) to measure $rCMR_{glc}$ (Table II). When comparing $rCMR_{glc}$ in 14 brain regions of 3- and 12-month-old rats, they observed a significant difference only in the frontal cortex, where rCBF at 12 months was 22% higher than the value at 3 months. At 24 months of age, rCBF was significantly lower than values at 12 months in the inferior colliculus, thalamus and hypothalamus, midbrain, pons, and medulla, but not in areas of the cerebral cortex, striatum, hippocampus, or superior colliculus. No significant differences between 24- and 34-month-old rats were observed in any brain regions.

These data suggest a discrepancy between the time courses of age effects on $rCMR_{glc}$ and rCBF in the Fischer-344 rat. Whereas $rCMR_{glc}$ is lower in 12-month-old rats than in 3-month-old rats, rCBF under the same conditions is not different or higher at 12 months than at 3 months. Furthermore, rCBF is reduced at 24 months of age as compared with 12 months, while $rCMR_{glc}$ shows no significant age differences between 12 and 24 months. Although the reason for this apparent dissociation between rCBF and $rCMR_{glc}$ is unknown, such a discrepancy could reflect an age-related change in the coupling relation between rCBF and $rCMR_{glc}$ or in the lumped constant in the operational equation for calculating $rCMR_{glc}$ by the $[^{14}C]$DG technique.

A similar lack of correlation was observed by Le Ponchin-Lafitte and Rapin (1980). These investigators did not measure $rCMR_{glc}$, but followed the brain accumulation of radioactivity from $[^{3}H]$deoxyglucose, in 3-, 8-, and 15-month-old Wistar rats; and assessed cerebral blood flow with ^{125}I-iodoantipyrine. They noted that, while uptake of radioactivity into the brain from $[^{3}H]$deoxyglucose declined with age, CBF was unaffected.

Despite the various discrepancies noted above, the studies indicated a constancy of CMR_{glc} in rats after the first year and in CBF after the second year of age. These generalizations imply that in the rat brain, indices of cerebral metabolism and blood flow do not reflect the morphological and neurochemical sequelae of aging, some of which are exacerbated after the first and the second years of life. Examples from morphological studies include the demonstration by Inukai (1928) that cerebellar Purkinje cells are lost after 12 months of age, and observations by Feldman and Dowd (1975) that dendritic spines in the cortex are lost with progressive aging, especially during the second and third postnatal years. Examples of senescence-associated chemical changes in the rat brain after the first year include observations in CD-F rats by Puri and Volicer, (1977) that phosphodiesterase activities at low substrate concentrations are significantly lower at 24 and 30 months, but not at 12 months, compared with 4 months; and that dopamine stimulates adenylate cyclase in striata from 12- and 24-month-old, but not 30-month-old, rats. Similarly, Maggi *et al.* (1979) reported that in the cerebellem and brain stem of Wistar rats, binding of $[^{3}H]$dihydroalprenolol, a β-adrenergic receptor agonist, is reduced significantly at 24 but not 12 months compared with 3 months. A continuous decrease in $[^{3}H]$dihydroalprenolol binding between 12 and 24 months of age also was seen in the rat cerebellum (Weiss *et al.*, 1979). In addition, activities of the neurotransmitter synthetic enzymes, choline acetyltransferase and glutamic acid decarboxylase in cerebral cortices from 26-month-old Sprague Dawley rats were 54% and 52% lower, respectively, than in 10-month-old rats (Strong *et al.*, 1980).

Rapoport and London (1982) hypothesized that the relative paucity of decrements in $rCMR_{glc}$ and rCBF during senescence of the rat brain may reflect compensatory changes that occur with minimal age-related pathology. In the aged rat brain, axonal terminal degeneration (Vaughan and Peters,

Table III. Regional Cerebral Metabolic Rates for Glucose in Beagle Dogs of Different Ages[a]

Brain region	Age: 3 Years (5)	6 Years (4)	10–12 Years (6)	14–16 Years (3)
Regional cerebral metabolic rate for glucose (μmol glucose/100 g per min)				
Cerebral cortex				
Superior frontal gyrus	44 ± 3	34 ± 0.9[b]	37 ± 2	28 ± 2[b]
Temporal pole	30 ± 2	26 ± 0.7	29 ± 1	23 ± 2[b]
Caudate nucleus, head	37 ± 0.7	32 ± 1[b]	29 ± 1	28 ± 2
Hippocampus	20 ± 0.6	15 ± 1[b]	17 ± 0.8	17 ± 0.9
Geniculate bodies	32 ± 2	27 ± 0.4[b]	24 ± 1	19 ± 3[b]
Superior colliculus	34 ± 2	28 ± 1[b]	32 ± 1	25 ± 2[b]
Inferior colliculus	57 ± 2	43 ± 0.5[b]	39 ± 2	34 ± 3[c]

[a]Data obtained from London *et al.* (1983). Each value is the mean ± S.E.M. for the number of dogs indicated in parentheses.
[b]Significant difference from previous age, $p \leq 0.05$.
[c]Significant difference from 6 years but not from 10–12 years.

1974), deterioration of pyramidal cell dendrites (Feldman, 1977; Vaughan, 1977) and dendritic spine loss (Feldman and Dowd, 1975) have been observed. However, pathological changes such as senile (neuritic) plaques, amyloidosis, neurofibrillary tangles, and granulovacuolar degeneration, some of which occur in the senescent human brain (Wiśniewski and Terry, 1976; Ball, 1978) have not been observed in the rat brain. In contrast, the aged canine brain exhibits cerebral amyloidosis (Wiśniewski *et al.*, 1970; Pauli and Luginbuhl, 1971) and occasional structures which resemble human neuritic plaques (Von Braunmuhl, 1956; Osetowska, 1966; Wiśniewski *et al.*, 1970) as well as histopathology which is comparable to that observed in the senescent human cortex (Scheibel *et al.*, 1975, 1976, 1977; Mervis, 1978). The presence of these neuropathologic findings in the senescent canine brain suggests that $rCMR_{glc}$ may not be sustained throughout senescence, as it is in the relatively pathology-free rat brain. Therefore, $rCMR_{glc}$ was measured in Beagle dogs of different ages (Garden, *et al.*, 1980; London *et al.*, 1983).

Mean rates of $rCMR_{glc}$ in representative brain regions of Beagle dogs at four ages are presented in Table III. More detailed accounts appear elsewhere (London *et al.*, 1983). Seven gray matter regions showed significant age-related decrements (23%–40%) between 3 and 14 years. The greatest decrements (36%–41%) occurred in the superior frontal gyrus, geniculate bodies, and inferior colliculus. Some regions, such as the head of the caudate nucleus and the hippocampus, showed lower rates of glucose utilization (15% and 25%, respectively) at 6 years compared with 3 years, with no differences between rates at 6 and 14–16 years. Other regions, such as the superior frontal gyrus, the geniculate bodies, and the superior and inferior colliculi, showed significantly lower $rCMR_{glc}$ (11%–30%) at 14–16 years compared with 6 years. In most cases, the significant differences between $rCMR_{glc}$ at 6 years compared with later ages occurred between 10–12 and 14–16 years. In the temporal cortex, $rCMR_{glc}$ was not significantly lower at 10–12 and 14–16 years compared with 3 years; but it was significantly lower at 14–16 years compared with 10–12 years. Thus, unlike the situation in the rat, where $rCMR_{glc}$ is unchanged in senescencence, $rCMR_{glc}$ is reduced during senescence of the Beagle.

Neuronal plasticity is evident in senescent rats in the olfactory bulb and occipital cortex, where dendritic arborization is augmented (Connor *et al.*, 1980; Hinds and McNally, 1980). Evidence for plasticity also has been obtained in the frontal cortices of two old dogs, where filamentous tufting along apical dendrites resembled the proliferation of new dendritic membrane (Mervis, 1978). However, the rarity of the phenomenon in dogs, taken with the findings of $rCMR_{glc}$ decrements in the senescent canine cortex, suggests that plasticity in the aged canine brain does not compensate adequately to prevent decrements in $rCMR_{glc}$.

6. AGE EFFECTS ON CEREBRAL METABOLISM: HUMAN STUDIES

Early studies of human aging, using the N_2O technique for simultaneous measurement of CBF and $CMRO_2$, have produced variable results, partially because of the lack of uniformity in selection of subjects. For example, Fazekas *et al.* (1952) did not exclude hypertensives, and found that CBF and $CMRO_2$ were significantly lower by 22% and 31% in 15 men above 50 years of age compared with nine younger men. Scheinberg *et al.* (1953), who studied hospitalized patients admitted for minor elective procedures, attempted to rule out clinical vascular disease. They separated 51 subjects into three age groups: young (18–36 years), middle-aged (38–55 years), and old (56–79 years). As in the previous study, CBF was lower in the older subjects; the 22% difference between the youngest and oldest group was statistically significant. $CMRO_2$ was not lower in the middle-aged group than in the young group. In contrast, Schieve and Wilson (1953) noted no significant differences in CBF or $CMRO_2$ among three groups of subjects with no history or evidence of vascular or hypertensive cardiovascular disease. However, although the oldest subjects (56–76 years of age) in this study were energetic working people essentially in good health, the two groups of younger subjects (21–35 years, 35–45 years) contained patients with schizophrenia. Because of evidence for decreased rCBF and $rCMR_{glc}$ in premotor and frontal regions of brains from schizophrenics (Lassen and Ingvar, 1972; Farkas *et al.*, 1980), it seems possible that schizophrenia in the young subjects could have masked age differences in CBF or $CMRO_2$.

Kety (1956) reviewed the results of 11 previous studies on CBF and $CMRO_2$ as a function of age. He noted that, except for five of the series, where the subjects were apparently normal nonhospitalized individuals, practically all of the studies included hospitalized patients. Kety noted that, at about the time of puberty, there was an apparently rapid fall in the overall cerebral circulation, which continued through adolescence. From the third decade onward, there was a gradual but continuous decline in the function through middle and old age. $CMRO_2$ showed a rapid decline followed by a more gradual decrease with advancing years. A criticism of this conclusion (Obrist, 1979; Sokoloff, 1979) is that the variability in the medical status of the subjects in several of the studies confounds the effects of age *per se* with sequelae of disease.

Some more recent studies, performed on normotensive subjects, which otherwise appeared normal, indicated an age-related decline in CBF or $CMRO_2$. In studies by Naritomi *et al.* (1979) as well as by Melamed *et al.* (1980), the ^{133}Xe inhalation technique was employed to study rCBF. These experiments demonstrated significant delines in rCBF with age. The data by Melamed *et al.* (1980) indicated that rCBF decrements were more marked in anterior regions than posterior regions of the left, but not the right, hemisphere. Naritomi *et al.* (1979) noted that rCBF was significantly, negatively correlated with age in all regions except the brain stem and cerebellar regions. An age-dependent decline in global CBF also was demonstrated by Frackowiak *et al.* (1980), who use ^{15}O with PET. In this study, $CMRO_2$ also was reduced with age.

In contrast, other studies have suggested that cerebral metabolism is slightly, if at all, affected by aging *per se*. For example, Lassen *et al.* (1960) noted that in five old normal men (average age, 72 years), mean CBF was nearly the same as in 11 young normal subjects (average age, 24 years); $CMRO_2$ was significantly lower than in young normals, but only by 9%. More recently, a lack of an age effect on CBF has been seen by the use of the ^{131}I-RISA method in 41 subjects between the ages of 21 and 76 years of age (Fujishima and Omae, 1980).

Dastur *et al.* (1963) made comparisons between a group of 26 healthy elderly men (mean age, 71 years) and 15 young subjects (mean age, 20.8 years). Although CBF and $CMRO_2$ did not differ between the two groups, CMR_{glc} was significantly lower, by 23%, in the elderly. As an explanation for the discrepancy between the results with $CMRO_2$ as compared with CMR_{glc}, the authors suggested that with aging, subtle changes in cerebral metabolism could lower CMR_{glc} before a reduction in $CMRO_2$ became apparent. Sokoloff (1979a) proposed that older subjects may have higher blood levels of ketone bodies, which could partially replace glucose as metabolic substrates. However, despite the demonstration by Owen *et al.* (1967) that ketone bodies can serve as substrates for brain metabolism, there is no evidence that ketone bodies are present in sufficiently high levels in old subjects to explain the age difference in CMR_{glc} observed by Dastur *et al.* (1963).

Dastur *et al.* (1963) stated alternatively that blood glucose determinations in their laboratory were far less accurate than blood oxygen measurements, casting doubt on the validity of the significant, age-related decrement in CMR_{glc}. In this regard, Lying-Tunell *et al.* (1980) failed to show a difference in CMR_{glc} between two groups consisting of 10 subjects each, with age ranges of 21–24 years and 55–65 years. No age differences were seen also in CBF and $CMRO_2$.

More recently, investigators have examined $rCMR_{glc}$ as a function of age by means of the $[^{18}F]$FDG technique. In one of these investigations, Kuhl *et al.* (1982) reported that overall CMR_{glc} in 40 normal volunteers between the ages of 18 and 78 years demonstrated a gradual decline, when plotted as a function of age. At age 78, whole brain mean CMR_{glc} was 26% less than at age 18. However, this alteration was of the same order as the variance among subjects at any age, and no statement was made about the statistical significance of this decline or of other results in the study. There was no selective decrease in $rCMR_{glc}$ by any particular brain region. However, the ratio of superior frontal cortex/superior parietal cortex CMR_{glc} declined with age, possibly reflecting a selective degeneration of the superior frontal cortex. In this regard, Brody (1976) found that progressive loss of human neuronal cell bodies with advancing age was more pronounced in the frontal and temporal cortices than elsewhere.

Duara *et al.* (1982) presented preliminary findings obtained with the $[^{18}F]$FDG technique in 20 healthy men, aged 21–83 years. As compared with the study by Kuhl *et al.* (1982), in which the subjects were exposed to ambient light and noise during the $[^{18}F]$DG procedure, subjects in the study by Duara *et al.* (1982) had their ears plugged and eyes covered. In most brain regions, $rCMR_{glc}$ tended to decline with age, but the trends did not reach statistical significance at $p \leq 0.05$. It was apparent that a much larger sample size was required to determine if $rCMR_{glc}$, as measured by the $[^{18}F]$FDG technique, declined in normal human aging.

A tabular summary of the studies discussed in this section of the chapter is shown in Table IV.

7. AGE EFFECTS ON CEREBRAL METABOLISM IN RESTING AND ACTIVATED STATES—FUNCTIONAL RELEVANCE

The results of the foregoing studies demonstrate that cerebral metabolism does not decline consistently in resting, healthy animals, and people during senescence. However, other indices of brain function, such as performance by rats in complex mazes (Goodrick, 1968; 1972) or human performance on the Benton Revised Visual Retention Test, show significant age-related decrements (Arenberg, 1978). Therefore, it appears that measurements of cerebral metabolism in the resting state may be insensitive to cerebral functional losses. Perhaps greater benefit could be derived from studies of cerebral metabolism during functional challenge of the brain. With the use of PET, one might be able to correlate simultaneous deficits in psychomotor performance with decreased metabolism in specific brain regions. Thus, the metabolic measurement might elucidate the anatomic loci associated with functional deficits.

Discrete age-related losses in the function of a particular neurotransmitter system may not affect regional metabolism in the resting state because each brain region comprises various cell types and neurotransmitter systems. By the same reasoning, regional losses in specific neuronal markers, such as neurotransmitter receptors, in the absence of other deficits, need not alter cerebral metabolism in resting animals or people. However, metabolic measurements can be used to study cerebral function in particular neurotransmitter pathways in activated states.

London *et al.* (1982) used oxotremorine, a cholinergic muscarinic agonist drug (Cho *et al.,* 1962; Lévy and Michel-Ber, 1965), to study muscarinic receptor function in aging rats. In previous studies, oxotremorine had stimulated $rCMR_{glc}$ in motor systems regions (Dow Edwards et al., 1981) as well as the cerebral cortex (Dam *et al.,* 1982) and Papez circuit of the rat brain (Dam and London, 1983).

In control rats (resting state), $rCMR_{glc}$ was lower at 12 months than at 3 months, but the same as at 24 months. However, in many brain regions, treatment of 12-month-old rats with 0.1 mg/kg of oxotremorine elevated $rCMR_{glc}$ to the same level as in 3 month old rats given 0.1 mg/kg of oxotremorine. This was not true of 24-month-old rats, which generally showed no significant response to this dose of oxotremorine. Thus, an age-related deficit which was not apparent in the resting state was manifested by pharmacologic activation with a direct receptor agonist. Data from a representative brain region (striatum) are shown in Table V.

In similar studies, Smith (1981) studied the $rCMR_{glc}$ responses to apomorphine, a dopamine agonist, in the striatum and substantia nigra of young

Table IV. Cerebral Blood Flow and Metabolic Rates for Oxygen and Glucose in Human Aging[a]

Literature reference	No. of subjects	Age in years (group means or ranges)	Technique employed	Health status	Age effect, level of statistical significance (p)		
					CBF	$CMRO_2$	CMR_{glc}
Fazekas *et al.* (1952)	9 Young 15 Old	23–42 50–91	N_2O	Hypertensives and elderly demented subjects included	↓ $p < 0.01$	↓ $p < 0.001$	Not measured
Scheinberg *et al.* (1953)	19 Young 32 Middle-aged 17 Old	18–36 38–55 56–79	N_2O	Normotensive hospital patients admitted for minor surgery. Evidence of vascular disease in some elderly subjects. Normal mental status	↓ Young vs. middle-aged $p < 0.01$ ↓ Middle-aged vs. old $p < 0.01$	No difference in young vs. middle-aged Middle-aged vs. old $p < 0.05$	Not measured
Schieve and Wilson (1952)	12 Young 10 Middle-aged 7 Old	21–35 35–45 56–76	N_2O	Normotensive. Schizophrenics included in young and middle-aged groups	No difference	No difference	Not measured
Lassen *et al.* (1960)	11 Young 5 Old	24 vs. 72	^{85}Kr	Normal performance on psychometric tests	No difference	↓ $p < 0.05$	Not measured
Dastur *et al.* (1963)	15 Young 26 Old	20.8 vs. 71	N_20	Normotensive. No arteriosclerosis	No change	No change	↓ $p < 0.05$

Naritomi *et al.* (1979)	28 Young 18 Older	21–40 40–63	^{133}Xe	Normotensive. No history of cerebrovascular disease	↓ $p < 0.001$	Not measured	Not measured
Frackowiak *et al.* (1980)	14	26–74	^{15}O, PET	Normotensive. No neurological or medical disease	↓ No statistics provided	↓ No statistics provided	Not measured
Melamed *et al.* (1980)	44	19–76	^{133}Xe	Normotensive. No neurological, pulmonary, or vascular disease	↓ $p < 0.001$	Not measured	Not measured
Lying-Tunell *et al.* (1980)	10 Young 10 Old	21–24 55–65	N_2O	No neurological or mental impairment	No change	No change	No change
Fujishima and Omae (1980)	41	21–76	^{131}I-RISA	No history or indication of hypertension or cerebral diseases	No difference	Not measured	Not measured
Kuhl *et al.* (1982)	40	18–78	[^{18}F]DG, PET	No evidence of dementia, hypertension, or cerebrovascular disease	Not measured	Not measured	Gradual↓. No statistics provided
Duara *et al.* (1982)	21	21–83	[^{18}F]DG, PET	Normotensive, carefully screened, healthy volunteers	Not measured	Not measured	Tendency to ↓ · $p > 0.05$.

[a]Modified from Duara *et al.* (1984).

Table V. Striatal $rCMR_{glc}$ Responses to Oxotremorine in Fischer-344 Rats of Different Ages[a]

Age (months)	Treatment	
	Control	Oxotremorine (0.1 mg/kg, i.p.)
	$rCMR_{glc}$ (μmol/100g per min)	
3	73 ± 5	77 ± 5
12	53 ± 5	78 ± 7[b]
24	47 ± 4[c]	63 ± 6[c]

[a]Data obtained from London, E. D., Rapoport, S. I., and Dam., M. (unpublished observations). Each value is the mean ± S.E.M. for 6–8 rats.
[b]Significant difference from $rCMR_{glc}$ in control rats, $p \leq 0.05$.
[c]Significant difference from $rCMR_{glc}$ in 3-month-old rats given the same treatment, $p \leq 0.05$.

adult (4–6 months) and aged (26–36 months) rats. She observed that in the striatum and substantia nigra of vehicle-injected aged rats, $rCMR_{glc}$ was significantly lower than in young rats. Preliminary results indicated that apomorphine significantly enhanced $rCMR_{glc}$ in the substantia nigra and striatum of young adult rats but not senescent rats. These studies provide another example of how metabolic studies involving pharmacologic activation may manifest age-associated functional defects, which are not apparent in the resting state.

Reduced $rCMR_{glc}$ responses to apomorphine and oxotremorine may reflect age-related defects in dopamine or muscarinic receptors, respectively. The studies described above performed on rats, may serve as prototypes for clinical studies of receptor function *in vivo.* Such studies are relevant to pharmacotherapy with direct receptor agonists. It has been suggested that the potential value of muscarinic agonist therapy for Alzheimer's disease warrants further investigation (Sitaram *et al.,* 1978). Perhaps assessment of $rCMR_{glc}$ responses to muscarinic agonist drugs, such as arecoline or oxotremorine, may help predict whether or not a patient with Alzheimer's disease might respond to direct receptor agonist therapy.

8. SUMMARY

CBF, $CMRO_2$, and CMR_{glc} are indices of cerebral energy metabolism which are closely related to one another. Although there are exceptions, these parameters frequently reflect cerebral function. Results of *in vitro* assays indicate that cerebral oxidative metabolism is reduced in aging. *In vivo* studies, performed on healthy, resting animals, and humans, do not consistently show declines in CBF, $CMRO_2$, and CMR_{glc} in senescence. This suggests an insensitivity of the parameters in the resting state to age-related functional decline.

It is suggested that future studies measure cerebral metabolism during pharmacological or physiological challenge.

REFERENCES

Alpert, N. M., Ackerman, R. H., Correia, J. A., Grotta, J. C., Chang, J. Y., and Taveras, J. M., 1979, Measurement of cerebral blood flow and oxygen metabolism in transverse section—Preliminary results, *Acta Neurol. Scand. Suppl. 72,* **60:**196–197.

Arenberg, D., 1978, Differences and changes with age in the Benton Visual Rentention Test, *J. Gerontol.* **33:**534–540.

Ball, M. J., 1978, Topographic distribution of neurofibrillary tangles and granulovacuolar degeneration in hippocampal cortex of aging and demented patients. A quantitative study, *Acta Neuropathol. (Berl.)* **42:**73–80.

Bessell, E. M., Foster, A. B., and Westwood, J. H., 1972, The use of deoxyfluoro-D-glucopyranoses and related compounds in a study of yeast hexokinase specificity, *Biochem. J.* **128:**199–204.

Brody, H., 1976, An examination of cerebral cortex and brainstem aging, in: *Neurobiology of Aging (Aging,* Vol. 13, R. D. Terry and S. Gershon, eds.), pp. 177–181, Raven Press, New York.

Chainy, G. B. N., and Kanungo, M. S., 1978, Induction and properties of pyruvate kinase of the cerebral hemisphere of rats of various ages, *J. Neurochem.* **30:**419–427.

Cho, A. K., Hasslett, W. L., and Jenden, D. J., 1962, The peripheral actions of oxotremorine, a metabolite of tremorine, *J. Pharmacol. Exp. Ther.* **138:**249–257.

Connor, J. R., Beban, S. E., Hansen, B., Hopper, P., and Diamond, M. C., 1980, Dendritic increases in the aged rat somatosensory cortex, *Abstr. Soc. Neurosci.* **6:**739.

Dam, M., and London, E. D., 1983, Glucose utilization in the Papez circuit: Effects of oxotremorine and scopolamine, Submitted.

Dam, M., and London, E. D., 1983, Glucose utilization in the Papez circuit: Effects of oxotremorine and scopolamine, *Brain Res.,* in press.

Dastur, D. K., Lane, M. H., Hansen, D. B., Kety, S. S., Butler, R. N., Perlin, S., and Sokoloff, L, 1963, Effects of aging on cerebral circulation and metabolism in man, in: *Human Aging: A Biological and Behavioral Study* (J. E. Birren, R. N. Butler, S. G. Greenhouse, L. Sokoloff, and M. R. Yarow, eds.), Public health Service Publication No. 986, pp. 57–76, U. S. Government Printing Office, Washington, D.C.

Des Rosiers, M. H., Kennedy, C., Patlak, C. S., Pettigrew, K. D., Sokoloff, L., and Reivich, M., 1974, Relationship between local cerebral blood flow and glucose utilization in the rat, *Neurology* **24:**389.

Dow-Edwards, D., Dam, M., Peterson, J. M., Rapoport, S. I., and London, E. D., 1981, Effect of oxotremorine on local cerebral glucose utilization in motor system regions of the rat brain, *Brain Res.* **226:**281–289.

Duara, R., Margolin, R. A., Robertson-Tchabo, E. A., London, E. D., Schwartz, M., Renfrew, J. W., Kessler, R., Sokoloff, L., Ingvar, D. H., and Rapoport, S. I., 1982, Regional cerebral glucose utilization in healthy men at different ages, *Neurology* **32:**A166–A167.

Duara, R., London, E. D., and Rapoport, S. I., 1984, Structural and metabolic changes in the aging brain, in: *Handbook of the Biology of Aging,* 2nd ed. (C. E. Finch and E. Schneider, eds.), in press, Van Nostrand-Reinhold Company, New York.

Duffy, T. E., Cavazzuti, M., Gregoire, N. M., Cruz, N. F., Kennedy, C., and Sokoloff, L, 1979, Regional cerebral glucose utilization in newborn Beagle dogs, *Trans. Am. Soc. Neurochem.* **10:**171.

Farkas, T., Reivich, M., Alavi, A., Greenberg, J. H., Fowler, J. S., MacGregor, R. R., Christman, D. R., and Wolf, A. P., 1980, The application of [^{18}F]2-deoxy-2-fluoro-D-glucose and positron emission tomography in the study of psychiatric conditions, in: *Cerebral Metabolism and Neural Function* (J. V. Passonneau, R. A. Hawkins, W. D. Lust, and F. A. Welsh, eds.), pp. 403–408, Waverly Press, Baltimore.

Fazekas, J. F., Alman, R. W., and Bessman, A. N., 1952, Cerebral physiology of the aged, *Am. J. Med. Sci.* **223:**245–257.

Feldman, M. L., and Dowd, C., 1975, Loss of dendritic spines in aging cerebral cortex, *Anat. Embryol (Berl.)* **148:**279–301.

Ferris, S. H., de Leon, M. J., Wolf, A. P., Farkas, T., Christman, D. R., Reisberg, B., Fowler, J. S., MacGregor, R., Goldman, A., George, A. E., and Rampal, S., 1980, Positron emission tomography in the study of aging and senile dementia, *Neurobiology of Aging* **1:** 127–131.

Fox, J. H., Parmacek, M. S., and Patel-Mandlik, K., 1975, Effect of aging on brain respiration and carbohydrate metabolism of Syrian hamsters, *Gerontologia* **21:**224–230.

Frackowiak, R. S. J., Lenzi, G. L., Jones, T., and Heather, J. D., 1980, Quantitative measurement of regional cerebral blood flow and oxygen metabolism in man using ^{15}0 and positron emission tomography: Theory, procedure and normal values, *J. Comput. Assist. Tomog.* **4:**727–736.

Freyhan, F. A., Woodford, R. B., and Kety, S. S., 1951, Cerebral blood flow and metabolism in psychoses of senility, *J. Nerv. Ment. Dis.* **113:**449–456.

Fujishima, M., 1967, Brain circulation studies in cases with cerebrovascular diseases by external counting of intravenously injected RISA, *Fukuoka Acta Med.* **58:**194–216.

Fujishima, M., and Omae, T., 1980, Brain flow and mean transit time as related to aging, *Gerontology* **26:**104–107.

Garbus, J., 1955, Respiration of brain homogenates of old and young rats, *Am. J. Physiol.* **183:**618–619.

Garden, A. S., Ohata, M., Rapoport, S. I., and London, E. D., 1980, Age-associated decrease in local cerebral glucose utilization (LCGU) in the Beagle, *Abst. Soc. Neurosci.* **6:**768.

Garfunkel, J. M., Baird, H. W., III, and Ziegler, J., 1954, The relationship of oxygen consumption to cerebral functional activity, *J. Pediat.* **44:**64–72.

Gibson, G. E., Peterson, C., and Jenden, D. J., 1981, Brain acetylcholine synthesis declines with senescence, *Science* **213:**674–676.

Goodrick, C. L., 1968, Learning, retention, and extinction of a complex maze habit for mature-young and senescent Wistar albino rats, *J. Gerontol.* **25:**298–304.

Goodrick, C. L., 1972, Learning by mature-young and aged Wistar albino rats as a function of test complexity, *J. Gerontol.* **27:**353–357.

Hagberg, B., and Ingvar, D. H., 1976, Cognitive reduction in presenile dementia related to abnormalities of the cerebral blood flow, *Br. J. Psychiat.* **128:**209–222.

Hinds, J. W., and McNelly, N. A., 1980, Correlation of aging changes in the olfactory epithelium and olfactory bulb of the rat, *Abstr. Soc. Neurosci.* **6:**739.

Ingvar, D. H., Lassen, N. A., 1979, Activity distribution in the cerebral cortex in organic dementia as revealed by measurements of regional cerebral blood flow, in: *Brain Function in Old Age* (F. Hochmeister and C. Müller, eds.), pp. 268–277, Springer Verlag, New York.

Inukai, T., 1928, On the loss of Purkinje cells with advancing age, from the cerebellar cortex of the albino rat, *J. Comp. Neurol.* **45:**1–31.

Kennedy, C., Sakurada, O., Shinohara, M., Jehle, J., and Sokoloff, L., 1978, Local cerebral glucose utilization in the normal conscious macaque monkey, *Ann. Neurol.* **4:**293–301.

Kety, S. S., 1956, Human cerebral blood flow and oxygen consumption as related to aging, *J. Chron. Dis.* **3:**478–486.

Kety, S. S., 1957, The general metabolism of the brain *in vivo*, in: *Metabolism of the Nervous System* (International Neurochemical Symposium II, Denmark, 1956, D. Richter, ed.), pp. 221–237, Pergamon Press, New York.

Kety, S. S., and Schmidt, C., 1945, The determination of cerebral blood flow in man by the use of nitrous oxide in low concentrations, *Am. J. Physiol.* **143**:53–66.

Kety, S. S., and Schmidt, C. F., 1948, The nitrous oxide method for the quantitative determination of cerebral blood flow in man: Theory, procedure and normal values, *J. Clin, Invest.* **27**:476–483.

Kuhl, D. E., Metter, E. J., Reige, W. H., and Phelps, M. E., 1982, Effects of human aging on patterns of local cerebral glucose utilization determined by the [^{18}F]fluorodeoxyglucose method, *J. Cerebral Blood Flow and Metab.* **2**:163–171.

Landau, W. M., Freygang, W. H., Rowland, L. P., Sokoloff, L., and Kety, S. S., 1955, The local circulation of the living brain; values in the unanesthetized and anesthetized rat, *Trans. Am. Neurol. Assoc.* **80**:125–129.

Lassen, N. A., 1959, Crebral blood flow and oxygen consumption in man, *Physiol. Rev.* **39**:183–238.

Lassen, N. A., 1974, Control of cerebral circulation in health and disease, *Circ. Res.* **34**:749–760.

Lassen, N. A., and Ingvar, D. H., 1972, Radioisotope assessment of regional cerebral blood flow, *Prog. Nucl. Med.* **1**:376–409.

Lassen, N. A., Munck, O., and Tottey, E. R., 1957, Mental function and cerebral oxygen consumption in organic dementia, *A.M.A. Arch. Neurol. Psychiat.* **77**:126–133.

Lassen, N. A., Feinberg, I., and Lane, M. H., 1960, Bilateral studies of cerebral oxygen uptake in young and aged normal subjects and in patients with organic dementia, *J. Clin. Invest.* **39**:491–500.

Lassen, N. A., Hoedt-Rasmussen, K., Sorensen, S. C., Skinhøj, E., Cronquist, S., Forss, B., and Ingvar, D. H., 1963, Regional cerebral blood flow in man determined by krypton,^{85} *Neurology* **13**:719–727.

Lassen, N. A., Ingvar, D. H., and Skinhøj, E., 1978, Brain function and blood flow, *Sci. Am.* **239**:62–71.

Le Ponchin-Lafitte, M., and Rapin, J. R, 1980, Age-associated changes in deoxyglucose uptake in whole brain, *Gerontology* **26**:265–269.

Lévy, J., and Michel-Ber, E., 1965, Sur le métabolite de la trémorine, l'oxotrémorine, *Therapie* **20**:265–267.

London, E. D., Nespor, S. M., Ohata, M., and Rapoport, S. I., 1981, Local cerebral glucose utilization during development and aging of the Fischer-344 rat, *J. Neurochem.* **37**:217–221.

London, E. E., Mahone, P., Rapoport, S. I., and Dam, M., 1982, Effect of age on oxotremorine-induced stimulation of local cerebral glucose utilization, *Fed. Proc.* **41**:1323.

London, E. D., Ohata, M., Takei, H., French, A. W., and Rapoport, S. I., 1983, Regional cerebral metabolic rate for glucose in Beagle dogs of different ages, *Neurobiol. Aging,* in press.

Lying-Tunell, U., Lindblad, B. S., Malmlund, H. O., and Persson, B., 1980, Cerebral blood flow and metabolic rate of oxygen, glucose, lactate, ketone bodies and amino acids. I. Young and old normal subjects, *Acta Neurol. Scandinav.* **62**:265–275.

Maggi, A., Schmidt, M. J., Ghetti, B., and Enna, S. J., 1979, Effect of aging on neurotransmitter receptor binding in rat and human brain, *Life Sci.* **24**:367–374.

Mallett, B. L., and Veall, N., 1963, Investigation of cerebral blood flow in hypertension, using radioactive-xenon inhalation and extracranial recording, *Lancet* **1**:1081–1082.

Marcus, M. L., Busija, D. N., Bischof, C. J., and Heistad, D. D., 1981, Methods for measurement of cerebral blood flow, *Fed. Proc.* **40**:2306–2310.

McIlwain, H., Anguiano, G., and Cheshire, J. D., 1951, Electrical stimulation *in vitro* of the metabolism of glucose by mammalian cerebral cortex, *Biochem. J.* **50:**12–18.

Meier-Ruge, W., Reichlmeier, K., and Iwangoff, P., 1976, Enzymatic and enzyme histochemical changes of the aging animal brain and consequences for experimental pharmacology on aging, in: *Neurobiology of Aging (Aging,* Vol. 3, R. D. Terry and S. Gershon, eds.), pp. 379–388, Raven Press, New York.

Melamed, E., Lavy, S., Bentin, S., Cooper, G., and Rinot, Y., 1980, Reduction in cerebral blood flow during normal aging in man, *Stroke* **11:**31–35.

Mervis, R. F., 1978, Structural alterations in neurons of aged canine neocortex: A Golgi study, *Exp. Neurol.* **62:**417–432.

Naritomi, H., Meyer, J. S., Sakai, F., Yamaguchi, F., and Shaw, T., 1979, Effects of advancing age on regional cerebral blood flow: Studies in normal subjects and subjects with risk factors for atherothrombotic stroke, *Arch. Neurol.* **36:**410–416.

Nilsson, B., Rehncrona, S., and Siesjö, B. K., 1978, Coupling of cerebral metabolism and blood flow in epileptic seizures, hypoxia and hypoglycaemia, *Ciba Found. Symp.* **56:**199–218.

Obrist, W. D., 1979, Cerebral circulatory changes in normal aging and dementia, in: *Bayer Symposium VII: Brain Function in Old Age* (F. Hoffmeister, C. Müller, and H. P. Krause, eds.), pp. 278–287, Springer-Verlag, Berlin.

Obrist, W. D., Thompson, H. K., King, C. H., and Wang, H. S., 1967, Determination of regional cerebral blood flow by inhalation of 133-xenon, *Circ. Res.* **20:**124–135.

Orbist, W. D., Chivian, E., Conqvist, S., and Ingvar, D. H., 1970, Regional cerebral blood flow in senile and presenile dementia, *Neurology* **20:**315–322.

Ohata, M., Sundaram, U., Fredericks, W. R., London, E. D., and Rapoport, S. I., 1981, Regional cerebral blood flow during development and ageing of the rat brain, *Brain* **104:**319–332.

Osetowska, E., 1966, Étude anatomorphathologique sur le cerveau de chiens seniles, in: *Proceedings of the Fifth International Congress of Neuropathology* (F. Luthy and A. Bischoff, eds.), pp. 497–502, Excerpta Medica Foundation, Amsterdam.

Owen, O. E., Morgan, A. P., Kemp, H. G., Sullivan, J. M., Herrera, M. G., and Cahill, G. F., 1967, Brain metabolism during fasting, *J. Clin. Invest.* **46:**1589–1595.

Parmacek, M. S., Fox, J. H., Harrison, W. H., Garron, D. C., and Swenie, D., 1979, Effect of aging on brain respiration and carbohydrate metabolism of CBF_1 mice, *Gerontology* **25:**185–191.

Patel, M. S., 1977, Age-dependent changes in the oxidative metabolism in the rat brain, *J. Gerontol.* **32:**643–646.

Patterson, J. L., Jr., Heyman, A., and Nichols, F. T., Jr., 1950, Cerebral blood flow and oxygen consumption in neurosyphills, *J. Clin. Invest.* **29:**1327–1334.

Pauli, B., and Luginbuhl, H., 1971, Fluorescenzmikroskopische Untersuchungen der cerebralen Amyloidose bei alten Hunden und senilen Menschen, *Acta Neuropathol. (Berl.)* **19:**121–128.

Peng, M.-T., Peng, Y.-I., and Chen, F.-N., 1977, Age-dependent changes in the oxygen consumption of the cerebral cortex, hypothalamus, hippocampus, and amygdaloid in rats, *J. Gerontol.* **32:**517–522.

Phelps, M. E., Hoffman, E. J., Huang, S. C., and Kuhl, D. E., 1978, ECAT: A new computerized tomographic imaging system for positron emitting radiopharmaceuticals, *J. Nucl. Med.* **19:**635–647.

Posner, J. P., 1973, Newer techniques of cerebral blood flow measurement, in: *Cerebral Vascular Diseases: Eighth Conference* (F. H. McDowell and R. W. Brennan, eds.), pp. 145–162, Grune and Stratton, New York.

Puri, S. K., and Volicer, L., 1977, Effect of aging on cyclic AMP levels and adenylate cyclase and phosphodiesterase activities in the rat corpus striatum, *Mech, Aging Dev.* **6:**53–58.

Raichle, M. E., Grubb, R., Gado, M. H., Eichling, J. O., and Ter-Pogossian, M. M., 1976, Correlation between regional cerebral blood flow and oxidative metabolism. *In vivo* studies in man, *Arch. Neurol.* **33**:523–526.

Rapoport, S. I., and London, E. D., 1982, Brain metabolism during aging of the rat and dog: Implications for brain function in man during aging and dementia, in: *Neural Aging and its Implications in Human Neurological Pathology (Aging,* Vol. 18, R. D. Terry, C. L. Bolis, and G. Toffano, eds.), pp. 79–88, Raven Press, New York.

Reiner, J. M., 1947, The effect of age on carbohydrate metabolism of tissue homogenates, *J. Gerontol.* **2**:315–320.

Reivich, M., Jehle, J. W., Sokoloff, L., and Kety, S. S., 1969, Measurement of regional cerebral blood flow with (^{14}C)-antipyrine in awake cats, *J. Appl. Physiol.* **27**:296–300.

Reivich, M., Kuhl, D., Wolf, A., Greenberg, J., Phelps, M., Ido, T., Casella, V., Fowler, J., Hoffman, E., Alavi, A., Som, P., and Sokoloff, L., 1979, The [^{18}F]Fluorodeoxyglucose method for the measurement of local cerebral glucose utilization in man, *Circ. Res.* **44**:127–137.

Roy, C. W., and Sherrington, C. S., 1980, On the regulation of the blood supply of the brain, *J. Physiol.* (London) **11**:85–108.

Ryder, E., and Izquierdo, P., 1979, Enzymatic profile of mitochondria isolated from selected brain regions of young adult and one year old rats, *Fed. Proc.* **38**:516.

Sakurada, O., Kennedy, C., Jehle, J., Brown, J. D., Carbin, G. L., and Sokoloff, L., 1978, Measurement of local cerebral blood flow with iodo[^{14}C]antipyrine, *Am. J. Physiol.* **234**:H59–H66.

Sanadi, D. R., 1977, Metabolic changes and their significance in aging, in *Handbook of the Biology of Aging* (C. E. Finch and L. Hayflick, eds.), pp. 73–98, Van Nostrand-Reinhold, New York.

Scheibel, M. E., Lindsay, R. D., Tomiyasu, U., and Scheibel, A. B., 1975 Progressive dendritic changes in aging human cortex, *Exp. Neurol.* **47**:392–403.

Scheinberg, P., Blackburn, I., Rich, M., and Saslaw, M., 1953, Effects of aging of cerebral circulation and metabolism, *Arch. Neurol. Psych.* **70**:77–85.

Schieve, J. F., and Wilson, W. P., 1953, The influence of age, anaesthesia and cerebral arteriosclerosis on cerebral vascular activity to Co_2, *Am. J. Med.* **15**:171–174.

Schwark, W. S., Singhal, R. L., and Ling, G. M., 1971, Metabolic control mechanisms in mammalian systems. Regulation of pyruvate kinase in the cerebral cortex, *J. Neurochem.* **18**:123–134.

Siesjö, B. K., 1978, *Brain Energy Metabolism,* John Wiley and Sons, Chichester.

Sitaram, N., Weingartner, H., and Gillin, J. C., 1978, Human serial learning: Enhancement with arecholine and choline and impairment with scopolamine, *Science* **201**:274–276.

Smith, C. B., 1981, Age-related changes in local rates of cerebral glucose utilization in the rat, in: *Brain Neurotransmitters and Receptors in Aging and Age-Related Disorders (Aging,* Vol. 17, S. J. Enna, T. Samorajski, and B. Beer, eds.), pp. 195–201, Raven Press, New York.

Smith, C. B., Goochee, C., Rapoport, S. I., and Sokoloff, L., 1980, Effects of ageing on local rates of cerebral glucose utilization in the rat, *Brain* **103**:351–365.

Sokoloff, L., 1960, The metabolism of the central nervous system *in vivo,* in: *Handbook of Physiology; Neurophysiology* (J. Field, H. W.Magoun, V. E. Hall, eds.), Vol. 3, pp. 1843–1864, American Physiological Society, Washington, D. C.

Sokoloff, L., 1972, Circulation and energy metabolism of the brain, in: *Basic Neurochemistry* (2nd edition, G. J. Siegel, R. W. Albers, R. Katzman, and B. W. Agranoff, eds.), pp. 338–413, Little Brown and Co., Boston.

Sokoloff, L., 1977, Relation between physiological function and energy metabolism in the central nervous system, *J. Neurochem.* **29**:13–26.

Sokoloff, L., 1978, Local cerebral energy metabolism: Its relationships to local functional activity and blood flow, in: *Cerebral Vascular Smooth Muscle and its Control* (Ciba Foundation Symposium 56), pp. 171–197, Elsevier–Excerpta Medica–North-Holland, Amsterdam.

Sokoloff, L., 1979a, Effects of normal aging on cerebral circulation and energy metabolism, in: *Brain Function in Old Age* (Bayer Symposium VII, F. Hoffmeister and C. Müller, eds.), pp. 367–380, Springer Verlag, New York.

Sokoloff, L., 1979b, The [^{14}C]deoxyglucose method: Four years later, in: *Cerebral Blood Flow and Metabolism* (Acta Neurologica Scandinavica, Suppl. 72, Vol. 60, International Symposium on Cerebral Blood Flow and Metabolism, Tokyo, 1979, F. Gotoh, H. Nagai, and Y. Tazaki, eds.), pp. 640–649, Munksgaard, Copenhagen.

Sokoloff, L., Reivich, M., Kennedy, C., Des Rosiers, M. H., Patlak, C. S., Pettigrew, K. D., Sakurada, O., and Shinohara, M., 1977, The ^{14}C-deoxyglucose method for the measurement of local cerebral glucose utilization: Theory, procedure, and normal values in the conscious and anesthetized albino rat, *J. Neurochem.* **28**:897–916.

Strong, R., Hicks, P., Hsu, L., Bartus, R. T., and Enna, S. J., 1980, Age-related alterations in the rodent brain cholinergic system and behavior, *Neurobiol. Aging* **1**:59–62.

Sveinsdottir, E., 1965, Clearance curves of Kr-85 or Xe-133 considered as sum of mono-exponential outwash functions. Description of a computer program for the simple case of only two compartments, *Acta Neurol. Scand.* **41**:*Suppl.* 14:69–71.

Takagaki, G., 1968, Control of aerobic glycolysis and pyruvate kinase activity in cerebral cortex slices, *J. Neurochem.* **15**:903–907.

Ter-Pogossian, M. M., Eichling, J. O., Davis, D. O., Welch, M. J., and Metzger, J. M., 1969, The determination of regional cerebral blood flow by means of water labeled with radioactive oxygen-15, *Radiology* **93**:31–40.

Toga, A. W., and Collins, R. C., 1981, Metabolic response of optic centers to visual stimuli in the albino rat: Anatomical and physiologic considerations, *J. Comp. Neurol.* **199**:443–464.

Vaughan, D. W., 1977, Age-related deterioration of pyramidal cell basal dendrites in rat auditory cortex, *J. Comp. Neurol.* **171**:501–516.

Vaughan, D., and Peters, A., 1974, Neuroglial cells in the cerebral cortex of rats from young adulthood to old age: An electron microscope study, *J. Neurocytol.* **3**:405–429.

Vitorica, J., Andrés, A.,Satrústegui, J., and Machado, A., 1981, Age-related quantitative changes in enzyme activities of rat brain, *Neurochem. Res.* **6**:127–136.

Von Braunmuhl, A. V., 1956, "Kongophile Angiopathie" und "Senile Plaques" bei greisen Hunden, *Arch. Psychiatr. Nervenkr.* **194**:396–414.

Weiss, B., Greenberg, L., and Cantor, E., 1979, Age-related alterations in the development of adrenergic denervation supersensitivity, *Fed. Proc.* **38**:1915–1921.

Wiśniewski, H., Johnson, A. B., Raine, C. S., Kay, W. J., and Terry, R. D., 1970, Senile plaques and cerebral amyloidosis in aged dogs. A histochemical and ultrastructural study, *Lab. Invest.* **23**:287–296.

4

Age-Related Alterations in β-Adrenergic Modulation of Cardiac Cell Function

CHARLES R. FILBURN and EDWARD G. LAKATTA

1. INTRODUCTION

The cardiovascular system's raison d'être is to transport oxygen, nutrients, blood cells, and other substances to body tissues, to remove the waste products of metabolism, and to maintain constant temperature. Since these are not fixed but can vary over at least a tenfold range (i.e., from sleep to maximum exercise) the cardiovascular system, to maintain optimal efficiency, adjusts its level of function to meet the body requirements.

There are several recognized mechanisms which control the level of cardiovascular function (i.e., cardiac output). These are illustrated in Fig. 1 and include: the volume of blood returned to the heart between beats (preload), the intrinsic capacity of the cardiac muscle to develop force and to shorten following excitation (contractility), the delivery of oxygen to support metabolism required for contractility (coronary flow), the resistance of blood flow by the peripheral vasculature (afterload), and the rate at which the heart beats (heart rate). These factors may be regarded as the immediate determinants of cardiac function and are interdependent so that a change in one due, for example, to disease or aging would elicit obligatory compensatory changes in the others.

In some instances the interaction among these control mechanisms is direct, while in others the interaction occurs via reflexes. Each of the regulatory factors depicted in Fig. 1 is subject to autonomic modulation. During stress, cardiac function must increase severalfold, and the change in the extent of

---

CHARLES R. FILBURN and EDWARD G. LAKATTA • Laboratory of Molecular Aging and Cardiovascular Section, National Institute on Aging, National Institutes of Health, Gerontology Research Center, Baltimore City Hospitals, Baltimore, Maryland 21224.

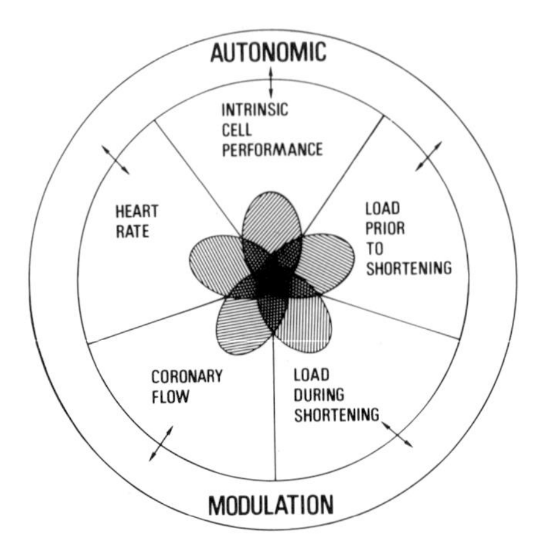

Figure 1. Factors that govern cardiac output. The overlap depicted in the center of the figure indicates the interdependence of these determinants of function. The bidirectional arrows indicate that each function not only is modulated by autonomic tone but also demonstrates a negative feedback on this modulation. Reprinted from Lakatta (1982).

autonomic modulation of each of these is a major determinant of the resultant level of cardiac output that is achieved. In particular, with a stress such as running, enhanced sympathetic and reduced vagal influence result in (1) an increase in heart rate, (2) an increase in amount of blood returning to the heart and, in particular, increase in the rate of this return by reduced venous capacitance, (3) a redistribution of blood flow to working organs, i.e., blood is shunted toward working muscles and away from the viscera, (4) an increase in coronary blood flow, (5) a reduction in some components of afterload, i.e., aortic input impedence, and (6) more efficient excitation-contraction coupling which results in greater extent of shortening of each cell, thereby permitting the heart to eject more blood and at a faster rate.

2. CARDIAC FUNCTION DURING SENESCENCE

It has been observed that with advancing age, the cardiovascular response to stress becomes deficient (Lakatta, 1980). There are many reasons which might explain this phenomenon. In many studies, variables other than age have often not been controlled. Two important variables in this regard are the presence of occult (clinically undiagnosed) coronary artery steneosis and physical "deconditioning," i.e., being "out of shape," both of which vary with age but are quite independent of the "aging process" per se. Intrinsic age-related structural or functional changes in the factors in Fig. 1 are a third type of mechanism that must be considered to explain the decline in maximal cardiovascular function. A diminution in the effectiveness of autonomic (sympathetic) modulation of these factors represent a fourth mechanism. The purpose of this chapter is to examine (1) our current understanding of the fundamental aspects of how β stimulation of cardiac tissues is affected, (2) what is known regarding age-related changes in postsynaptic β-adrenergic modulation of cardiac muscle, and (3) what types of studies are required to further this understanding.

Several types of evidence suggest a diminished β-adrenergic response in cardiovascular tissues with advancing age (Lakatta, 1980). The effect of age on the response to dynamic exercise has recently been studied in a physically active subset of the participants of the Baltimore Longitudinal Study of Aging who are screened for the presence of occult coronary disease with exercise ECG monitoring and radionuclide cardiac imaging (Rodeheffer *et al.*, 1981). In this study, cardiac volumes were measured during upright bicycle exercise with gated blood pool scan technique. The results showed that in elderly subjects during vigorous exercise (1) the increment in heart rate was less than in younger adults; (2) while heart size at rest was not age related, the heart became enlarged both at the end of the filling period and at the end of the contraction period, compared to younger subjects; (3) the ejection fraction, i.e., the proportion of the filling volume pumped with each stroke, increased less than in younger subjects. While these age-related deficiencies in the response to stress could be due to intrinsic age changes in multiple factors in Fig. 1, a decline in autonomic modulation of these factors as a single explanation could account for the entire pattern of altered hemodynamics observed to occur with advancing age.

Data from additional studies in man support the notion that the effectiveness of autonomic modulation declines with age. Age differences in cardiac output during vigorous exercise were lessened when the exercise was performed subsequently during β-adrenergic blockade (Conway *et al.*, 1971). During vigorous exercise, serum catecholamines increase by an order of magnitude and provide an estimate of the extent of catecholamine secretion by the neuroendocrine system. At maximum exercise in healthy fit subjects (Tzankoff *et al.*, 1980), the age-related diminution in heart rate was accompanied by an *increase* in serum catecholamines in elderly subjects relative to that in younger subjects (Fig. 2). Thus failure to elaborate and release catecholamines during stress cannot explain the apparent diminution in adrenergic responsiveness. Rather, these results suggest a decrement in target organ response. This notion is supported by the observation that rapid injections of isoproterenol, a beta agonist, elicit less of an increase in heart rate and cardiac index in elderly versus younger adult men (London *et al.*, 1976; Yin *et al.*, 1976; Kuramoto *et al.*, 1978; Vestal *et al.*, 1979; Bertel *et al.*, 1980). A decrease in maximum heart rate response induced by isoproterenol in senescent versus adult beagle dogs has also been observed, yet the heart rate response to an extrinsic pacemaker was not altered with aging (Yin *et al.*, 1979). Also, during exercise, senescent beagles exhibit an increase in aortic input impedance which would tend to reduce blood flow from the heart, while their younger adult counterparts did not exhibit this increased impedance to blood flow (Yin *et al.*, 1981). When the exercise was repeated during β-adrenergic blockade, both senescent (12 yr old) and younger adult (1–3 yr) dogs exhibited an increase in aortic impedance. These results suggest that in the absence of β-blockade, aortic input impedance did not rise during exercise in the younger dogs because beta autonomic stimulation served to relax the aorta and that this response was different or did not occur in the older cohort. The direct vasodilatory response of β-adrenergic agonists studied directly in aorta removed from rats and rabbits decreases sharply with advancing age (Fleisch, 1981). Finally, in cardiac muscle isolated from rats, the direct response to β-adrenergic stimulation decreases with age (Fig. 3).

Taken together, the results of all these studies in diverse preparations from a variety of species including man suggest that many facets of autonomic modulation (Fig. 1) become less effective with advancing age. More recent studies have attempted to determine what different specific cellular responses to β-adrenergic stimulation occur in cardiac muscle with advancing age. Ultimately, β-adrenergic modulation of excitation-contraction in cardiac muscle, as in each of the other modules in Fig. 1, results from changes in cellular Ca^{2+} metabolism.

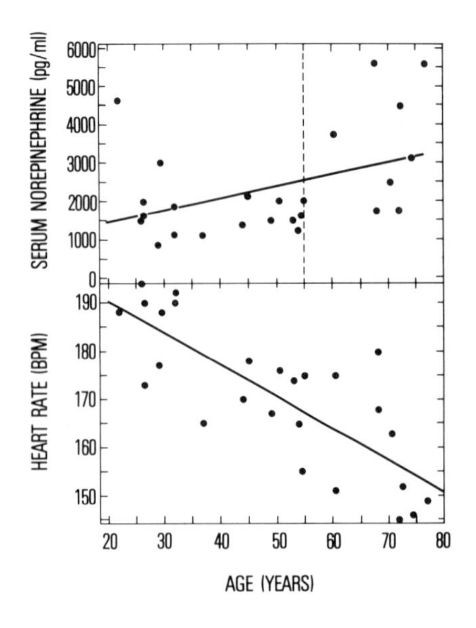

Figure 2. The effect of advanced age on heart rate and serum norepinephrine levels at maximum treadmill exercise. Subjects were participants of the Baltimore Longitudinal Study of Aging who were judged free from occult coronary artery disease by a thorough examination which included prior stress testing with ECG monitoring. Heart rate and serum norepinephrine at rest were not age related. From Tzankoff *et al.* (1980).

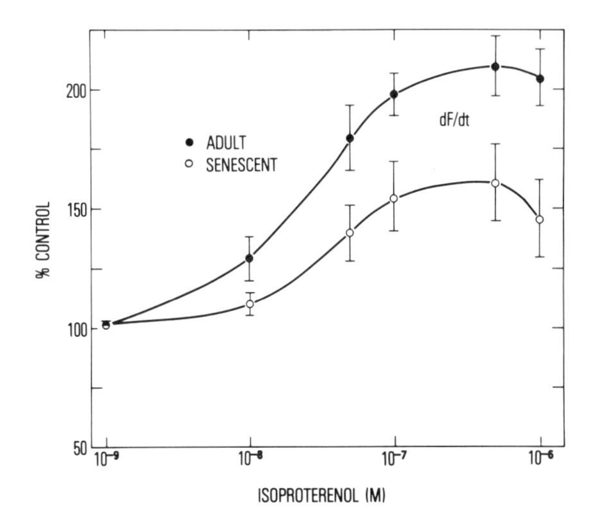

Figure 3. The effect of isoproterenol on the maximum rate of force development (dF/dt) in perfused interventricular septa from adult (7–9 months) and senescent (24–26 months) rats. Prior to isoproterenol, no age difference in dF/dt was present. From Guarnieri *et al.* (1980).

3. MOLECULAR ASPECTS OF β-ADRENERGIC REGULATION DURING SENESCENCE

A scheme of the cellular biochemical reactions controlled by or controlling intracellular Ca^{2+} metabolism and β-adrenergic mediated enhancement of contractility in the cardiac cell is presented in Fig. 4. Interaction of catecholamines with the β-receptor stimulates adenylate cyclase, causing a rise in intracellular cyclic AMP and a consequent activation of cyclic AMP-dependent protein kinase. Elevated protein kinase activity in turn enhances the state of phosphorylation of the various endogenous substrates of this enzyme. These substrates include enzymes of glycogen and intermediary metabolism, contractile proteins, and proteins of sarcolemmal and sarcoplasmic reticulum membranes. Various modulators, in the forms of receptor-mediated reduction of adenylate cyclase, cyclic nucleotide phosphodiesterase, protein kinase inhibitor, or phosphoprotein phosphatase may act to attenuate cyclic AMP-mediated protein phosphorylation. The net effect of these phosphorylations is to provide the heart with additional energy and Ca^{2+} and to enable it to effectively utilize the Ca^{2+} in increasing the rate and/or the strength of contractions. Each of these sites of phosphorylation will be discussed in more detail below, for their respective roles and the influence of other, cyclic AMP-independent, Ca^{2+}-dependent protein kinases are currently under intensive investigation and are not fully understood.

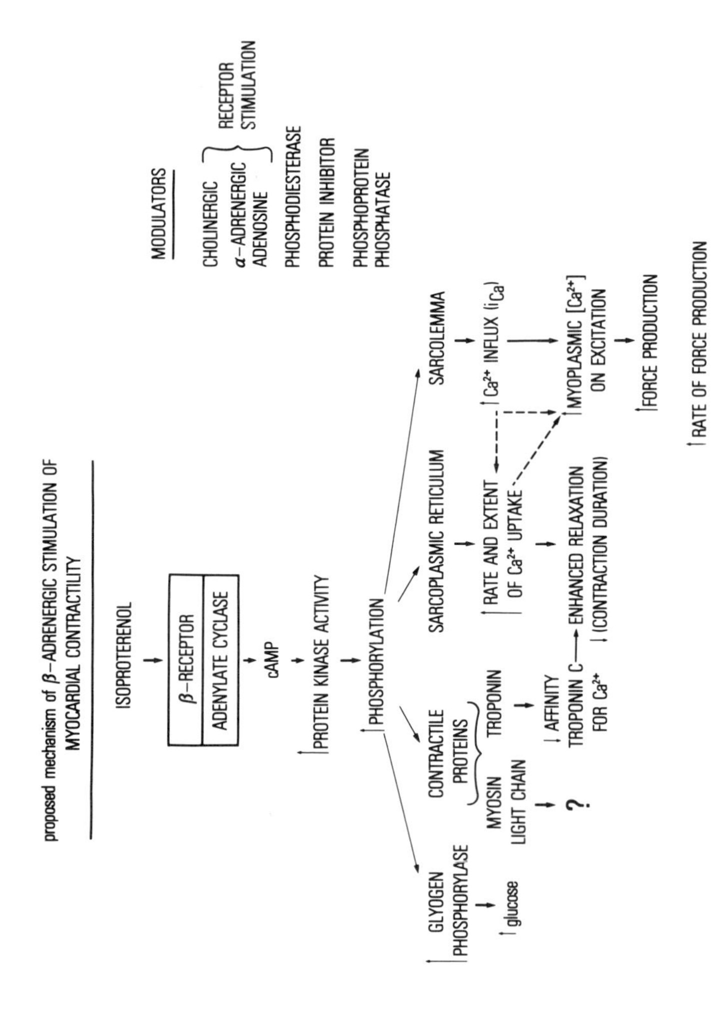

Figure 4. Relationship of β-adrenergically mediated changes in cellular biochemical reactions to enhancement of contractility in cardiac muscle.

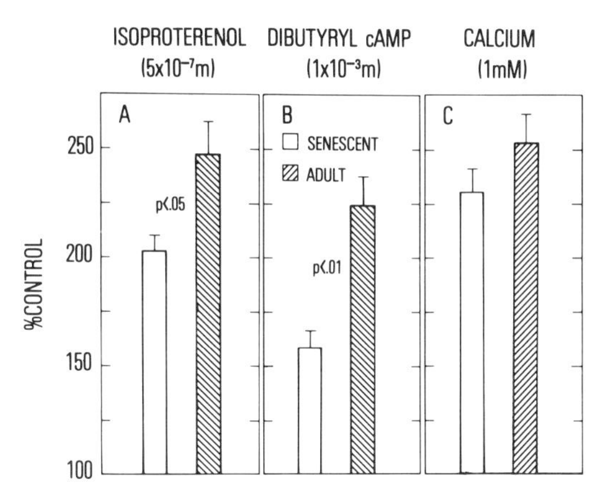

Figure 5. The effect of isoproterenol, 5×10^{-7} M (A), dibutyryl cAMP, 10^{-3} M (B), and enhanced [CA] in the perfusate to 1 mM (C) in perfused interventricular septa from adult and senescent rats. Redrawn from Guarnieri *et al.* (1980).

Identifying the various sites of phosphorylation and understanding their roles in modulating intracellular Ca^{2+} compartmentalization is essential for explaining the diminished response of the aged myocardium to β-adrenergic stimulation. Guarnieri *et al.* (1980) showed that, following stimulation by isoproterenol, interventricular septa from the senescent rat which demonstrated a reduced contractile response compared to those from younger adult rats (Figs. 3, 5) contained the same elevated levels of cyclic AMP and cyclic AMP-dependent protein kinase (Table I) as did septa from adult rats. The finding of comparable, elevated levels of cyclic AMP in both age groups appears at odds with recent reports of an age-associated decrease in isoproterenol stimulation of adenylate cyclase in rat heart membrane preparations (O'Connor *et al.,* 1981; Narayanan and Derby, 1982). In both of these studies, 3–4 month, i.e., matur-

Table I. Effect of Isoproterenol (5×10^{-8} M) on Cyclic AMP and Cyclic AMP-Dependent Protein Kinase in Adult and Senescent Rat Myocardium[a]

	Cyclic AMP pmol/mg wet wt.		Protein kinase activity ratio (−cyclic AMP/+cyclic AMP)	
	Control	Isoproterenol	Control	Isoproterenol
Adult (7–9 months)	0.343 ± 0.02 (8)	0.617 ± 0.03 (9)	0.191 ± 0.011 (8)	0.439 ± 0.02 (9)
Senescent (22–25 months)	0.356 ± 0.01 (9)	0.635 ± 0.01 (6)	0.193 ± 0.017 (8)	0.434 ± 0.02 (6)

[a]Isoproterenol was continuously infused into isolated perfused intraventricular septa and, at the peak of the contractile response, each septum was quick frozen. The septa were powdered and samples taken for analysis of cyclic AMP or cyclic AMP-dependent protein kinase. From Guarnieri *et al.* (1980).

ing, rats were compared to 24 month, senescent rats. However, O'Connor *et al.* (1981) also examined adult, 12-month rats, and observed only a minimal change compared to 24-month rats. In addition, Narayanan and Derby (1982) noted the similarity of the changes observed in the heart in β-receptor affinity, and guanine nucleotide regulation of both affinity and coupling to adenylate cyclase closely resembled the changes observed in maturation of the rat reticulocyte to an erythrocyte (Bilezikian *et al.*, 1977). Thus, while an effect of "aging" on cyclic AMP generation is evident in the heart, it appears to be a maturational change, rather than a postmaturational change, and therefore may be unrelated to the diminished response to catecholamines in the senescent compared to adult myocardium.

This conclusion is supported by additional studies of Guarnieri *et al.* (1980) which showed that perfusion with dibutyryl cyclic AMP stimulated contractility in adult septa to the same extent as isoproterenol, but produced approximately one-half this response in senescent septa (Fig. 5). It appears, then, that the senescent heart is unable, despite apparently adequate protein kinase activity, to elevate sufficiently the state(s) of phosphorylation at one or more of the sites in Fig. 4 to levels comparable to that of mature septa. Alternatively, the density of one or more of these sites may be reduced during senescence and thus limit the response to β-adrenergic stimulation. In contrast, an equal extent of phosphorylation might produce less change in Ca^{2+} ion transport or binding in the senescent vs. adult heart. Thus, the problem of identifying the mechanism underlying the defect(s) in response to β-adrenergic stimulation in the senescent heart appears to be reduced to measuring the density of the possible sites of phosphorylation, the increase in phosphorylation upon β-adrenergic stimulation, and the resultant change in Ca^{2+} metabolism.

Recent studies on the heart, skeletal muscle, and other tissues indicate considerable complexity in the interactions of protein kinases, protein kinase inhibitors, phosphoprotein phosphatases, and phosphoprotein phosphatase inhibitors in controlling phosphorylation-dependent regulatory proteins, with

Table II. Correlation of Cyclic AMP-Dependent Protein Kinase Activity Ratio with Isoproterenol-Induced Changes in Contractile Parameters[a]

	$+dF/dt$	DF	TPF	$-dF/dt$	RF½
r	0.98	0.91	0.92	0.98	0.99
p	0.003	0.034	0.028	0.003	0.001
n	5	5	5	5	5

[a]Values are derived from the means of protein kinase activity ratios and relative in contractile parameters in 3–5 septa at each of five different concentrations of isoproterenol ranging from 5×10^{-9} M to 10^{-6} M. Data from Filburn and Lakatta, in preparation. $+dF/dt$, maximum rate of force development; DF, developed force; TPF, time to peak force; $-dF/dt$, maximum rate of force decline; $RF½$, time of 50% decrease.

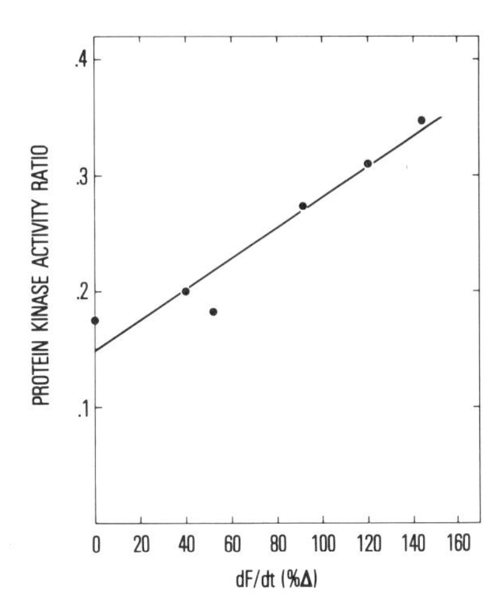

Figure 6. Relationship of protein kinase activity ratio to $+dF/dt$ in isolated perfused rat septa. Data were obtained as described in Table I and were used to obtain the correlation shown in Table II.

an important role for intracellular localization of both enzymes and substrates, particularly in the heart (Cohen, 1981). Consequently, the sequence of events in β-adrenergic regulation in the heart depicted in Fig. 4 will be discussed with regard to the evidence implicating cyclic AMP-mediated phosphorylation of the putative sites, the importance of compartmentation, and current information on the effect of senescence. In addition, since adrenergic regulation of myocardial contractility appears to involve both α- and β-receptors, and may be influenced by several other neurohumoral and/or hormonal factors, these interactions and possible studies of the effect of senescence will be discussed.

4. MECHANISM OF β-ADRENERGIC REGULATION

4.1. Activation of Cyclic AMP-Dependent and Protein Kinase

The temporal relationship between catecholamine stimulated levels of cyclic AMP and myocardial contractility from a large number of studies, plus the ability of the analog dibutyryl cyclic AMP to mimic β-adrenergic effects, strongly suggests that cyclic AMP mediates β-adrenergic stimulation of contractility (for reviews see Tsein, 1978; Scholz, 1980) as indicated in Fig. 4. The dose–response relationship between β-adrenergic activation of cyclic AMP-dependent protein kinase and contractility has been demonstrated in working heart preparations (Keely and Corbin, 1977) and is shown for rat septa in Table II and Fig. 6.

Elevation of cyclic AMP and activation of protein kinase in the heart, however, does not always result in enhanced contractility. Prostaglandin E_1 elevates both parameters but fails to stimulate contractility, to activate phosphorylase kinase, to inhibit glycogen phosphorylase, or to increase phosphorylation of troponin I and a number of other proteins, all of which occur upon exposure to isoproterenol (Keely, 1979; Hayes *et al.*, 1979; 1981). Even with isoproterenol stimulation the relationships are not simple and straightforward. Isoproterenol immobilized on glass beads, and presumably acting locally on only a portion of a papillary muscle, produces an inotropic effect with no detectable changes in cyclic AMP level (Ventner *et al.*, 1975). In addition, isoproterenol stimulates phosphorylation of a 28,000 dalton protein in isolated myocardial cells at concentrations which produce no detectable change in cyclic AMP (Onorato and Rudolph, 1981). Similar apparent dissociations of a physiological response from changes in cyclic AMP levels induced by adenylate cyclase-stimulating agonists have been observed in other tissues (Cooke *et al.*, 1976; Ling and Marsh, 1977) but in these cases a relationship between a response and protein kinase activity ratio was observed. Cumulatively, the data can be interpreted to indicate that compartmentation of cyclic AMP, protein kinase, and/or protein substrates exists in cells, especially myocytes, and that phosphorylation within these subcellular compartments and their interaction in modulating cellular [Ca^{2+}] plays a crucial role in determining responsiveness to catecholamines.

In the heart two types of cyclic AMP-dependent protein kinase which exist and are shown to be differentially distributed between cytosolic and particulate fractions have been isolated (Corbin *et al.*, 1977). Hayes *et al.* (1980) found that perfusion of rabbit hearts with isoproterenol or PGE_1 increased the soluble protein kinase activity ratio, but only isoproterenol caused accumulation of membrane bound cyclic AMP, translocation of catalytic C subunit to the cytosol, and phosphorylase activation, all of which gave similar dose–response curves to a range of concentrations of isoproterenol (Brunton *et al.*, 1981). The only detectable high affinity binding sites of bound cyclic AMP were the regulatory subunits of cyclic AMP-dependent protein kinases, with saturation of particulate binding sites following perfusion with isoproterenol but not PGE_1. Thus, the compartmentation and activation of protein kinases in particulate and soluble pools appears to have functional significance for contractile effects.

To date the only data available on compartmentation of protein kinases in senescent hearts show no change during senescence in the particulate/cytosolic distribution, the level of cyclic AMP-dependent protein kinase activity, or the responsiveness of either cytosolic or particulate protein kinase to cyclic AMP (Guarnieri *et al.*, 1980). The effect of aging on changes in activation of particulate activity after exposure to isoproterenol remains to be established.

4.2. Activation of Glycogen Phosphorylase and Intermediary Metabolism

Mobilization of glucose occurs following adrenergic stimulation of the heart as a consequence of phosphorylation-dependent activation of glycogen phosphorylase via phosphorylase kinase and inhibition of glycogen synthase (Mitchell *et al.,* 1980). In addition, pyruvate dehydrogenase appears to be stimulated indirectly by β-adrenergic stimulation (Hiraoka *et al.,* 1980), while phosphofructokinase is stimulated by α-receptor stimulation (Clark and Patten, 1981). Thus, normal catecholamine stimulation is associated with greater mobilization of and capacity to use a source of energy to support the demands of enhanced contractility. Since decreases in specific mitochondrial enzymes have been observed in the senescent heart (see Hansford, 1980; Hansford and Castro, 1982), the capacity to elicit some of these metabolic responses would appear to be an important factor in the reduced β-adrenergic response observed during senescence. However, as shown in Fig. 5, senescent isolated rat septa exhibit the same increase in contractility as mature septa following an increase in perfusate Ca^{2+} from a submaximal to a maximal dose. Since this increase approximates that seen in mature septa following isoproterenol, and both groups exhibit the same response to increasing concentrations of perfusate Ca^{2+} (Guaranieri *et al.,* 1980), the inability to mobilize sufficient energy to mediate the enhanced contractility, at least on an acute basis, in not involved in the diminished response to β-adrenergic stimulation observed in senescent hearts.

4.3. Stimulation of Ca^{2+} Flux at the Sarcolemmal Membrane

An increase in contractile strength following β-adrenergic stimulation results in part from an increase in cytosolic [Ca^{2+}] during systole (Fig. 7), which occurs in part from increased influx of Ca^{2+} across the sarcolemmal membrane following depolarization. An increased capacity of the sarcoplasmic reticulum to remove Ca^{2+} from the cytosol and to subsequently release more upon depolarization, also occurs. This removal of Ca^{2+} from the cytosol also functions to regulate the duration of contraction, which is reduced following catecholamines. An increased efflux of Ca^{2+} is required to prevent overloading the cell with Ca^{2+}. Thus, both direct and indirect evidence points to the sarcolemma as a major site of action during β-adrenergic stimulation.

Movement of Ca^{2+} across the sarcolemmal membrane is the net result of flux through voltage-dependent channels, a Na^{+}/Ca^{2+} antiporter, an outwardly directed Ca^{2+} pump associated with a Ca^{2+}-dependent ATPase activity, and an inward passive leak (Fig. 8). Cyclic AMP-dependent phosphorylation of sarcolemmal membrane proteins has been demonstrated both *in vivo*

(Walsh *et al.*, 1979) and in isolated membranes (for a review see Tada and Katz, 1982; also Rinaldi *et al.*, 1981). In addition, Ca^{2+}-calmodulin dependent phosphorylation has been observed (Lamers *et al.*, 1981), as well as phosphorylation by phosphorylase β-kinase (Sulakhe and St. Louis, 1977; Caroni and Carafoli, 1981). Some variation in the number, molecular weights, and relative amounts of phosphorylation of sarcolemmal proteins is evident in the studies performed to date, due partly to differences in isolation methods, to contamination by sarcoplasmic reticulum, and to use or lack of use in heat denaturation of samples prior to electrophoresis. Thus, at present, it is difficult to definitively associate phosphorylation of specific sarcolemmal proteins with a given function involving Ca^{2+} transport.

Studies comparing effects of phosphorylation on sarcolemmal Ca^{2+} transport properties, however, do indicate that such relationships are likely to exist. Following β-adrenergic stimulation the slow inward current, evident during the plateau phase of the action potential and attributable in large part to an ion channel for Ca^{2+} (Fig. 7), is increased. Using sarcolemmal vesicles Rinaldi *et*

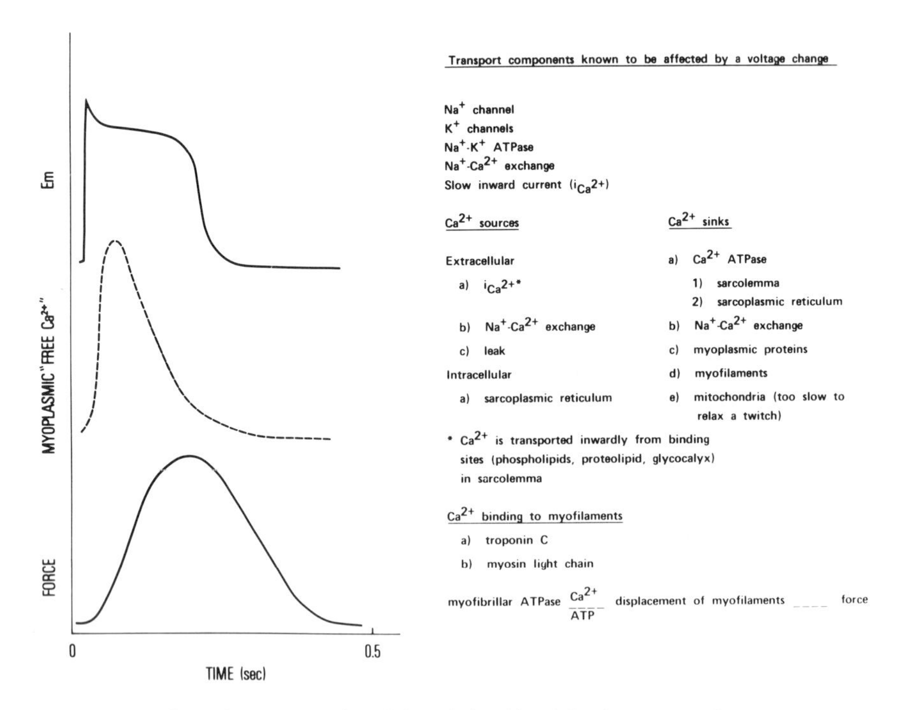

Figure 7. A schematic representation of the relationship of the time course of excitation (action potential), the increase in myoplasmic [Ca], and contraction in a cardiac cell. Each of these is modulated by autonomic stimulation.

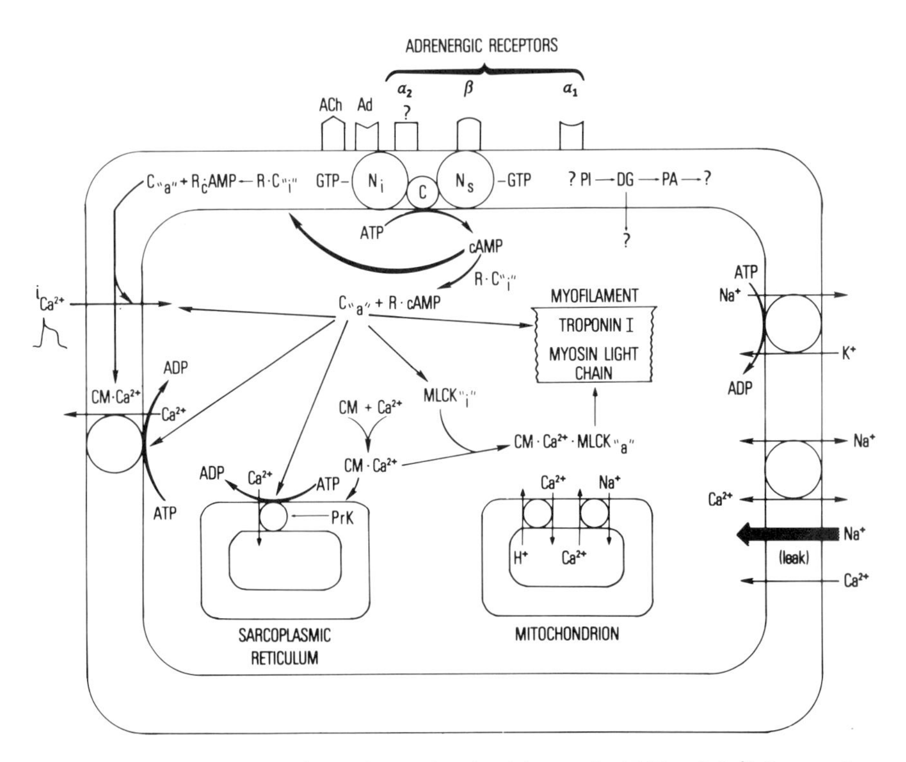

Figure 8. Scheme of the regulatory interactions involving cyclic AMP and Ca^{2+} in a cardiac myocyte. Agonists binding to specific receptors stimulates or attenuates cyclic AMP synthesis or stimulates breakdown of phosphatidylinositol. Cyclic AMP activates an inactive protein kinase, present both in the sarcolemma and cytosol, which then catalyzes phosphorylation of various proteins involved in the metabolism or action of intracellular Ca^{2+}. Intracellular Ca^{2+} interacts with calmodulin to form a complex which interacts with various enzymes involved in metabolism and action of both Ca^{2+} and cyclic AMP. The abbreviations used are as follows: Ad, adenosine receptor; ACh, acetylcholine receptor; α_1 and α_2, alpha-1 and alpha-2 adrenergic receptors; β, β-adrenergic receptor; C, catalytic subunit of adenylate cyclase; N_s and N_i, stimulatory and inhibitory GTP binding proteins associated with catalytic subunit; GTP, guanosine triphosphate; ATP, adenosine triphosphate; ADP, adenosine disphosphate; cAMP, cyclic AMP; PDE_i and PDE_a, inactive and active forms of phosphodiesterase; CM, calmodulin; AMP, adenosine monophosphate; C_i and C_a, inactive and active protein kinase catalytic subunit; R, regulatory cyclic AMP-binding subunit of protein kinase; $MLCK_i$ and $MLCK_a$, inactive and active forms of myosin light chain kinase; PK_i and PK_a, inactive and active forms of protein kinase; $i_{Ca^{2+}}$, inward Ca^{2+} current; PI, phosphatidylinositol; DG, diacylglycerol; PA, phosphatidic acid.

al. (1981) showed that pretreatment under phosphorylating conditions with the catalytic subunit of cyclic AMP-dependent protein kinase caused a two-fold increase in depolarization-induced Ca^{2+} uptake, possibly a manifestation of an *in vitro* equivalent of the slow Ca^{2+}-channel. Under these conditions most of the phosphorylation occurred in proteins of molecular weight 23,000 and 11,500; addition of only cyclic AMP produced a similar effect, presumably mediated by endogenous protein kinase. These proteins appear to be identical with those demonstrated by Jones *et al.* (1979) to be interconvertible by heating, i.e., the 23,000 M_r protein is a dimer of the 11,500 protein, and also appear very similar to those demonstrated in sarcoplasmic reticulum (see Tada and Katz, 1982).

Regulation of the Ca^{2+}-pumping ATPase, which serves to pump Ca^{2+} in an outward direction, by a phosphorylation–dephosphorylation process has been shown by Caroni and Carafoli (1981a). Pretreatment with phosphorylase phosphatase of sarcolemmal vesicles isolated from the dog heart inhibited Ca^{2+}-ATPase and Ca^{2+}-pumping, an effect which was reversed by ATP. In addition, Ca^{2+}-pumping activity was stimulated by phosphorylase β-kinase and inhibited by the protein inhibitor of the cyclic AMP-dependent protein kinase. Both the inhibitor and the Ca^{2+}-chelator EGTA prevented rephosphorylation of vesicles in a nonadditive manner, implying dual regulation by cyclic AMP and Ca^{2+} of distinct endogenous protein kinases (see Lamers *et al.,* 1981). Since the Ca^{2+} dependence of Ca^{2+} uptake and Ca^{2+}-ATPase was similar in phospho- and dephosphorylated vesicles, phosphorylation appears to affect only the efficiency, not the affinity for Ca^{2+}, of the Ca^{2+}-transporting system. It should be emphasized that both the Ca^{2+}-ATPase (Caroni and Carafoli, 1981b) and ATP-dependent Ca^{2+} uptake (Kuwayama and Kanagama, 1982) are activated by calmodulin, which appears to be an intrinsic part of the sarcolemmal membrane (Caroni and Carafoli, 1981b). Thus, calmodulin regulates the ATP-dependent Ca^{2+} uptake both directly and indirectly via a protein kinase. The importance of the membraneous-cytosolic distribution of calmodulin in this regulation remains to be determined.

In a related study on dog heart sarcolemma, Lamers *et al.* (1981) showed cAMP-dependent phosphorylation of a 27,000 M_r protein which was unaffected by heat prior to SDS-PAGE, and a 24,000 M_r protein which reduced to 9,000 after heating. Addition of Ca^{2+} + calmodulin resulted in the majority of the increased phosphorylation appearing in the same 9000 M_r band. This same 9000 M_r band and a 5000 M_r band were also observed in rat heart sarcolemma.

Thus, a state of phosphorylation of specific sarcolemmal membrane proteins is associated with changes in ATP-dependent or membrane potential-dependent Ca^{2+}-transporting properties. The fact that the most consistently observed phosphoprotein corresponds to one also present in sarcoplasmic reticulum is troublesome, since even the best preparations of sarcolemmal membranes appear to contain some contaminating sarcoplasmic reticulum. Alternatively, it is possible that one protein serves a similar role in both membranes, i.e., regulating distinctly different Ca^{2+}-ATPases which play an integral role in transducing energy into uphill Ca^{2+} transport. Further study is needed to firmly establish a functional role for phosphorylation of this and other sarcolemmal proteins.

Movement of Ca^{2+} across the sarcolemma also occurs via an antiporting carrier in an apparently electrogenic process in which Na^+ is exchanged for Ca^{2+} (for a review see Langer, 1982; Philipson and Nishimoto, 1982). This carrier-mediated exchange may play a role in inward movement of Ca^{2+} during depolarization and with repolarization may reverse direction to transport Ca^{2+} outward in exchange for Na^+. At present there is no direct evidence associating protein phosphorylation with a change in the kinetic properties of the Na^+/Ca^{2+} exchanger. However, an increase in affinity of the carrier for Na^+ does occur upon pretreatment of sarcolemmal vesicles with ATP (Kadoma *et al.*, 1982), an effect produced by 1 μM ATP (Reinlib *et al.*, 1981). Since a similar effect involving both Na^+ and Ca^{2+} occurs in barnacle muscle (DiPolo and Caputo, 1977) and axonal membranes (Blaustein, 1977; DiPolo, 1977b) but is not produced by a nonhydrolyzable ATP analog (DiPolo, 1977), regulation by phosphorylation may also be involved in this component of sarcolemmal Ca^{2+} transport.

From the foregoing discussion of sarcolemmal Na^+/Ca^{2+} exchange it should be apparent that any change in $[Na^+]_i$ will lead, at least transiently, to a change in $[Ca^{2+}]_i$. The sarcolemmal Na^+/K^+ ATPase pump, by modulating $[Na^+]_i$ will affect contractile activity in muscle, as seen in the effect of cardiac glycosides. More recently, in rat heart slices, Metzger and Lindner (1982) observed an inhibition of this enzyme by forskolin, a receptor-independent activator of adenylate cyclase, with a consequent increase in contractility. The associated rise in cyclic AMP and protein kinase activity suggested that this inhibition could be a result of cyclic AMP-mediated protein phosphorylation. If a similar effect occurs upon β-adrenergic stimulation, an additional mechanism must be pursued in the study of the reduced contractile response to β-adrenergic stimulation in senescent myocardium.

While the contractile response to cardiac glycosides is markedly depressed in the senescent vs. the adult heart (Gerstenblith *et al.*, 1979a; Guarnieri *et al.*, 1979), very little is known about the effect of senescence on the properties of the various components of sarcolemmal Na^+ or Ca^{2+} transport. Recently Narayanan (1981), in a very thorough study, compared ATP-dependent Ca^{2+} transport of sarcolemmal preparations from 3–4-month-old and 24–25-month-old rats. While the possibility exists that the effects seen were in part maturational (i.e., adult animals were not studied), it was shown that aged sarcolemma had a two-fold greater V_{max} for Ca^{2+} accumulation, with no change in concentration of Ca^{2+} required for half-maximal binding. No comparisons of the effects of cyclic AMP, cyclic AMP-dependent protein kinase, or calmodulin on Ca^{2+} transport activity have been made on mature vs. senescent sarcolemmal preparations. In addition, nothing is known about the kinetic parameters of Na^{+}/Ca^{2+} exchange in senescent sarcolemmal vesicles.

4.4. Stimulation of Ca^{2+} Transport by Sarcoplasmic Reticulum

In contrast to the relatively recent and incomplete characterization of sarcolemmal Ca^{2+} transport, somewhat more is known about regulation of Ca^{2+} transport into sarcoplasmic reticulum. Initially cyclic AMP and cyclic AMP-dependent protein kinase were shown to increase both Ca^{2+}-uptake and Ca^{2+}-ATPase activity, with a high correlation between the two effects (for a review see Tada and Katz, 1982). An ~22,000 M_r protein, designated phospholamban, was identified as the major site of phosphorylation, again with a good correlation between phosphate incorporation and Ca^{2+}-uptake and -ATPase activities. Since then, an appreciation of the effect of heat denaturation, detergents, and the importance of electrophoretic conditions has led to the realization that the ~22,000 M_r protein is a dimer of a smaller ~11,000 M_r protein, which itself appears to be a dimer of an ~5,500 M_r protein (see Kirchberger and Antonetz, 1982a). These *in vitro* studies plus the demonstration of *in vivo* isoproterenol stimulation of one or more of these subunits (LePeuch *et al.*, 1980; Onorato and Rudolph, 1981; Lindemann *et al.*, 1983) indicate a likely role for increased sarcoplasmic reticulum Ca^{2+}-pump activity in the β-adrenergic cardiac response. Lindemann *et al.* (1983) also showed in the guinea pig ventricle that, in addition to isoproterenol, histamine, and dibutyryl cyclic AMP increase phosphorylation of phospholamban and stimulated Ca^{2+}-ATPase activity of sarcoplasmic reticulum. Good correlations were observed with respect to both time of exposure and dose of these agents between phosphorylation and phospholamban and acceleration of relaxation.

Thus, an increased rate of removal of Ca^{2+} from the cytosol and, as a consequence, from the myofilaments, resulting in an increased rate of relaxation, may be due in part at least to the enhanced rate of Ca^{2+} transport by sarcoplasmic reticulum. Subsequent depolarization of the heart cell causes an increase in myoplasmic Ca^{2+}, resulting in a more rapid increase to a higher cytosolic level and a consequent increase in both contractile force and rate of force development. This enhanced Ca^{2+} level is likely to be due in part to enhanced Ca^{2+} release from sarcoplasmic reticulum, as indicated by the effects of cyclic AMP on contractility in mechanically skinned myocardial cells (Fabiato and Fabiato, 1975).

The Ca^{2+}-transporting activity of sarcoplasmic reticulum is also, unequivocally, regulated by a Ca^{2+} + calmodulin-dependent protein kinase. LePeuch *et al.* (1979) first demonstrated a stimulatory effect of Ca^{2+} + calmodulin on phosphate incorporation into phospholamban by an endogenous protein kinase, accompanied by an increased rate of Ca^{2+} transport. The suggestion that cyclic AMP regulation of Ca^{2+} transport depends on the level of prior calmodulin-dependent phosphorylation (LePeuch *et al.,* 1979) has not been borne out by other studies, but the dual, additive regulation by cyclic AMP and Ca^{2+} + calmodulin-dependent protein kinases has been confirmed in *in vitro* studies in several laboratories (see Tada and Katz, 1982; also Bilezikjian *et al.,* 1981; Kirchberger and Antonetz, 1982b). In both forms of regulation, the stimulatory effect on Ca^{2+}-transport is associated with a comparable increase in Ca^{2+}-ATPase activity, indicating an increase in the turnover rate of the ATPase with no change in efficiency of coupling to Ca^{2+} transport (Tada *et al.,* 1979; Kirchberger and Antonetz, 1982b). An actual physiological role for calmodulin-mediated phosphorylation remains to be established. In this regard it is noteworthy that using oubain, high perfusate Ca^{2+}, and the ionophore A23187, all treatments which elevate intracellular Ca^{2+}, Lindemann *et al.* (1983) failed to detect a change in phosphorylation of phospholamban in guinea pig ventricles.

The role of sarcoplasmic reticulum Ca^{2+} uptake in the diminished response of the senescent heart to β-adrenergic stimulation has been and remains an area of fruitful investigation. Froehlich *et al.* (1978) first demonstrated a significant decrease in oxalate-facilitated Ca^{2+} accumulation in microsomal fractions from senescent, 24–25-month Wistar rat hearts compared to 6–8-month, adult hearts. Mechanical studies on muscles isolated from the same hearts showed a prolongation in both time to peak tension and half relaxation time, resulting in an overall increase in contraction duration, as previously shown (Lakatta *et al.,* 1975). As noted above, Narayanan (1981) also observed an age-related decrease in sarcoplasmic reticulum Ca^{2+} pump function regardless as to whether this change was maturational or postmaturational in nature. The decrement observed may serve at least partially to explain the prolongation of contraction in the absence of exogenously added catecholamines.

Further studies by Kadoma *et al.* (1980) have shown that sarcoplasmic reticulum preparations for both adult (10–12 month) and senescent (23–24 month) rats respond to preincubation with cyclic AMP plus cyclic AMP-dependent protein kinase with a 50% stimulation of transport activity and a corresponding phosphorylation of phospholamban. The level of a Ca^{2+}-dependent, acid-stable but alkali-labile phosphoenzyme, assumed to be a measure of the Ca^{2+}-ATPase content, was similar in both age groups. Thus, the data suggest that cyclic AMP-dependent regulation of sarcoplasmic reticulum is unaltered with age, but that a reduced rate of turnover of a constant number of ATPase molecules, which persist even after stimulation by cyclic AMP-dependent protein kinase, underlies the lower transport activity. The lack of an age difference in cAMP dependent stimulation of Ca^{2+} transport in isolated sarcoplasmic reticulum fits well with the observation that the relaxant effect of β-adrenergic stimulation is unaltered with age (Lakatta et al., 1975; Guarnieri *et al.,* 1980). To date no studies have been made of the effect of Ca^{2+} and calmodulin on sarcoplasmic reticulum from senescent hearts.

While many studies on SR have focused on Ca^{2+}-uptake, the release of Ca^{2+} from this storage site has received less attention. Katz (1979) reported a slight enhancement of Ca^{2+} efflux from SR by prior phosphorylation with cyclic AMP-dependent protein kinase, but argued that under the conditions used, an effect of external $[Ca^{2+}]$ probably obscured a much greater stimulatory effect. Kirchberger and Wong (1978) observed a phosphorylation-dependent shift in the half-maximal concentration of external $[Ca^{2+}]$ for stimulation of efflux from SR vesicles. No further studies have been reported, nor is it known whether calmodulin acting by itself or through stimulation of phosphorylation, plays a role in release.

For some time it has been hypothesized that release of SR Ca^{2+} is triggered by a small amount of Ca^{2+} entering the cell following depolarization. Recent studies on single, "mechanically skinned" cardiac cells confirm this mechanism and show release to be affected by the level of preload of the SR with Ca^{2+}, the rate of change of $[Ca^{2+}]_{free}$, and the level of $[Ca^{2+}]_{free}$ used as a trigger (Fabiato and Fabiato, 1979; Fabiato, 1981, 1982). Recent evidence utilizing laser spectroscopic techniques suggests that Ca^{2+}-induced Ca^{2+} release also occurs in intact cardiac muscle (Lakatta and Lappe, 1981; Stern *et al.,* 1981; Kort and Lakatta, 1981).

Preliminary data suggest that the release properties of SR from senescent rats are different from preparations from younger animals, resulting in a requirement for a slightly higher level of Ca^{2+} to trigger release (Fabiato, 1982). Since the comparison made was 28 months versus 2–4 months, further study is required to confirm the change as being associated with senescence. It is of interest, however, that the transmembrane action potential is substantially lengthened in senescent cardiac muscle (Wei *et al.,* 1980), a finding consistent with a greater transsarcolemmal Ca^{2+} movement to trigger release.

4.5. Effects of β-Adrenergic Stimulation on Contractile Proteins

The regulation of myocardial muscle contraction by Ca^{2+} is believed due to binding to troponin C and, through a subsequent change in interaction between troponin C and I, relief of the inhibitory effect of troponin I on interaction between actin and myosin and actomyosin ATPase activity. A well-documented effect of β-adrenergic stimulation is phosphorylation of troponin I, with a resulting increase in the half-maximal $[Ca^{2+}]$ for activation of Mg^{2+}-dependent ATPase activity of isolated myofibrils or actomyosin (Bailin, 1979;

Holroyde *et al.*, 1979; Perry, 1979) and possibly of absolute force developed by hyperpermeable muscle preparations (Mope *et al.*, 1980). These effects appear to be due to a reduction in the Ca^{2+} binding of troponin C at peak saturation of Ca^{2+}-specific sites and an increase in rate of dissociation from this site (Robertson *et al.*, 1982). This latter effect could play a role in the increased relaxation rate in response to β-adrenergic stimulation discussed above.

Although the relaxant effects of catecholamines are not age related, measurements of the state of phosphorylation of troponin I in senescent and adult hearts under basal conditions and following stimulation with isoproterenol are of importance. To date there are no data available on this point. However, the Ca^{2+} dependence of ATPase activity in purified myofibrils and of force in detergent-treated cardiac preparations does not vary with adult aging, i.e., from 8 to 24 months of age (Bhatnager *et al.*, 1982).

An additional site of contractile protein phosphorylation that is currently being investigated is one of the low-molecular-weight subunits, the so-called P light chain, of myosin. Interest in this subunit stems in part from studies in smooth muscle in which phosphorylation of this subunit was shown to regulate myosin ATPase activity (for reviews see Perry, 1979; Adelstein and Eisenberg, 1980) and contraction (Kerrick *et al.*, 1980; Stull, 1980; Walsh *et al.*, 1982). The demonstration of phosphorylation and inhibition of cyclic AMP-dependent protein kinase of the Ca^{2+} + calmodulin-dependent protein kinase catalyzing this phosphorylation (Conti and Adelstein, 1980; Adelstein *et al.*, 1981a,b; Walsh, 1981) implies a role for this control mechanism in β-adrenergic regulation of smooth muscle contractility and raises the question of its presence in the heart. A Ca^{2+} + calmodulin-dependent myosin light chain kinase is present in the heart and is phosphorylated and inhibited by cyclic AMP-dependent protein kinase (Wolf and Hofmann, 1980; Hofmann and Wolf, 1981; Walsh and Guilleux, 1981). At present, however, the effect of phosphorylation of the light chain on myosin ATPase activity and the extent to which β-adrenergic stimulation alters the state of light chain phosphorylation remains controversial.

Unlike smooth muscle myosin ATPase, the activity of which is highly dependent upon phosphorylation of the P light chain, the myosin ATPase of striated muscle is highly active with unphosphorylated light chain (see Perry, 1979). Any effect of phosphorylation on activity is therefore likely to be modulatory in nature. Although some studies failed to detect effects on kinetic parameters (Morgan *et al.*, 1976; Perry, 1979; Stull, 1980), Pemrick (1980) showed an increase in skeletal muscle myosin ATPase activity that was independent of Ca^{2+} concentration, while Resnik and Gevers (1981) observed an increase in the V_{max} of myosin Ca^{2+}-ATPase activity. The facts that light chain phosphorylation alters the conformation of skeletal muscle myosin (Ritz-Gold *et al.*, 1980) and that comparable levels of myosin light chain kinase are present in smooth and striated muscle (see Walsh, 1982) further support a possible role for phosphorylation of light chains in regulation of contraction in striated muscle.

In vivo studies attempting to relate changes in light chain phosphorylation to changes in myocardial contractile activity also leave an unclear picture. Reports of little or no effect of epinephrine on light chain phosphorylation (Holroyde *et al.,* 1979; Jeacocke and England, 1980; High and Stull, 1980; Westwood and Perry, 1981) are balanced by demonstrations of a correlation between phosphorylation and active tension (Kopp and Barany, 1979) and between light chain phosphate content and V_{max} for *myosin* Ca^{2+}-ATPase activity (Resnik and Gevers, 1981). Indirect support for the *in vivo* functioning of the apparent dual, antagonistic regulation of light chain phosphorylation by cyclic AMP and Ca^{2+}, implied from the cyclic AMP-mediated inhibition of the light chain kinase discussed above, was obtained by Walsh and Guilleux (1981) by the use of β-receptor and calmodulin antagonists.

The mixed, disparate results in these reports appear to be due in part to the difficulty of eliminating contaminants in accurately measuring either ^{32}P or inorganic phosphate in myosin light chains (see Resnik and Gevers, 1981). The very low basal values observed by Kopp and Barany (1979) and Resnik and Gevers (1981), plus the very careful comparison of alternate methods performed in the latter study, suggests that, in fact, myosin light chain phosphorylation does occur upon inotropic stimulation and that cytosolic Ca^{2+} is the principal regulator. Since β-adrenergic stimulation increases intracellular Ca^{2+}, at least on a time-averaged or transient basis, there is no reason *a priori* to assume that negative cyclic AMP regulation of the light chain kinase will predominate over the positive regulation by Ca^{2+} + calmodulin that should accompany this change in Ca^{2+}.

It is apparent that much remains to be learned about the role of myosin light chain phosphorylation in the regulation of myocardial contractility. To date no studies have been made of it in senescent hearts. It may be relevant to note that the procedure of increasing perfusate [Ca^{2+}], and consequently intracellular Ca^{2+}, produced similar changes in contractility in adult and senscent septa (Guarnieri *et al.*, 1980), implying no major change during senescence in whatever role Ca^{2+} + calmodulin-mediated phosphorylation of the myosin P light chain plays in the contraction cycle.

4.6. Temporal Relationship of Protein Phosphorylation to the Contractile Response

Much of the work on cyclic AMP and Ca^{2+} + calmodulin regulation of protein phosphorylation in the heart has been performed *in vitro* with isolated enzymes, membranes, and contractile proteins. While these studies have shown

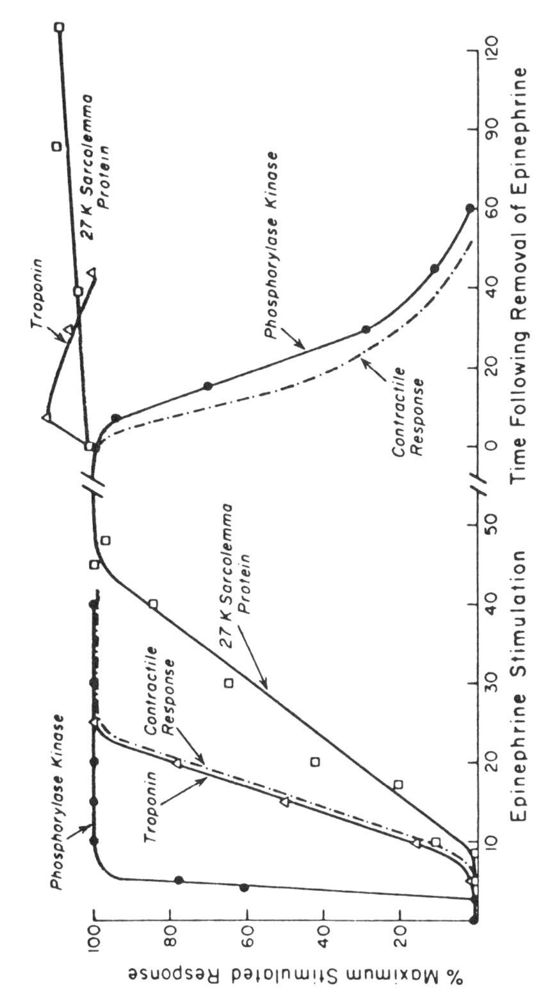

Figure 9. Time course of changes in protein phosphorylation and contractility in the rat heart upon addition and removal of epinephrine. From Sul *et al.* (1981).

some correlations between levels of phosphorylation of given proteins and certain functional parameters, definitively establishing a role for these phosphoproteins in the regulation of contraction requires, as a minimum, the demonstration *in vivo* of correlations, in both degree and time, between levels of phosphorylation of given proteins and specific changes in contractility. Sul *et al.* (1980) have pointed out that the temporal relationships between changes in phosphorylation of three proteins in the rat heart and the contractile response, examined following exposure to and then removal of epinephrine, is distinctively different for each of the proteins. The increase in phosphorylation of phosphorylase kinase, troponin I, and the 27,000 M_r sarcolemmal protein precede, coincide with, and follow, respectively, the contractile response (Fig. 9). Upon removal of epinephrine dephosphorylation of phosphorylase kinase follows close behind the drop in the contractile response, while both troponin I and the sarcolemmal protein remain maximally phosphorylated even after the heart has returned to a basal contractile state. These data suggest that the state of phosphorylation of neither of these latter two proteins is rate limiting in the descent phase of the response, and that possibly troponin I is important in the initial response.

These dissociations should not be too surprising, since other sites of regulatory phosphorylation, not yet measured in this paradigm, are also present and likely to influence the course of the response. Thus the distinct patterns of phosphorylation and dephosphorylation so far observed imply a complex interaction within the heart between protein substrates, protein kinases, and phosphoprotein phosphatase(s). This complexity is a reflection of several factors: multiple forms of the enzymes themselves, each with a characteristic substrate specificity; multiple effectors of the enzymes, including both activators and inhibitors; and differential subcellular localization of both enzymes and protein substrates, permitting certain interactions while precluding or minimizing others. Some of these factors have already been discussed in the preceding sections, and more will be mentioned below. The multiplicity of enzymes and effectors and the compartmentation within which they must function simply increase the difficulty of fully analyzing and explaining changes occurring during senescence, and make simple solutions very unlikely.

5. MODULATION OF β-ADRENERGIC STIMULATION

5.1. Regulation of Protein Kinase and Phosphoprotein Phosphatase Inhibitors

While studies to date indicate that total cyclic AMP-dependent protein kinase activity in adult and senescent hearts are comparable, studies on the rat septum indicate a moderate increase (18%) in the inhibitor as a consequence of senescence (Guarnieri *et al.*, 1980). The protein inhibitor of the catalytic subunit of cyclic AMP-dependent protein kinase appears to be present in the heart at a level sufficient to modulate the kinase activity under basal conditions (Walsh *et al.*, 1971), and changes in its level in the heart are known to be associated with nutritional status (Walsh and Ashby, 1973). Studies in other tissues suggest that hormonal status may also play a role (Kuo, 1975). Recently, Szmigielski (1981) showed that isoproterenol treatment of rat hearts caused a decrease in the activity level of this inhibitor, but failed to change the level of a different inhibitor active against cyclic nucleotide-dependent and independent protein kinases. Assessment of the extent of such changes in adult and senescent rat hearts may shed some light on the diminished contractile response to β-adrenergic stimulation observed by Guarnieri *et al.* (1980). As better techniques become available for measuring absolute levels (i.e., a radioimmunoassay), changes in inhibitor levels observed during senescence that have been observed need to be reassessed.

Analogous inhibitors of phosphoprotein phosphatases have also been identified (for a review see Cohen, 1982), one of which is known to be regulated by phosphorylation by cyclic AMP-dependent protein kinase. In addition, it is now apparent that multiple forms of phosphoprotein phosphatase exist and that, as with protein kinases, they are found in both the cytosol and particulate, subcellular fractions. A further, highly relevant complication is the fact that Ca^{2+} + calmodulin also regulates at least one of these phosphatases (Stewart *et al.*, 1982). Much of our current knowledge on phosphoprotein phosphatases comes from studies on liver and skeletal muscle, with very little information available on cardiac muscle. If the levels of the endogenous protein kinase substrates involved in regulation of contractile activity are found to be stable during senescence, but their states of phosphorylation change, a clear delineation of phosphoprotein phosphatase(s) and their regulation in the heart will be required to fully understand the age-related difference in response to β-adrenergic stimulation. Preliminary studies on phosphoprotein phosphatase in subcellular fractions of rat septa indicate a slight increase during senescence in activity associated with a crude membrane fraction (Guarnieri *et al.*, 1980), suggesting a partial role in the diminished contractile respone to β-adrenergic stimulation. Further study, using purified membrane fractions and assessing the effect(s) of inhibitors on levels of endogenous protein phosphorylation, should help elucidate the role of the enzyme(s) in modulating the contractile response.

5.2. Catecholamine Interaction through α and β Receptors

Recent studies on adrenergic receptors indicate that the heart contains both α and β receptors (Watanabe *et al.*, 1982; Benfey, 1982). Positive inotropic effects of catecholamines on the heart have been associated with α-receptor activation (Scholz, 1980; Benfey, 1982), but the necessity of low-frequency of stimulation in order to detect these effects suggests that the effects of β-adrenergic stimulation predominate *in vivo*. Nevertheless, some of the effects produced by α-receptor activation may influence some of the regulatory processes affected by β-receptor stimulation and thus merit some attention. Stimulation of α-receptors causes increases in phosphofructokinase activity (Clark and Patten, 1981; Patten *et al.*, 1982) and respiration-dependent mitochondrial Ca^{2+} uptake, with an associated increase in the sensitivity of intramitochondrial α-ketoglutarate dehydrogenase to extramitochondrial Ca^{2+} (Kessar and Crompton, 1981). Since mitochondria possess a large capacity to store Ca^{2+}, and release it through a Na^+/Ca^{2+} antiporter (Fig. 8), this increase in Ca^{2+} uptake may augment whatever role the mitochondrion plays in regulating cytosolic [Ca^{2+}]. At present this role remains controversial, with claims for mitochondrial buffering of cytosolic Ca^{2+} (Carafoli and Crompton, 1978; Bygrave, 1978; Nicholls, 1978) opposed by convincing data indicating that *in vivo* regulation of the Krebs cycle by intramitochondrial [Ca^{2+}] requires intramitochondrial total and free Ca^{2+} to be too low to permit substantial buffering (Hansford and Castro, 1982a; Coll *et al.*, 1982). It is noteworthy that both the rate of uptake of Ca^{2+} into and the rate of Ca^{2+} release from rat heart mitochondria is known to be diminished in senescence (Hansford and Castro, 1982b). At present no assessment of α-adrenergic action on any of these parameters has been performed in senescent hearts.

Activity of α_1-adrenergic receptors has recently been shown to produce the classic "phosphatidylinositol effect," i.e., increased ^{32}P-labeling of phosphatidic acid and phosphatidylinositol (Uchida *et al.*, 1982). In many tissues where this effect has been closely analyzed, the first step in this process is breakdown of phosphatidylinositol to phosphatidic acid and diacylglycerol. Diacylglycerol has been shown to markedly increase the sensitivity of a Ca^{2+} + phospholipid-dependent protein kinase to Ca^{2+} (Kaibuchi *et al.*, 1981). This protein kinase is present in the heart and catalyzes the phosphorylation of troponin-T (Wise *et al.*, 1982; Katoh *et al.*, 1982). The susceptibility of troponin-T to phosphorylation has been known for some time, but regulation of this reaction and its role in cardiac troponin function remain to be established (see Perry, 1979; Barany and Barany, 1980). In the event that α-adrenergic regulation of this protein kinase is established and linked to changes in troponin-T phosphorylation, another component of myofibrillar protein phosphorylation must be a target for future studies in the senescent heart.

Another effect of α-adrenergic activation, in the case of the α_2 subclass of receptor, that has been observed now in several tissues and appears to occur in the heart is inhibition of hormonally stimulated adenylate cyclase (Rodbell, 1980; Jakobs *et al.*, 1981). In assessing α-adrenergic involvement in heart metabolism, Keely *et al.* (1977) observed that perfusion of hearts with epinephrine plus the α-antagonist phentolamine potentiated the effects of epinephrine on activation of both cyclic AMP-dependent protein kinase and glycogen phosphorylase. A stimulatory effect on glucose uptake was also blocked by this antagonist but not by the β-blocker propranolol. In addition, α-receptor stimulation increased the level of cyclic GMP (also Amer and Byrne, 1975). Thus, α-receptor activation may modulate β-receptor activation through antagonism at the level of receptor-cyclase coupling, or indirectly through the action of cyclic AMP (see discussion of cholinergic antagonism below). The data on perfused senescent rat septa showing β-adrenergic elevation to comparable levels of cyclic AMP (Guarnieri *et al.,* 1980) sheds little on this point, since isoproterenol has very little α-agonist activity. However, the contractile response to norepinephrine, which has potent α as well as β activity, was also markedly diminished in cardiac muscle from senescent versus adult rats (Lakatta *et al.,* 1975b). Studies with the natural catecholamines in the presence and absence of the appropriate blockers is necessary to assess these α–β interactions.

5.3. Acetylcholine Interaction through Muscarinic Receptors

In marked contrast to β-adrenergic stimulation, acetylcholine interaction with muscarinic receptors produces a negative inotropic effect and a closely correlated increase in cyclic GMP. In addition, the analog 8-bromo cyclic GMP by itself reduces contractile force in cardiac fibers and atrial muscle (Nawrath, 1976; Singh and Flitney, 1981). When present along with a β-adrenergic agonist, acetylcholine attenuates the inotropic response normally observed (Watanabe and Besch, 1975; Gardner and Allen, 1976) while again increasing levels of cyclic GMP. Thus it appears that cyclic GMP mediates the action of acetylcholine in the heart.

The attenuating effect of cholinergic stimulation on the β-adrenergic response, however, is also due in part to a direct inhibitory effect on β-receptor stimulated adenylate cyclase. The effect is observed in membrane preparations, requires GTP, and involves Na^+ (Watanabe *et al.,* 1978; Jakobs *et al.,* 1979), and is independent of cyclic GMP. Furthermore, large increases in cyclic GMP have been observed with agents such as nitroprusside, but with no effect on contractile force (Diamond *et al.,* 1977; Katsuki *et al.,* 1977; Keely and Lincoln, 1978). At concentrations which produced comparable increases in cyclic GMP, acetylcholine antagonized the effect of epinephrine on contractile force while, again, nitroprusside had no effect (Keely and Lincoln, 1978). Finally, Brooker (1977) observed a negative inotropic effect of the cholinergic agonist carbachol while failing to detect a significant change in cyclic GMP.

This apparent paradox, with 8-bromo cyclic GMP reproducing the effect of acetylcholine but the latter, at a submaximal dose, failing to produce a measurable change in cyclic GMP, appears now to be due to compartmentation analagous to that shown for cyclic AMP. Upon rapidly measuring the activity ratio of cyclic GMP-dependent protein kinase in hearts perfused with acetylcholine or nitroprusside, Lincoln and Keely (1981) observed a stimulation by acetylcholine along with a two- to threefold increase in cyclic GMP, but no effect with nitroprusside despite an eight- to tenfold increase in the level of cyclic GMP. Combined, the two agents produced an additive effect on cyclic GMP but no further stimulation of the protein kinase above that produced by acetylcholine alone. These results certainly suggest some form of compartmentation, but in the absence of data on isolated cells it is impossible to determine whether the compartmentation is intra- or extracellular.

At present very little data exist supporting a mechanism of action of cyclic GMP in the heart involving protein phosphorylation. Wrenn and Kuo (1981) recently demonstrated cyclic GMP-dependent phosphorylation of a 70,000 M_r protein in the cytosol of the rat heart. The effect of cyclic GMP or of added cyclic GMP-dependent protein kinase on sarcolemmal preparations has not been reported. In view of the known inhibiting effect of acetylcholine and of 8-bromo cyclic GMP on Ca^{2+} influx and slow inward current (Nawrath, 1977), the sarcolemma appears to be a likely site of action and a good source for future studies.

To date the only studies concerning the action of acetylcholine in the heart during senescence are those of Frolkis *et al.* (1973, 1979), Kulchitskii (1980), and Kelliker and Conahan (1980). Frolkis and colleagues claim a reduction in old animals in the threshold of the negative chronotropic effects of the vagus nerve and a decrease in the levels of acetylcholine required to cause changes in contractility. Kulchitskii (1980) linked these observations to changes in cyclic GMP metabolism by demonstrating significantly greater increases in cyclic GMP in response to submaximal doses of acetylcholine in hearts of 24–26-month rats compared to 6–8-month rats. This difference persisted in the presence of acetylcholinesterase blockade, suggesting a change at or distal to the receptor. In contrast, Kelliher *et al.* (1980) demonstrated a decrease in the heart rate response in old rats to both vagal nerve stimulation and to bolus injections of methacholine. No studies have been performed by these or other investigators assessing β-adrenergic-cholinergic interactions with respect to contractility or cyclic nucleotide levels. It is evident that much remains to be done to resolve these discrepancies and to fully describe the effect of senescence on cholinergic regulation.

5.4. Adenosine Interaction through Adenylate Cyclase

Another important factor in catecholamine regulation of myocardial contractility is the nucleoside adenosine. By itself adenosine depresses the force of contraction and reduces both the plateau phase and the duration of the action potential (Johnson and McKinnon, 1956; DeGubouff and Sleator, 1965). These effects appear to be due to a reduction in cell membrane permeability to Ca^{2+}, as seen in a reduction in the rate of rise and amplitude of the slow action potential of K^+-depolarized guinea pig atria treated with norepinephrine (Schrader *et al.,* 1975). Further studies with catecholamines and adenosine have shown that adenosine attenuation of β-adrenergic stimulation of contractility is associated with a reduction of elevated levels of cyclic AMP (Schrader *et al.,* 1977; Urthaler *et al.,* 1981; Baumann *et al.,* 1981). This reduction in cyclic AMP has also been observed in isolated heart cells (Hazeki and Ui, 1981) and is explained by a direct adenosine inhibition of both basal and β-adrenergically stimulated adenylate cyclase (Schrader *et al.,* 1977; Baumann *et al.,* 1981; Webster and Olsson, 1981). Some of this adenosine antagonism of the contractile responses to catecholamines is also related to inhibition of adrenergic neurotransmission (Hedquist and Fredholm, 1979).

That regulation of adenosine actually occurs *in vivo* is suggested by the observations that adenosine is normally released by the myocardium (Rubio and Berne, 1969) and that nerve stimulation increases this release (Fredholm and Hedquist, 1978). Interestingly, the adenosine released under normoxic conditions is derived from a relatively small adenosine nucleotide compartment that also appears to be the source of cyclic AMP (Schrader and Gerlach, 1976). The relationship of this compartment to those in which cyclic AMP-dependent and cyclic GMP-dependent protein kinase are involved remains to be determined.

At present no studies have been made of the effect of senescence on adenosine regulation. Since isoproterenol stimulation of adult and senescent rat hearts produces the same elevated level of cyclic AMP (Guarnieri *et al.,* 1980), any adenosine antagonism present under the conditions of perfusion and stimulation must have been minimal or neutralized by compensating changes in β-adrenergic regulation. Further studies employing adenosine and various adenosine analogs are needed to fully assess effects of senescence.

6. SUMMARY

It is evident from the foregoing discussion that within the last decade there has been an explosion of new information regarding mechanisms of ion channels transport, and binding by membranes and contractile proteins, i.e., mechanisms that regulate excitation, contraction, and relaxation in the cardiac cell. In spite of this plethora of new information, a full understanding of these mechanisms in themselves and of their interactions, which govern cell function at any instant, is not at hand. Autonomic stimulation modulates cardiac cell performance through its influence on these mechanisms. Although much new information regarding specific details of autonomic modulation has recently emerged, it will require substantial time and additional study before these mechanisms and their interactions will fully be understood. Given the alterations in autonomic modulation of excitation–contraction that have been described to occur with age in the rat, this aging model appears to be a potentially valuable tool for further study of these fundamental mechanisms at the cellular level. In the process, the precise steps which limit the autonomic responsiveness of senescent tissue will be elucidated.

REFERENCES

Adelstein, R. S., and Eisenberg, E., 1980, Regulation and kinetics of the actin-myoxin-ATP interaction, *Ann. Rev. Biochem* **49:**921–956.

Adelstein, R. S., Pato, M. D., and Conti, M. A. 1981b, The role of phosphorylation in regulatory contractile proteins, *Adv. Cyclic Nucleotide Res.* **14:**361–373.

Adelstein, R. S., Pato, M. D., Selless, J. R., Conti, M. A., and Eaton, C. R., 1981a, Regulation of contraction by reversible phosphorylation of myosin and myosin kinase, in: *Cold Spring Harbor Conferences on Cell Proliferation,* Vol. 8, *Protein Phosphorylation* (O. M. Rosen and E. G. Krebs, eds.), pp. 811–822, Cold Spring Harbor Laboratory.

Amer, M. S., and Byrne, J. E., 1975, Interchange of adenyl and guanyl cyclases as an explanation for transformation of β- and α-adrenergic responses in the rat striatum, *Nature* **256:**421–424.

Bailin, G., 1979, Phosphorylation of a bovine cardiac actin complex, *Am. J. Physiol.* **236:**C41–C46.

Barany, M., and Barany, K., 1980, Phosphorylation of myofibrillar proteins, *Ann. Rev. Physiol.* **42:**275–292.

Barany, M., Barany, K., Barron, J. T., Kopp, S. J., Doyle, D. D., Hager, S. R., Schlesinger, D. H., Hama, F., Sayers, S. T., and Janis, R. A., 1981, Protein phosphorylation in live muscle, in: *Cold Spring Harbor Conference on Cell Proliferation,* Vol. 8, *Protein Phosphorylation* (O. M. Rosen and E. G. Krebs, eds.), pp. 869–886, Cold Spring Harbor Laboratory.

Baumann, G., Schrader, J., and Gerlach, E., 1981, Inhibitory action of adenosine on histamine- and dopamine-stimulated cardiac contractility and adenylate cyclase in guinea pigs, *Circ. Res.* **48:**259–266.

Benfey, B. G., 1982, Function of myocardial α-adrenoreceptors, *Life Sci.* **31:**101–112.

Bertel, O., Buhler, F. R., Kiowski, W., and Lutold, B. E., 1980, Decreased beta adrenoreceptor responsiveness as related to age, blood pressure, and plasma catecholamines in patients with essential hypertension, *Hypertension* **2:**130–138.

Bhatnagar, G. M., Walford, G. D., Beard, E. S., and Lakatta, E. G., 1982, Dissociation of time to peak force and myofibrillar ATPase activity with aging of the myocardium, *Fed. Proc.* **41:**1513.

Bilezikjian, L. M., Kranias, E. G., Potter, J. D., and Schwartz, A., 1981, Studies on phosphorylation of canine cardiac sarcoplasmic reticulum by calmodulin-dependent protein kinase, *Circ. Res.* **49:**1356–1362.

Blaustein, M. P., 1977, Effects of internal and external cations and of ATP on sodium-calcium and calcium-calcium exchange in squid axons, *Biophys. J.* **20:**70–111.

Brooker, G., 1977, Dissociation of cyclic GMP from the negative inotropic action of carbachol in guinea pig atria, *J. Cyclic Nucleotide Res.* **3:**407–413.

Brunton, L. L., Hayes, J. S., and Moyer, S. E., 1981, Compartments of cyclic AMP and protein kinase in heart: data supporting their existence and speculations on their subcellular basis, in: *Cold Spring Harbor Conference on Cell Proliferation,* Vol. 8, *Protein Phosphorylation* (O. M. Rosen and E. G. Krebs, eds.), pp. 225–235, Cold Spring Harbor Laboratory.

Bygrave, F. L., 1978, Mitochondria and the control of intracellular calcium, *Biol. Rev.* **53:**43–79.

Carafoli, E., and Crompton, M., 1978, The regulation of intracellular calcium by mitochondria, *Ann. NY Acad. Sci.* **307:**269–284.

Caroni, P., and Carafoli, E., 1981a, Regulation of Ca^{2+}-pumping ATPase of heart sarcolemma by a phosphorylation-dephosphorylation process, *J. Biol. Chem.* **256:**9371–9373.

Caroni, P., and Carafoli, E., 1981b, The Ca^{2+}-pumping ATPase of heart sarcolemma. Characterization, calmodulin dependence, and partial purification, *J. Biol. Chem.* **256:**3263–3270.

Clark, M. G., and Patten, G. S., 1981, Adrenaline activation of phosphofructokinase in rat heart mediated by α-receptor mechanism independent of cyclic AMP, *Nature* **292:**461–463.

Cohen, P., 1982, The role of protein phosphorylation in neural and hormonal control of cellular activity, *Nature* **296:**613–620.

Coll, K. E., Suresh, K. J., Corkey, B. E., and Williamson, J. R., 1982, Determination of the matrix free Ca^{2+} concentration and kinetics of Ca^{2+} efflux in liver and heart mitochondria, *J. Biol. Chem.* **257:**8696–8704.

Conti, M. A., and Adelstein, R. S., 1980, Phosphorylation by cyclic adenosine 3′:5′-monophosphate-dependent protein kinase regulates myosin light chain kinase, *Fed. Proc.* **39:**1569–1573.

Conway, J., Wheeler, R., and Sannerstedt, R., 1971, Sympathetic nervous activity during exercise in relation to age, *Cardiovasc. Res.* **5:**577–581.

Cooke, B. A., Lindh, M. L., and Janszen, F. H. A., 1976, Correlation of protein kinase activation and testosterone production after stimulation of Leydig cells with luteinizing hormone, *Biochem. J.* **160:**439–446.

Cooper, R. H., Sul, H. S., and Walsh, D. A., 1981, Phosphorylation and activation of the cardiac isoenzyme of phosphorylase kinase by the cAMP-dependent protein kinase, *J. Biol. Chem.* **256:**8030–8038.

Corbin, J. D., Sugden, P. H., Lincoln, T. M., and Keely, S. M., 1977, Compartmentalization of adenosine 3′,5′-monophosphate and adenosine 3′,5′-monophosphate-dependent protein kinase in heart tissue, *J. Biol. Chem.* **253:**3854–3861.

DeGubareff, T., and Sleator, W., 1965, Effects of caffeine on mammalian atrial muscle and its interaction with adenosine and calcium, *J. Pharmacol. Exp. Ther.* **148:**202–214.

Diamond, J., Ten Eich, R. E., and Tropani, A. M., 1977, Are increases in cyclic GMP levels responsible for the negative inotropic effects of acetylcholine in the heart? *Biochem. Biophys. Res. Commun.* **79:**912–918.

Dipolo, R., 1977, Characterization of the ATP-dependent calcium efflux in dialyzed squid giant axons, *J. Gen. Physiol.* **69:**795–813.

Dipolo, R., and Caputo, C., 1977, The effect of ATP on calcium efflux in dialyzed barnacle muscle fibres, *Biochem. Biophys. Acta* **470:**389–394.

Fabiato, A., 1981, Myoplasmic free calcium concentration reached during the twitch of an intact isolated cardiac cell and during calcium-induced release of calcium from the sarcoplasmic reticulum of a skinned cardiac cell from the adult rat or rabbit, *J. Gen. Physiol.* **78:**457–497.

Fabiato, A., 1982, Calcium release in skinned cardiac cells: Variations with species, tissues, and development, *Fed. Proc.* **41:**2238–2244.

Fabiato, A., and Fabiato, F., 1975, Relaxing and inotropic effects of cyclic AMP on skinned cardiac cells, *Nature* **253**:556–558.

Fabiato, A., and Fabiato, F., 1979, Calcium and cardiac excitation-contraction coupling, *Ann. Rev. Physiol.* **41**:473–484.

Fleisch, J. H., 1981, Age-related decrease in beta adrenoreceptor activity of the cardiovascular system, *Trends Pharmacol. Sci.* **2**:337–339.

Fredholm, B. B., and Hedquist, P., 1978, Adenosine—A transsynaptic modulator of norepinephrine release? in: *Catecholamines: Basic and Clinical Frontiers* (E. Usdin, ed.), pp. 1146–1148, Pergamon Press, New York.

Froehlich, J. P., Lakatta, E. G., Beard, E., Spurgeon, H. A., Weisfeldt, M. L., and Gerstenblith, G., 1978, Studies of sarcoplasmic reticulum function and contraction duration in young adult and aged rat myocardium, *J. Mol. Cell. Cardiol.* **10**:427–438.

Frolkis, V. V., Bezrukov, V. V., Duplenko, Y. K., Shchegoleva, I. V., Schevtchuk, V. G., and Verkhratsky, N. S., 1973, Acetylcholine metabolism and cholinergic regulation of functions in aging, *Gerontologia* **19**:45–57.

Frolkis, V. V., Schevtchuk, V. G., Verkhratsky, N. S., Stupina, A. S., Karpova, S. M., and Lakiza, T. Y., 1979, Mechanisms of neurohormonal regulation of heart function in aging, *Exp. Aging Res.* **5**:441–477.

Gardner, R. M., and Allen, D. O., 1976, Effect of acetylcholine and glycogen phosphorylase activity and cyclic nucleotide content in isolated perfused rat hearts, *J. Cyclic Nucleotide Res.* **2**:171–178.

Gerstenblith, G., Spurgeon, H., Froehlich, J. P., Weisfeldt, M. L., and Lakatta, E. G., 1979, Diminished inotropic responsiveness of ouabain in aged rat myocardium, *Circ. Res.* **44**:517–523.

Guarnieri, T., Spurgeon, H., Froehlich, J. P., Weisfeldt, M. L., and Lakatta, E. G., 1979, Diminished inotropic response but unaltered toxicity to acetyltrophanthidin in the senescent beagle, *Circ. Res.* **60**:1548–1554.

Guarnieri, T., Filburn, C. R., Zitnick, G., Roth, G. S., and Lakatta, E. G., 1980, Contractile and biochemical correlates of β-adrenergic stimulation of the aged heart, *Am. J. Physiol.* **239**:H501–H508.

Hansford, R. G., 1980, Metabolism and energy production, in: *The Aging Heart, Its Function and Response to Stress* (M. L. Weisfeldt, ed.), Vol. 12, pp. 25–76, Raven Press, New York.

Hansford, R. G., and Castro, F., 1982a, Intramitochondrial and extramitochondrial free calcium ion concentrations of suspensions of heart mitochondria with very low plausibly physiological, contents of total calcium, *J. Bioenerg. Biomembr.* **14**:171–186.

Hansford, R. G., and Castro, F., 1982b, Effect of senescence on Ca^{2+}-ion transport by heart mitochondria, *Mech. Ageing Dev.* **19**:5–13.

Hansford, R. G., and Castro, F., 1982c, Age-linked changes in the activity of enzymes of the tricarboxylate cycle and lipid oxidation, and of carnitine content, in muscles of the rat, *Mech. Ageing Dev.* **19**:191–201.

Hayes, J. S., Brunton, L. L., Brown, J. H., Reese, J. B., and Mayer, S. E., 1979, Hormonally specific expression of cardiac protein kinase activity, *Proc. Natl. Acad. Sci. USA* **76**:1570–1574.

Hayes, J. S., Brunton, L. L., and Mayer, S. E., 1980, Protein phosphorylation catalyzed by cyclic AMP-dependent protein kinase by isoproterenol and prostaglandin E_1, *J. Biol. Chem.* **255**:5113–5119.

Hayes, J. S., Bowling, N., King, K. L., and Bader, G. B., 1982, Evidence for selective regulation of the phosphorylation of myocyte proteins by isoproterenol and prostaglandin E_1, *Biochim. Biophys. Acta* **714**:136–142.

Hazeki, O., and Ui, M., 1981, Modification of islet-activating protein of receptor-mediated regulation of cyclic AMP accumulation in isolated rat heart cells, *J. Biol. Chem.* **256**:2856–2862.

Hedquist, P., and Fredholm, B. B., 1979, Inhibitory effect of adenosine on adrenergic neuroeffector transmission in the rabbit heart, *Acta Physiol. Scand.* **105**:120–122.

High, C. W., and Stull, J. T., 1980, Phosphorylation of myosin in perfused rabbit and rat hearts, *Am. J. Physiol.* **239**:H756–H764.

Hiraoka, T., DeBuysere, M., and Olson, M. S., 1980, Studies on the effects of β-adrenergic agonists on the regulation of pyruvate dehydrogenase in the perfused rat heart, *J. Biol. Chem.* **255**:7604–7609.

Hofmann, F., and Wolf, H., 1981, Basic properties of myosin light-chain kinase from bovine cardiac muscle, in: *Cold Spring Harbor Conference on Cell Proliferation,* Vol. 8, *Protein Phosphorylation* (O. M. Rosen and E. G. Krebs, eds.), pp. 841–854, Cold Spring Harbor Laboratory.

Holroyde, M. J., Howe, E., and Solaro, R. J., 1979, Modification of calcium requirements for activation of cardiac myofibrillar ATPase by cyclic AMP dependent phosphorylation, *Biochim. Biophys. Acta* **586**:63–69.

Jakobs, K. H., Aktories, K., and Schultz, G., 1979, GTP-dependent inhibition of cardiac adenylate cyclase by muscarinic cholinergic agonists, *N. S. Arch. Pharmacol.* **310**:113–119.

Jakobs, K. H., Aktories, K., and Schultz, G., 1981, Inhibition of adenylate cyclase by hormones and neurotransmitters, *Adv. Cyclic Nucleotide Res.* **14**:173–187.

Jeacocke, S. A., and England, P. J., 1980, Phosphorylation of myosin light chains in perfused rat heart. Effect of adrenaline and increased cytoplasmic calcium ions, *Biochem. J.* **188**:763–768.

Johnson, E. A., and McKinnon, M. G., 1956, Effect of acetylcholine and adenosine on cardiac cellular potentials, *Nature* **178**:1174–1175.

Jones, L. R., Besch, H. R., Jr., Fleming, J. W., McConnaughey, M. M., and Watanabe, A. M., 1979, Separation of vesicles of cardiac sarcolemma from vesicles of cardiac sarcoplasmic reticulum, *J. Biol. Chem.* **254**:530–539.

Kadoma, M., Sacktor, B., and Froehlich, J. P., 1980, Stimulation by cAMP and protein kinase of calcium transport in sarcoplasmic reticulum from senescent rat myocardium, *Fed. Proc.* **39**:2040.

Kadoma, M., Froehlich, J., Reeves, J., and Sutko, J., 1982, Kinetics of sodium ion induced calcium ion release in calcium ion loaded cardiac sarcolemmal vesicles: Determination of initial velocities by stopped-flow spectrophotometry, *Biochemistry* **21**:1914–1918.

Kaibuchi, K., Takai, Y., and Nishizuka, Y., 1981, Cooperative roles of various membrane phospholipids in the activation of calcium-activated, phospholipid-dependent protein kinase, *J. Biol. Chem.* **256**:7146–7149.

Katoh, N., Wise, B. C., and Kuo, J. F., 1982, Phosphorylation of cardiac troponin-I and troponin-T by cardiac phospholipid-sensitive Ca^{2+}-dependent protein kinase and inhibition of their phosphorylation by agents, *Fed. Proc.* **41**:1538.

Katsuki, S., Arnold, W., and Murad, F., 1977, Effects of sodium nitroprusside, nitroglycerin, and sodium azide on levels of cyclic nucleotides and mechanical activity of various tissues, *J. Cyclic Nucleotide Res.* **3**:239–248.

Katz, A. M., 1978, Role of the contractile proteins and sarcoplasmic reticulum in the response of the heart to catecholamines: An historical review, *Adv. Cyclic Nucleotide Res.* **11**:303–343.

Keely, S. L., 1979, Prostaglandin E_1 activation of heart cyclic AMP-dependent protein kinase: Apparent dissociation of protein kinase activation from increases in phosphorylase activity and contractile force, *Mol. Pharmacol.* **15**:235–245.

Keely, S. L., and Corbin, J. D., 1977, Involvement of cyclic AMP-dependent protein kinase in the regulation of heart contractile force, *Am. J. Physiol.* **233:**H269–H275.

Keely, S. L., and Lincoln, T. M., 1978, On the question of cyclic GMP as the mediator of the effects of acetylcholine on the heart, *Biochim. Biophys. Acta* **543:**251–257.

Keely, S. L., Corbin, J. D., and Lincoln, T., 1977, Alpha adrenergic involvement in heart metabolism. Effects on adenosine cyclic 3′5′-monophosphate, adenosine 3′,5′-monophosphate-dependent protein kinase, guanisine cyclic 3′,5′-monophosphate, and glucose transport, *Mol. Pharmacol.* **13:**965–975.

Kelliher, G. J., and Conahan, T., 1980, Changes in vagal activity and response to muscarinic receptor agonists with age, *J. Gerontol.* **35:**842–849.

Kerrick, W. G. L., Hoar, P. E., and Cassidy, P. S., 1980, Calcium-activated tension: The role of myosin light chain phosphorylation, *Fed. Proc.* **39:**1558–1563.

Kessar, P., and Crompton, M., 1981, The α-adrenergic-mediated activation of Ca^{2+} influx into cardiac mitochondria, *Biochem. J.* **200:**379–388.

Kirchberger, M. A., and Antonetz, T., 1982a, Phospholamban: Dissociation of the 22,000 molecular weight protein of cardiac sarcoplasmic reticulum into 11,000 and 5,000 molecular weight form, *Biochem. Biophys. Res. Commun.* **105:**152–156.

Kirchberger, M. A., and Antonetz, T., 1982b, Calmodulin-mediated regulation of calcium transport and (Ca^{2+} + Mg^{2+})-activated ATPase activity in isolated cardiac sarcoplasmic reticulum, *J. Biol. Chem.* **257:**5685–5691.

Kirchberger, M. A., and Wong, D., 1978, Calcium efflux from isolated cardiac sarcoplasmic reticulum, *J. Biol. Chem.* **253:**6941–6945.

Kopp, S. J., and Barany, M., 1979, Phosphorylation of the 19,000 dalton light chain of myosin in perfused rat heart under the influence of negative and positive inotropic agents, *J. Biol. Chem.* **254:**12,007–12,012.

Kort, A. A., and Lakatta, E. G., 1981, Light scattering identifies diastolic myoplasmic Ca^{2+} oscillation in diverse mammalian cardiac tissues, *Circulation* **64:**, Part II, IV-62.

Kulchitskii, O. K., 1980, Effect of acetylcholine on the cyclic GMP level in the rat heart at different ages, *Bull. Exp. Biol. Med.* **90:**1237–1239.

Kuo, J. F., 1975, Changes in activities of modulator of cyclic AMP-dependent and cyclic GMP-dependent protein kinases in pancreas and adipose tissue from alloxan-induced diabetic rats, *Biochem. Biophys. Res. Commun.* **65:**1214–1220.

Kuramoto, K., Matsushita, S., Mifune, J., Sakai, M., and Murakami, M., 1978, Electrocardiographic and hemodynamic evaluation of isoproterenol test in elderly ischemic heart disease, *Jpn, Circ. J.* **42:**955–960.

Kuwayama, H., and Kanazawa, T., 1982, Purification of cardiac sarcolemmal vesicles: High sodium pump content and ATP-dependent, calmodulin-activated calcium uptake, *J. Biochem.* **91:**1419–1426.

Lakatta, E. G., 1980, Age-related alterations in the cardiovascular response to adrenergic mediated stress, *Fed. Proc.* **39:**3173–3177.

Lakatta, E. G., 1983, Determinants of cardiovascular performance: Modifications due to aging, *J. Chronic Dis.* **36:**15–30.

Lakatta, E. G., and Lappe, D. L., 1981, Diastolic scattered light fluctuation, resting force and twitch force in mammalian cardiac muscle, *J. Physiol. (Lond.)* **315:**369–394.

Lakatta, E. G., Gerstenblith, G., Angell, C. S., Shock, N. W., and Weisfeldt, M. L., 1975a, Prolonged contraction duration in aged myocardium, *J. Clin, Invest.* **55:**61–68.

Lakatta, E. G., Gerstenblith, G., Angell, C. S., Shock, N. W. and Weisfeldt, M. L., 1975b, Diminished inotropic response of aged myocardium to catecholamine, *Circ. Res.* **36:**262–269.

Lamers, J. M. J., Stinis, H. T., and DeJonge, H. R., 1981, On the role of cyclic AMP and Ca^{2+}-

calmodulin-dependent phosphorylation in the control of (Ca^{2+} + Mg^{2+})-ATPase of cardiac sarcolemmal, *FEBS Lett.* **127:**139–143.

Langer, G. A., 1982, Sodium-calcium exchange in the heart, *Ann. Rev. Physiol.* **44:**435–449.

LePeuch, C. J., Harich, J., and DeMaille, J. G., 1979, Concerted regulation of cardiac sarcoplasmic reticulum calcium transport by cyclic adenosine monophosphate dependent and calcium-calmodulin-dependent phosphorylation, *Biochemistry* **18:**5150–5157.

LePeuch, C. J., Guilleux, J. C., and DeMaille, J. G., 1980, Phospholamban phosphorylation in the perfused rat heart is not solely dependent on β-adrenergic stimulation, *FEBS Lett.* **114:**165–168.

Lincoln, T. M., and Keely, S. L., 1981, Regulation of cardiac cyclic AMP-dependent protein kinase, *Biochim. Biophys. Acta* **676:**230–244.

Ling, W. Y., and Marsh, J. M., 1977, Re-evaluation of the role of cyclic adenosine 3′,5′-monophosphate and protein kinase in the stimulation of steroidogenesis by luteinizing hormone in bovine corpus luteum slices, *Endocrinology* **100:**1571–1578.

London, G. M., Safar, M. E. E., Weiss, Y. A., and Milliez, P. L., 1976, Isoproterenol sensitivity and total body clearance of propranolol in hypertensive patients, *J. Clin. Pharmacol.* **16:**174–182.

Metzger, H., and Lindner, E., 1982, Forskolin-dependent activation of an adenylate cyclase of rat heart membranes leads to an inhibition of a membrane-bound Na,K-ATPase, *H. Z. Physiol. Chem.* **363:**466–467.

Mitchell, J. W., Mellgren, R. L., and Thomas, J. A., 1980, Phosphorylation of heart glycogen synthase by cAMP-dependent protein kinase. Regulatory effects of ATP, *J. Biol. Chem.* **255:**10368–10374.

Mope, L., McClellan, G. B., and Winegrad, S., 1980, Calcium sensitivity of the contractile system and phosphorylation of troponin in hyperpermeable cardiac cells, *J. Gen. Physiol.* **75:**271–282.

Morgan, M., Perry, S. V., and Ottaway, J., 1976, Myosin light-chain phosphatase, *Biochem. J.* **157:**687–697.

Narayanan, N., 1981, Differential alterations in ATP-supported calcium transport activities of sarcoplasmic reticulum and sarcolemma of aging myocardium, *Biochim. Biophys. Acta* **678:**442–459.

Narayanan, N., and Derby, J., 1982, Alterations in the properties of β-adrenergic receptors of myocardial membranes in aging: Impairments in agonist-receptor interactions and guanine nucleotide regulation accompany diminished catecholamine-responsiveness of adenylate cyclase, *Mech. Ageing Dev.* **19:**127–139.

Nawrath, H., 1976, Does cyclic GMP mediate the negative inotropic effect of acetylcholine in the heart? *Nature* **267:**72–74.

Nicholls, D. G., 1978, The regulation of extramitochondrial free calcium ion concentration by rat liver mitochondria, *Biochem. J.* **176:**463–474.

O'Connor, S. W., Scarpace, P. J., and Abrass, I. B., 1981, Age-associated decrease of adenylate cyclase activity in rat myocardium, *Mech. Ageing Dev.* **16:**91–95.

Onorato, J. J., and Rudolph, S. A., 1981, Regulation of protein phosphorylation by inotropic agents in isolated myocardial cells, *J. Biol. Chem.* **256:**10697–10703.

Patten, G. S., Filsell, O. H., and Clark, M. G., 1982, Epinephrine regulation of phosphofructokinase in perfused rat heart. A calcium ion-dependent mechanism mediated via α-receptors, *J. Biol. Chem.* **257:**9480–9486.

Pemrick, S. M., 1980, The phosphorylated L_2 light chain of skeletal myosin is a modifier of the actomyosin ATPase, *J. Biol. Chem.* **255:**8836–8841.

Perry, S. V., 1979, The regulation of contractile activity in muscle, *Biochem. Soc. Trans.* **7:**596–617.

Philipson, K. D., and Nishimoto, A. Y., 1982, Na^+-Ca^{2+} exchange in inside-out cardiac sarcolemmal vesicles, *J. Biol. Chem.* **257**:5111–5117.

Reinlib, L., Caroni, P., and Carafoli, E., 1981, Studies on heart sarcolemma: Vesicles of opposite orientation and the effect of ATP on the Na^+/Ca^{2+} exchanger, *FEBS Lett.* **126**:74–76.

Resink, T. J., and Gevers, W., 1981, Altered adenosine triphosphatase activities of natural actomyosin from rat heart perfused with isoprenaline and ouabain, *Cell Calcium* **2**:105–123.

Rinaldi, M. L., LePeuch, C. J., and DeMaille, J. G., 1981, The epinephrine induced activation of the cardiac slow Ca^{2+} channel is mediated by the cAMP-dependent phosphorylation of calciductin, a 23,000 M_r sarcolemmal protein, *FEBS Lett.* **129**:277–281.

Ritz-Gold, C. J., Cooke, R., Blumenthal, D. K., and Stull, J. T., 1980, Light chain phosphorylation alters the confirmation of skeletal muscle myosin, *Biochem. Biophys. Res. Commun.* **93**:209–214.

Robertson, S. P., Johnson, J. D., Holroyde, M. J., Kranias, E. G., Potter, J. D., and Solaro, R. J., 1982, The effect of troponin I phosphorylation on the Ca^{2+}-binding properties of the Ca^{2+}-regulatory site of bovine cardiac troponin, *J. Biol. Chem.* **257**:260–263.

Rodbell, M., 1980, The role of hormone receptors and GTP-regulatory proteins in membrane transduction, *Nature* **284**:17–22.

Rodeheffer, R., Gerstenblith, G., Fleg, J. L., Lakatta, E. G., Clulow, J., Kollman, C. H., Weisfeldt, M. L., and Becker, L. C., 1981, The impact of age on gated blood pool scan (GBPS) measurements of LV volumes during exercise, *Circulation* **64,** Part II, 243.

Rubio, R., and Berne, R. M., 1969, Release of adenosine by the normal myocardium in dogs and its relationship to the regulation of coronary resistance, *Circ. Res.* **25**:407–415.

Scholz, H., 1980, Effects of beta- and alpha-adrenoreceptor activators and adrenergic transmitter releasing agents on the mechanical activity of the heart, in: *Handbook of Experimental Pharmacology,* Vol. 54, Part I: *Adrenergic Activators and Inhibitors* (L. Szckeres, ed.), pp. 651–733, Springer-Verlag, Berlin.

Schrader, J., and Gerlach, E., 1976, Compartmentation of cardiac adenine nucleotides and formation of adenosine, *Pfluegers Arch.* **367**:129–135.

Schrader, J., Rubio, R., and Berne, R. M., 1975, Inhibition of slow action potentials of guinea pig atrial muscle by adenosine: A possible effect on Ca^{2+} influx, *J. Mol. Cell. Cardiol.* **7**:427–433.

Schrader, J., Bauman, J., and Gerlach, E., 1977, Adenosine as inhibitor of myocardial effects of catecholamines, *Pfluegers Arch.* **372**:29–35.

Singh, J., and Flitney, F. W., 1981, Inotropic responses of the frog ventricle to dibutyryl cyclic AMP and 8-bromo cyclic GMP and related changes in endogenous cyclic nucleotide levels, *Biochem. Pharmacol.* **30**:1475–1481.

Stern, M. D., Kort, A. A., Bhatnagar, G. M., and Lakatta, E. G., 1981, Laser scattering fluctuations in non-beating rat myocardium related to spontaneous myoplasmic calcium oscillations, *Biophys. J.* **33**:284a.

Stewart, A. A., Ingebritsen, T. S., Manalan, A., Klee, C. D., and Cohen, P., 1982, Discovery of a Ca^{2+}- and calmodulin-dependent protein phosphatase, *FEBS Lett.* **137**:80–84.

Stull, J. T., 1980, Phosphorylation of contractile proteins in relation to muscle function, *Adv. Cyclic Nucleotide Res.* **13**:39–94.

Sul, H. S., Cooper, R. H., McCullough, T. E., Pickett-Geis, C. A., Angelos, K. L., and Walsh, D. A., 1981, Regulation of cardiac phosphorylase kinase, in: *Cold Spring Harbor Conference on Cell Proliferation,* Vol. 8, *Protein Phosphorylation* (O. M. Rosen and E. G. Krebs, eds.), pp. 343–356, Cold Spring Harbor Laboratory.

Sulakhe, P. V., and St. Louis, P. J., 1977, Stimulation of calcium accumulation in cardiac sarcolemma by phosphorylase kinase, *Biochem. J.* **164**:457–459.

Szmigielski, A., 1981, Modulation of the activity of endogenous protein kinase inhibitors in rat heart by the beta adrenergic receptor, *Arch. Int. Pharmacol.* **249**:64–71.

Tada, M., and Katz, A. M., 1982, Phosphorylation of the sarcoplasmic reticulum and sarcolemma, *Ann. Rev. Physiol.* **44**:401–423.

Tsien, R. W., 1978, Cyclic AMP and contractile activity in heart, *Adv. Cyclic Nucleotide Res.* **8**:363–420.

Tzankoff, S. T., Fleg, J. L., Norris, A. H., and Lakatta, E. G., 1980, Age-related increase in serum catecholamine levels during exercise in healthy adult men, *Physiologist* **23**:50.

Uchida, T., Bhatnagar, G. M., Lakatta, E. G., and Filburn, C. R., 1982, Alpha-adrenergic stimulation of $^{32}PO_4$ labelling of phosphatidyl-inositol and phosphatidic acid in cultured rat heart cells, *Fed. Proc.* **41**:1523.

Urthaler, F., Woods, W. T., Jones, T. N., and Walker, A. A., 1981, Effects of adenosine on mechanical performance and electrical activity in the canine heart, *J. Pharmacol. Exp. Ther.* **216**:254–260.

Ventner, J. C., Ross, J., and Kaplan, N. O., 1975, Lack of detectable change in cyclic AMP during the cardiac inotropic response to isoproterenol immobilized on glass beads, *Proc. Natl. Acad. Sci. USA* **72**:824–828.

Vestal, R. E., Wood, A. J. J., and Shand, D. G., 1979, Reduced beta-adrenoreceptor sensitivity in the elderly, *Clin. Pharmacol. Ther.* **26**:181–186.

Walsh, D. A., and Ashby, D. C., 1973, Protein kinases: Aspects of their regulation and diversity, *Recent Prog. Horm. Res.* **29**:329–359.

Walsh, D. A., Ashby, D. C., Gonzabs, C., Calkins, D., Fisher, E. H., and Krebs, E. G., 1971, Purification and characterization of a protein inhibitor of adenosine 3′,5′-monophosphate dependent protein kinase, *J. Biol. Chem.* **246**:1977–1985.

Walsh, D. A., Clippinger, M. S., Sivaramakrishnan, S., and McCullough, T. E., 1979, Cyclic adenosine monophosphate dependent and independent phosphorylation of sarcolemmal membrane proteins in perfused rat heart, *Biochemistry* **18**:871–877.

Walsh, M. P., 1981, Calmodulin-dependent myosin light chain kinases, *Cell Calcium* **2**:333–352.

Walsh, M. P., and Guilleux, J. C., 1981, Calcium and cyclic AMP-dependent regulation of myofibrillar calmodulin-dependent myosin light chain kinases from cardiac and skeletal muscles, *Adv. Cyclic Nucleotides Res.* **14**:375–390.

Walsh, M. P., Bridenbaugh, R., Hartshorne, D. J., and Kerrick, W. G. L., 1982, Phosphorylation-dependent activated tension in skinned gizzard muscle fibers in the absence of Ca^{2+}, *J. Biol. Chem.* **257**:5987–5990.

Walsh, M. P., Vallet, B., Autric, F., and DeMaille, J G., 1979, Purification and characterization of bovine cardiac calmodulin-dependent myosin light chain kinase, *J. Biol. Chem.* **254**:12136–12144.

Watanabe, A. M., and Besch, H. R., 1975, Interaction between cyclic adenosine monophosphate and cyclic guanosine monophosphate in guinea pig ventricular myocardium, *Circ. Res.* **37**:309–317.

Watanabe, A. M., McConnaughey, M. M., Strawbridge, R. A., Fleming, J. W., Jones, L. R., and Besch, H. R., 1978, Musarinic cholinergic receptor modulation of β-adrenergic receptor affinity for catecholamines, *J. Biol. Chem.* **253**:4833–4836.

Watanabe, A. M., Jones, L. R., Monalan, A. S., and Besch, H. R., 1982, Cardiac autonomic receptors. Recent concepts from radiolabeled ligand-binding studies, *Circ. Res.* **50**:161–174.

Webster, S., and Olsson, R. A., 1981, Adenosine regulation of canine cardiac adenylate cyclase, *Biochem. Pharmacol.* **30**:369–373.

Wei, J. Y., Spurgeon, H. A., and Lakatta, E. G., 1980, Transmembrane action potential duration and contractile activation are lengthened in cardiac muscle of senescent rats, *Clin. Res.* **28:**619A.

Westwood, S. A., and Perry, S. V., 1981, The effect of adrenaline on the phosphorylation of the P light chain of myosin and troponin I in the perfused rabbit heart, *Biochem. J.* **197:**185–193.

Wise, B. C., Glass, D. B., Chou, C. H. J., Raynor, R. L., Katoh, N., Schatzman, R. C., Turner, R. S., Kibler, R. F., and Kuo, J. F., 1982b, Phospholipid-sensitive Ca^{2+}-dependent protein kinase from heart. I. Purification and general properties, *J. Biol. Chem.* **257:**8481–8488.

Wolff, H., and Hofmann, F., 1980, Purification of myosin light chain kinase from bovine cardiac muscle, *Proc. Natl. Acad. Sci. USA* **77:**5852–5855.

Wrenn, R. W., and Kuo, J. F., 1981, Cyclic GMP-dependent phosphorylation of an endogenous protein from rat heart, *Biochem. Biophys. Res. Commun.* **101:**1274–1280.

Yin, F. C. P., Spurgeon, H. A., Raizes, G. S., Greene, H. L., Weisfeldt, M. L., and Shock, N. W., 1976, Age-associated decrease in chronotropic response to isoproterenol, *Circulation* **54:**Suppl. II, II-167.

Yin, F. C. P., Spurgeon, H. A., Greene, H. L., Lakatta, E. G., and Weisfeldt, M. L., 1979, Age-associated decrease in heart rate response to isoproterenol in dogs, *Mech. Ageing Dev.* **10:**17–25.

Yin, F. C. P., Weisfeldt, M. L., and Milnor, W. R. 1981, Role of aortic input impedance in the decreased cardiovascular response to exercise with aging in dogs, *J. Clin. Invest.* **68:**28–38.

5

A Systems Analysis–Thermodynamic View of Cellular and Organismic Aging

JAIME MIQUEL, ANGELOS C. ECONOMOS, and JOHN E. JOHNSON, JR.

Some scientists cling to the reductionist faith—that the way to know everything is to concentrate on the investigation of the lower-most level, which is consequently styled the "fundamental" level. A wiser and sounder strategy of scientific research is to gain an understanding of the phenomena and regularities on all integration levels.
Theodosius Dobzhansky (1964)

1. INTRODUCTION

Major theories of aging usually focus on senescent failure in important physiological systems (neuroendocrine, immunological, etc.) or specific molecular mechanisms such as DNA loss, enzyme inactivation, or macromolecular cross-linkage. Seldom is an attempt made to integrate the data using a systems analysis approach (Miller, 1978), and even more rarely is aging viewed as a ther-

JAIME MIQUEL • Biomedical Research Division, NASA Ames Research Center, Moffett Field, California 94035. **ANGELOS C. ECONOMOS** • Genetics Laboratory, Catholic University of Louvain, Louvain-la-Neuve, Belgium. **JOHN E. JOHNSON, JR.** • Department of Neurology, Johns Hopkins University School of Medicine, Baltimore, Maryland 21205, and National Institute on Aging, National Institutes of Health, Baltimore City Hospitals, Baltimore, Maryland 21224.

modynamic process resulting from the interaction of an organism with the energy and matter of its environment (see Strehler, 1967; Sacher, 1967; Economos and Miquel, 1977; Miquel *et al.,* 1979; Economos, 1982).

Although the prevalent reductionistic studies have provided much essential information, only a systems analysis approach can offer an integrated picture of a phenomenon as complex and multifaceted as aging. Furthermore, a systems-thermodynamic viewpoint may be useful for theory testing, since the adequacy of a postulated mechanism of aging is best judged by its compatibility with senescent changes found in various animal models at *all* levels of biological organization.

Previous publications from our laboratory have dealt with a preliminary systems analysis of senescent disorganization (entropy gain) at several levels, namely, population (Economos, 1979; Economos and Miquel, 1977, 1979, 1980; Economos *et al.,* 1979), cell (Miquel *et al.,* 1979), and subcellular organelle (Economos *et al.,* 1980; Miquel *et al.,* 1980; Fleming *et al.,* 1982). The present chapter summarizes our concepts on the most salient aging changes found at these key levels of biological organization and also considers briefly the role of "pace-setting" physiological systems. Finally, we integrate the reviewed data in a general hypothesis of metazoan aging. This hypothesis states that the senescent changes found at the organism and cell levels are rooted in the inability of the mitochondria of fixed postmitotic cells to dissipate completely the entropy which arises as a by-product of the oxidative reactions involved in the production of ATP.

2. A BLACK BOX VIEW OF THE AGING PROCESS: MORTALITY KINETICS AND INCREASE IN INTERINDIVIDUAL DIFFERENCES

In contrast to most technical artifacts, that are comparatively simple in their structure and function, multicellular organisms show a bewildering array of physiological systems (Fig. 1), organs and cells (Figs. 2–4), and subcellular organelles (Fig. 5). All of these systems and subsystems should be analyzed in depth in order to understand aging. However, a preliminary analysis of the thermodynamics of aging can be obtained by disregarding the components of an organism and focusing instead on the interactions of the organism with the environment. When an organism is thus viewed, in engineering terms, as a "black box" (Fig. 6), the quantitative relationships among its inputs and outputs can be determined without paying much attention to internal processes. Among the questions to be answered by this thermodynamic-systems approach are which of the functions or "useful output" of an organism show senescent changes and which among the many environmental factors are responsible for these changes.

The most important integrated function of an organism is its ability to remain alive by succesfully resisting environmental challenges to its organization. Therefore, it is proper to start our analysis of aging by focusing on the fact that, in contrast to certain unicellular protozoa that are able to avoid aging (through the process of continuous growth followed by replication), most metazoans such as nematodes, insects, and mammals have a limited life span even when kept in the most favorable and sheltered environments. As illustrated in Fig. 7, the number of individuals of a population decreases with the passage of time, as the population ages, and none of the members of the population lives beyond a certain age that is characteristic for each species. The dynamics of this process, which are called "mortality kinetics," have been analyzed in detail by many authors and have stimulated the development of a number of theories of mortality (for a recent review see Economos, 1982). Over one and a half centuries ago, a British actuarian, Benjamin Gompertz, hypothesized that the force of mortality (ratio of the number of individuals dying in a short age interval, over the number of survivors at the beginning of that interval) increased exponentially with age in human populations (Gompertz, 1825). This hypothesis was subsequently accepted as a universal law of mortality kinetics. However, doubts as to the general applicability of Gompertz's law have been occasionally expressed (Beard, 1959) and a reexamination of the mortality kinetics of a considerable number of metazoan species, including humans, indicates that Gompertz's law is only an approximation, not valid over the terminal part of the life span, during which force of mortality levels off (Economos and Miquel, 1978; Economos, 1979, 1980, 1982).

As pointed out elsewhere (Economos, 1980), the often repeated statement that aging necessarily results in a continuously increasing force of mortality would be valid only if a population consisted of identical individuals, whose weighted sum of capacities of physiological functions or health state index ("vitality") in early maturity was the same. However, every population has a certain, usually nonuniform, frequency distribution of vitalities of its individuals; this distribution influences the form of the frequency distribution of deaths and consequently population mortality kinetics. This assumption is supported by a study (Economos and Miquel, 1979) which showed that if the initial frequency distribution of vitality of a human population (military aviators) is changed by eliminating from the population the weak, sick, or unfit individuals by means of a severe but noninvasive physical examination, the mortality kinetics is affected accordingly. That the members of a population may differ considerably in their biological properties is illustrated by Fig. 8, showing the frequency distribution of body weight in a population of mice. As discussed in detail elsewhere (Economos and Miquel, 1980), it is apparent that measurement of body weight, like any other physiological, structural, or biochemical characteristic, may yield a different value for each individual member of an animal population. This variability is still present, though smaller, in inbred animal populations or when only individuals of the same age are considered.

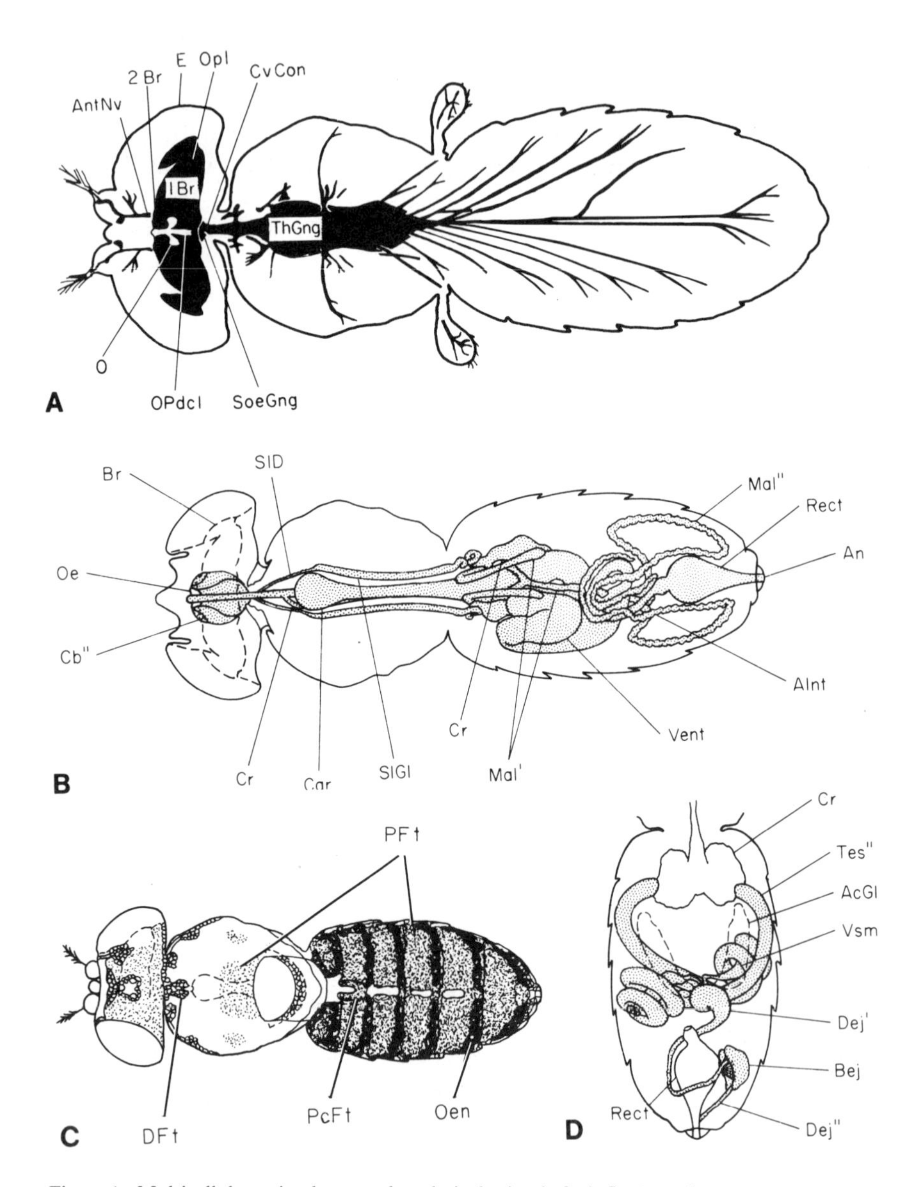

Figure 1. Multicellular animals, even the relatively simple fruit fly *D. melanogaster* represented in this figure, contain a number of interacting systems such as the nervous system (A), the digestive system (B), the adipose tissue (C), and the reproductive (male) system (D). (Redrawn after Miller, 1965.) This makes advisable a *systems analysis* approach for clarification of the mechanisms of aging.

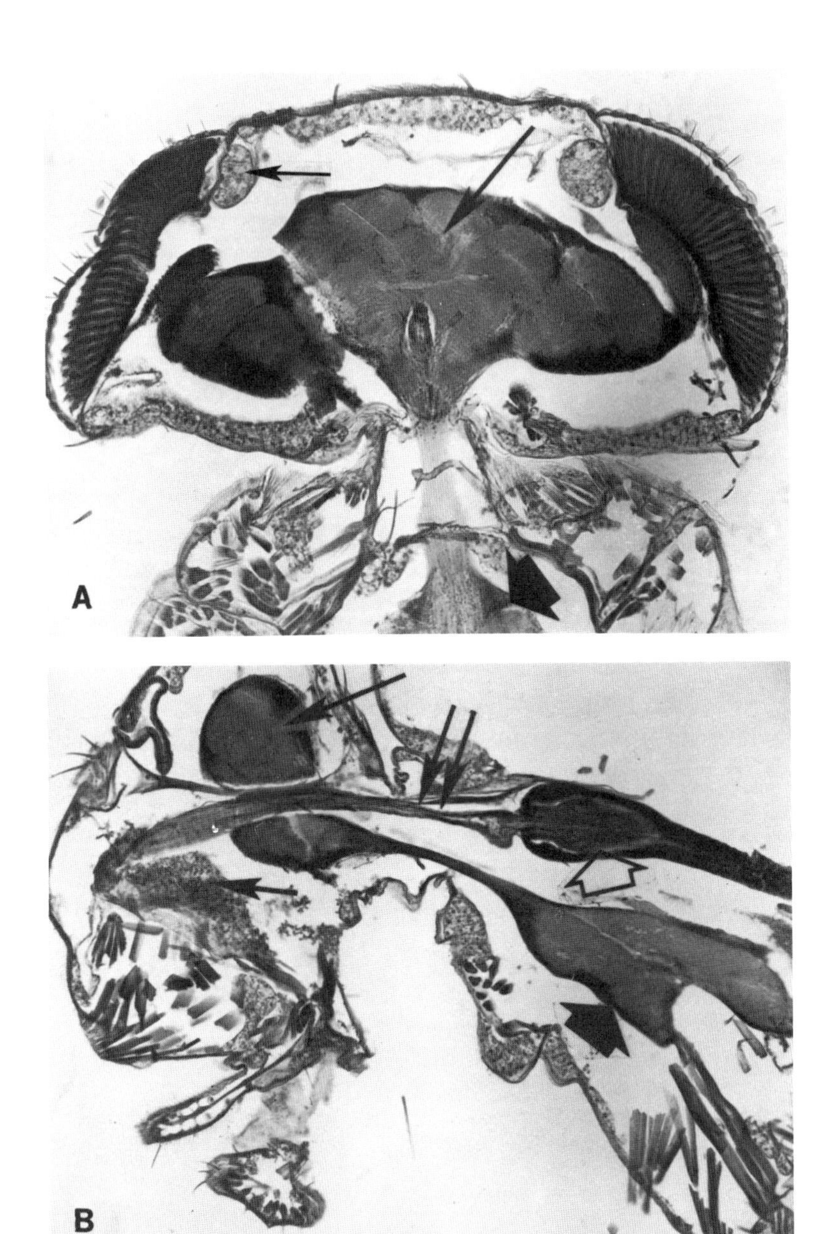

Figure 2. At a lower level of organization than that illustrated in the preceding figure, light microscopy shows that horizontal (A) and lateral (B) sections of the fly head, neck, and anterior thorax contain a number of structures such as brain (long arrow), adipose tissue (short arrow), and thoracic ganglion (thick arrow). The lateral section (B) also shows the organization of the esophagus (double arrows) and cardia (hollow arrow). [Paraffin sections stained by the FAN-method of Miquel *et al.*, 1968 (×150).] The considerable histological complexity of animal tissues illustrated by this figure renders difficult the interpretation of gerontological data obtained by standard biochemical analysis of whole organs.

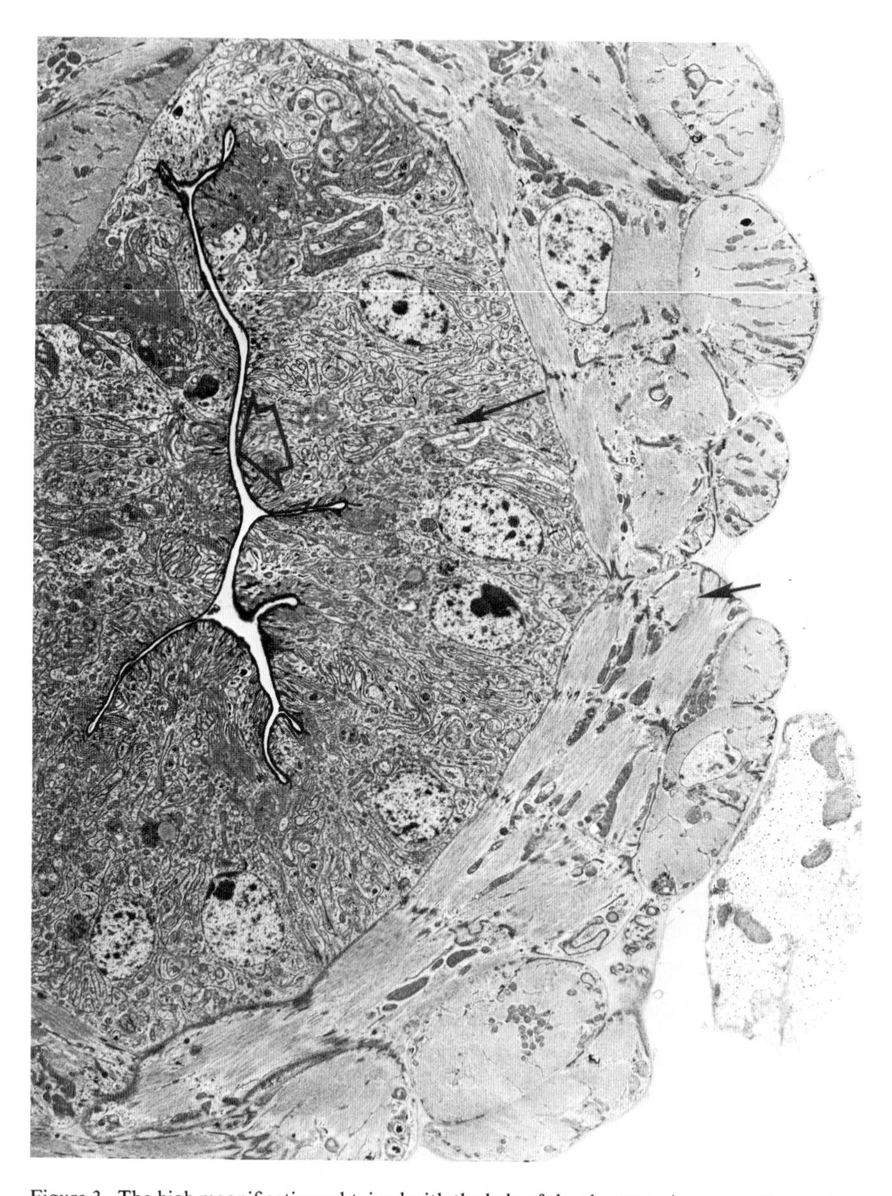

Figure 3. The high magnifications obtained with the help of the electron microscope allow a clear demonstration of the cells, that are the building blocks of tissues and organs. This electron micrograph shows that a simple organ such as the *Drosophila* rectum depends for its function on the "shield" or "sieve"-like effect of an intraluminal interface of chitin (hollow arrow), the absorptive properties of the epithelial cells (long arrow), and the contractibility of muscle (short arrow). × 7000.

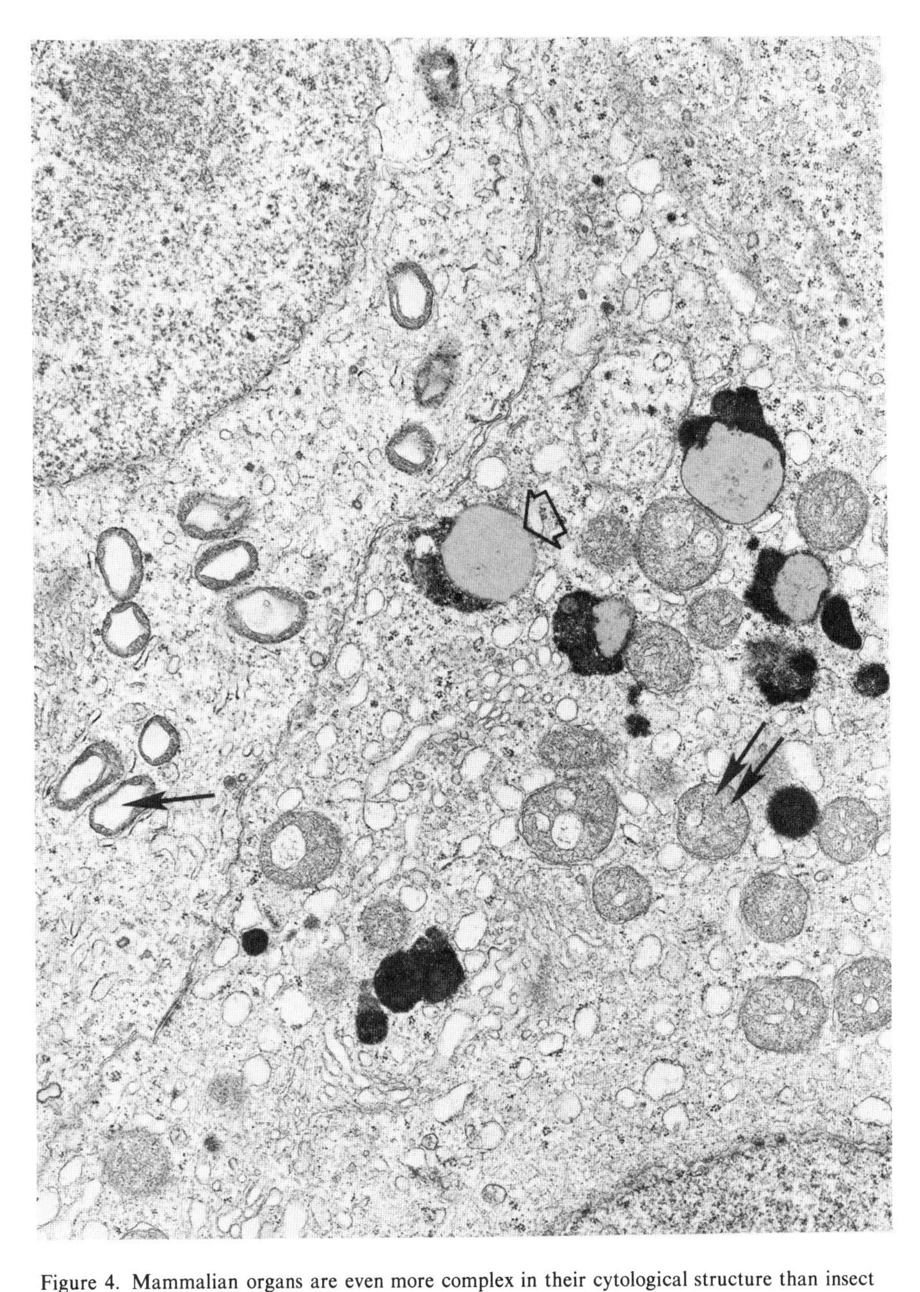

Figure 4. Mammalian organs are even more complex in their cytological structure than insect organs. This is exemplified by this electron micrograph of the testis of a 4-month-old mouse, showing a spermatogonium on the left and a Sertoli cell on the right. Whereas the spermatogonia replicate very rapidly, even in old animals, the Sertoli cells are irreversibly differentiated and do not divide in the adult animal. Notice the "empty" appearance of the mitochondria of the spermatogonium (single arrow) in contrast to the more electron-dense configuration of the Sertoli-cell mitochondria (double arrows). This last cell also contains lipofuscin granules (hollow arrow) similar to those found in old nerve cells. ×18,200.

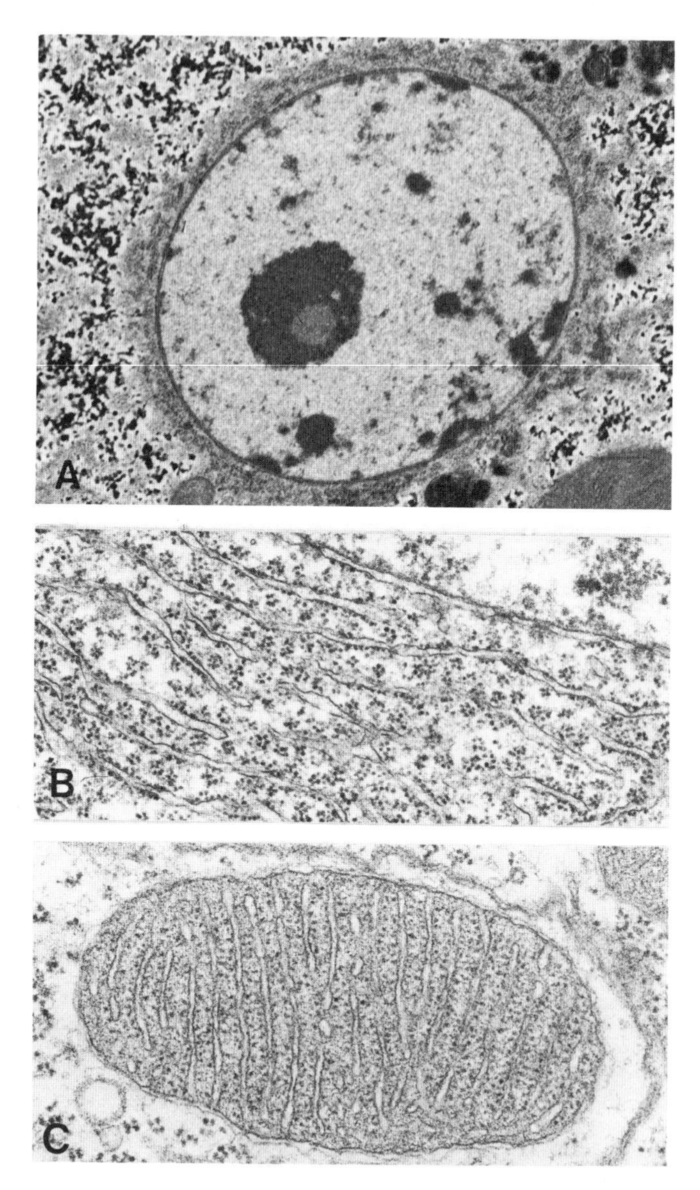

Figure 5. Among the most important subsystems of the cell are the nucleus (A), that is the main repository of the cell genome and the source of the informational macromolecules that control most synthetic activities of the cytoplasm, the protein-synthesizing granular endoplasmic reticulum (B), and the mitochondrion (C), a semiautonomous energy-producing organelle containing its own genome. [The cell source and magnification are as follows: A, *Drosophila* fat body, $\times$ 45,000; B, neuron from the lateral vestibular nucleus of the rat, $\times$30,780; C, neuron from the cephalic ganglionic center ("brain") of *Drosophila,* $\times$38,500.]

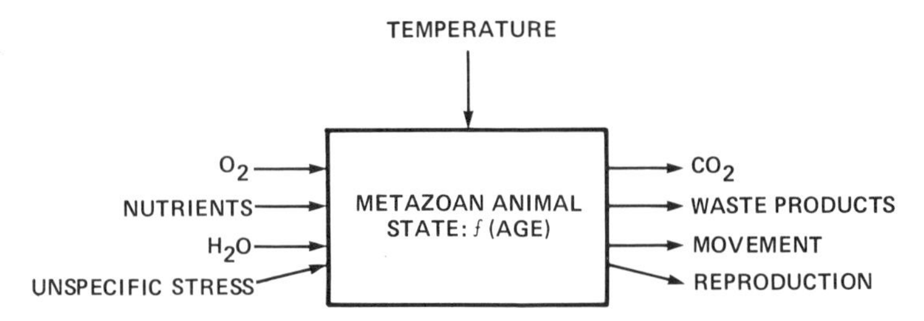

Figure 6. A simplified diagram of an animal as a black box in thermodynamic interaction with the energy and matter of the environment. According to the rate-of-living concept of aging, the structure and function of the organism deteriorate with age at a rate depending on the rate of exchange with several parameters of the environment. Moreover, in agreement with the principles of entropy accumulation in open (or dissipative) structures, the flow of energy and matter through the organism may result in irreversible disorder (i.e., structural disorganization) in the subsystems indicated in the above figures, namely, organs, cells, and subcellular organelles (Miquel *et al.*, 1979).

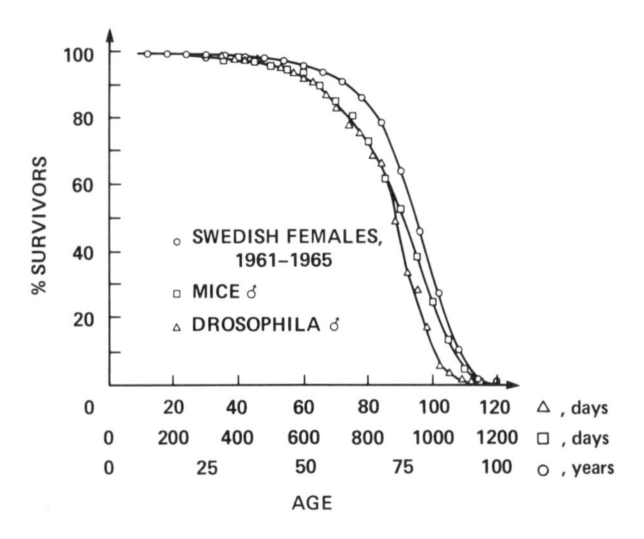

Figure 7. There is a remarkable similarity in the mortality curves of different species for populations maintained in optimum environments. It is interesting that, despite the considerable differences in median and maximum life span, the force of mortality accelerates in all species with the passage of time up to an inflection point, after which there is a progressive slowing-down of the force of mortality.

In contrast to body weight, which is one of the most easily measured physiological variables, the "inherent vitality" of a particular member of a population (that determines its life span in an optimum environment) cannot be assessed accurately by current gerontological methods. This is unfortunate, since it is likely that animals differ in their vitality at least as much as they differ in their body weight. As shown in Fig. 9, when the frequency of deaths of a mammalian population is plotted in percent of deaths per time interval, the histogram obtained shows an asymmetric distribution similar to that of the body weights (Fig. 8). Thus, we may tentatively conclude that mortality kinetics and distribution of body weight, as well as other physiological parameters, reflect the initial heterogeneity of the population, which is composed of a majority of individuals whose parameters cluster around a central value and of progressively less numerous subgroups which increasingly deviate from the central value. Thus, in contrast to previous concepts that stress the influence of random environmental factors on *aging adults* as the main determinant of mortality kinetics, we propose that mortality kinetics reflect, at least in part, the phenotypic diversity resulting from random environmental influences which modulate the genome-programmed developmental process. In other words, the main factors responsible for the striking differences in life span shown by the members of a particular animal population living in an optimum environment exert their influence on the very maleable immature individuals rather than on the adults.

In a systems-thermodynamic context, both the mortality kinetics of an aging population and the age-related shifts in the frequency distribution of various physiological parameters indicate that the flow of energy and matter, schematically represented by Fig. 6, results in a degradation of the physiological mechanisms controlling homeostasis, leading to a degree of disorganization incompatible with maintenance of the living state, i.e., death—the final entropic event.

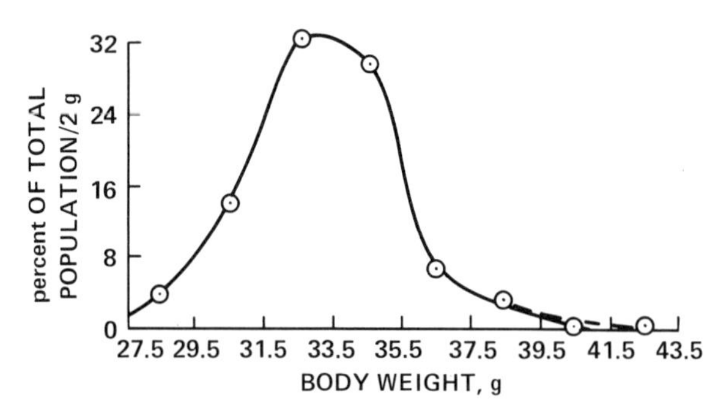

Figure 8. Frequency distribution of body weight of 126 male mice showing the heterogeneous character of the animals as regards this important physiological parameter (Economos and Miquel, 1980).

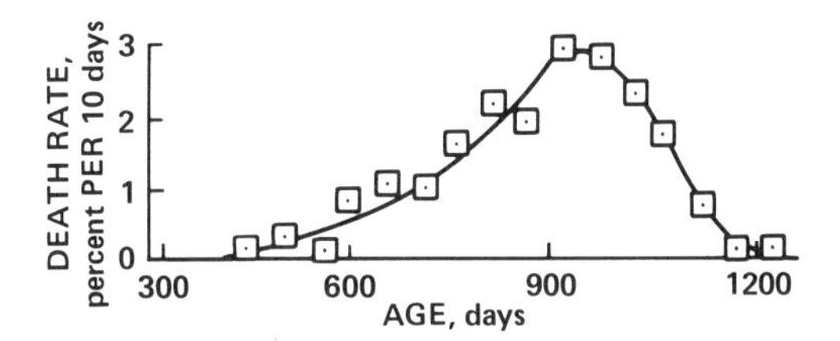

Figure 9. When mortality kinetics is plotted as percent of mice dying per time interval instead of the usual procedure (Fig. 7), the similarity with the body weight distribution for mice of the same C57BL strain becomes apparent. This is compatible with our hypothesis that mortality kinetics is to a considerable degree a reflection of the heterogeneity of populations.

3. ORGANISM–ENVIRONMENT INTERACTION CAUSING SENESCENCE

With some notable exceptions (Harman, 1956; Gerschman, 1964; Strehler, 1967; Sacher, 1967), the current concern with "intrinsic" molecular mechanisms of senescence has resulted in a relative disregard for the essential role of the environment in triggering age-related biological disorganization. Although the expression "intrinsic aging" is often used as synonymous with "programmed" or "fundamental" aging (as opposed to "environmentally induced life shortening"), the present authors feel that fundamental aging changes seen in biological organisms are the consequence of organism–environment interaction and, therefore, there is no "intrinsic aging." According to this environmental–biology view of aging (Miquel *et al.*, 1979), the unraveling of the specific mechanisms through which the environment causes senescent disorganization may be a more rational approach than the search for "intrinsic aging clocks" and "self-destruction devices." Thus, it seems appropriate to focus on the main environmental factors which are essential for the preservation of the living state and simultaneously take a toll, gradually disorganizing the systems of the adult animal until life can no longer be maintained.

Among the factors listed in Fig. 6, temperature, oxygen, and nutrients are known to influence the rate of metazoan aging. The most extensive studies, starting with the classic research by Loeb and Northrop (1917), Pearl (1928), and Alpatov and Pearl (1929), have been performed on the poikilotherm *Drosophila melanogaster.* Although some contradictory results have been reported, the majority of the recent studies on both *Drosophila* (Miquel *et al.*, 1976) and house flies (Sohal, 1982) suggest that, within the normal temperature range to which the flies are adapted, life span is inversely proportional to ambient temperature. Moreover, the longer survival of poikilothermic vertebrates such as fish at lower temperatures is also due to a decreased rate of aging, as in the above-mentioned insect experiments (see review on experimental life span prolongation by Sacher, 1977).

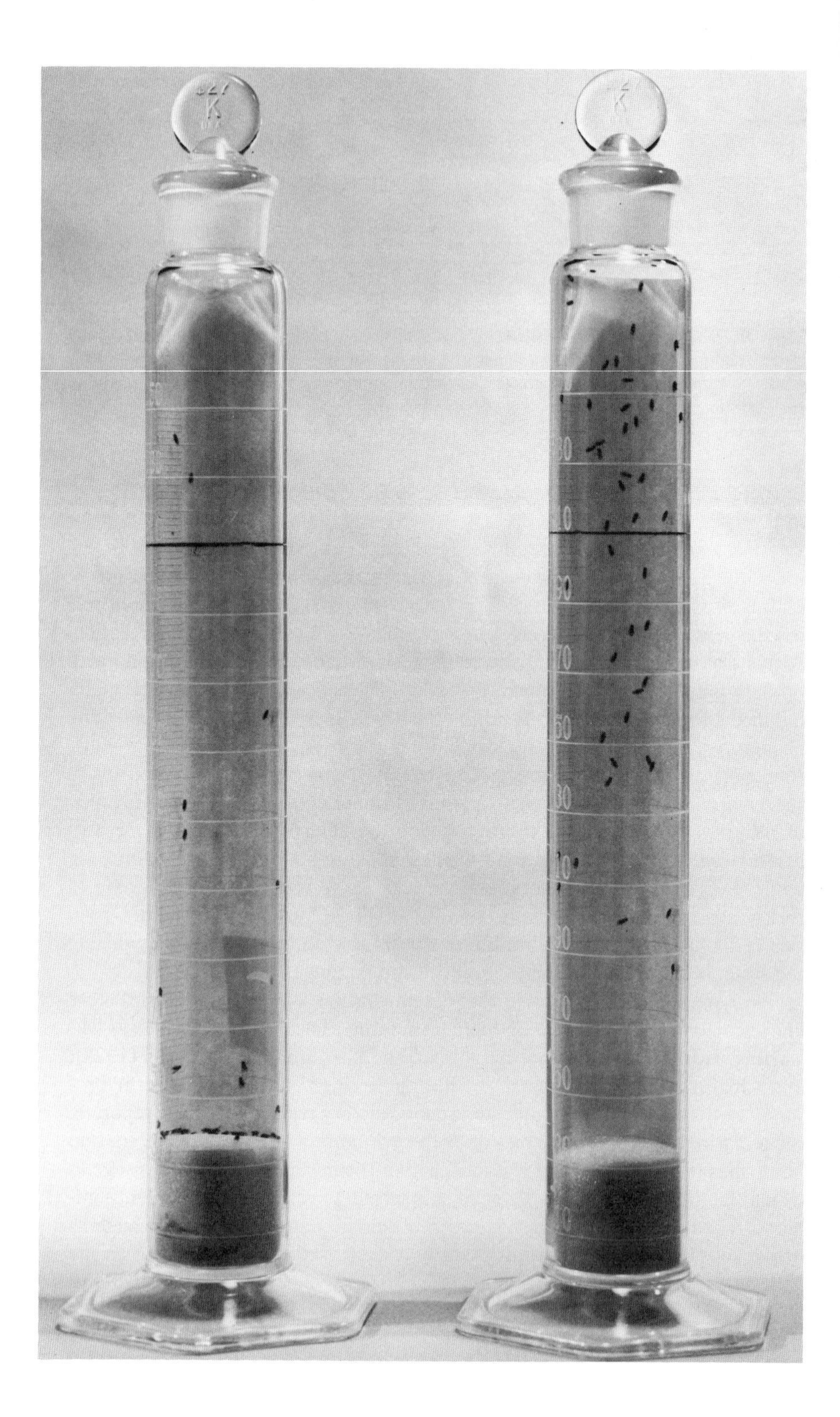

Although it has been suggested that thermal denaturation of proteins may be the cause of the accelerated aging of animals maintained at higher temperatures and of organismic aging in general (Rosenberg *et al.*, 1973; Atlan *et al.*, 1976), thermal denaturation reactions, because of their high activation energies, are not likely to play a main role in the genesis of senescence (Strehler, 1977). As already shown by Loeb and Northrop in 1917, the temperature dependence of the aging process is more comparable to that of enzyme-catalyzed reactions (an activation energy of about 12–21 kcal) than to protein denaturation reactions (60 kcal activation energy).

In agreement with the above, our *Drosophila* research suggests that the rate of aerobic metabolism rather than temperature per se is the main pacemaker of poikilotherm aging. In flies exposed to different temperatures (Miquel *et al.*, 1976) or fed antioxidants (Economos *et al.*, 1980, 1982; Miquel *et al*, 1982) there was a negative linear correlation between mean life span and metabolic rate. Interestingly, despite the considerable differences in the rate of respiration of flies living at different temperatures, we found that flies tended to use an approximately constant amount (4–5 ml per fly) of O_2 throughout their adult life (Miquel *et al.*, 1976).

The data on the effects of caloric restriction on the aging process of fish and mammals (see review by Sacher, 1977) are also consistent with the concept that metabolic rate influences the rate of aging and that, in agreement with the classic studies by Rubner (1908), life span correlates negatively with the rate of energy expenditure.

The specific mechanisms by which metabolic rate or, more specifically, oxygen consumption, may induce aging will be discussed in the following sections, which deal with senescence at progressively lower levels of biological organization.

4. PHYSIOLOGICAL DECLINE AND CELL AGING

It is well known that numerous physiological functions of experimental animals (Figs. 10–12) and humans show age-related losses before the final functional failure of the organism resulting in death. A detailed description of these manifestations of aging at the physiological-system level can be found in the review by Shock (1970) and a monograph edited by Masoro (1981).

It is interesting from a systems-gerontological viewpoint (that is concerned with the search for rate-limiting senescent processes) that parameters

←

Figure 10. Determination of "vitality" (physiological competence) in fruit flies by the negative geotaxis test (Miquel *et al.*, 1976). When excited by repeated shaking of the cylinder against a rubber mat placed on the table, most old flies (left) stay on the bottom, while young flies (right) rapidly reach the top of the cylinder.

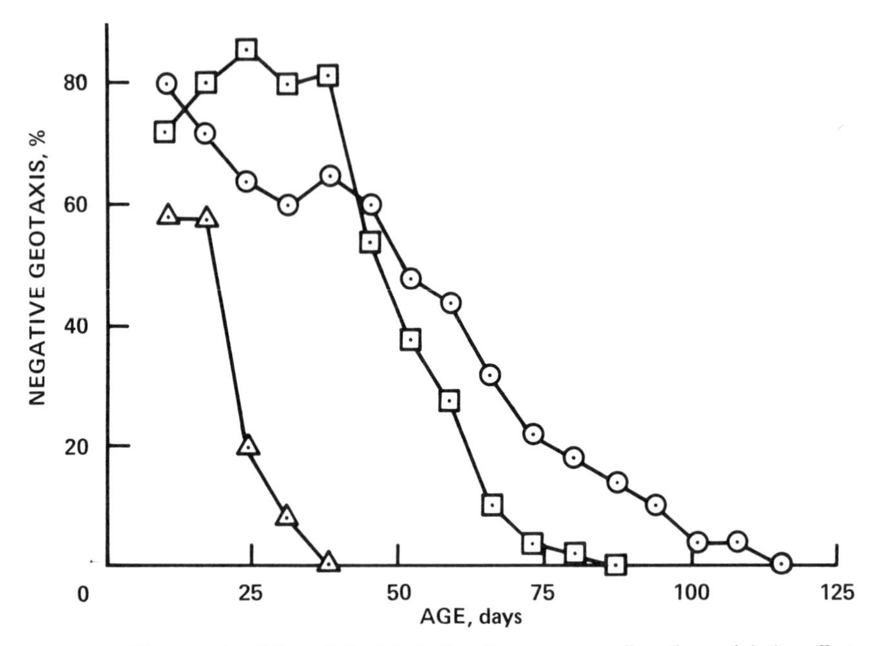

Figure 11. The age-related loss of physiological performance as well as the modulating effects of the environment on this loss are illustrated in this study of negative geotaxis in fruit flies kept at several ambient temperatures. The points represent the percentage of flies that reached the 250-ml line in a volumetric cylinder 20 sec after the flies were shaken to the bottom. Note the slightly higher vitality of the 21°C flies (○) as compared with the 18°C flies (⊙), up to approximately 40 days of age, and the earlier fall to zero values in the 21°C flies. On the other hand, exposure to 27°C (△) resulted in drastically reduced vitality over the whole life span. These results are in agreement with the predictions of the rate-of-living theory proposed by Pearl (1928) and with the concept that the rate of flow of energy (heat) or matter (ultimately oxygen, since, in flies and other poikilotherms, higher ambient temperatures result in higher oxygen utilization) control the rate of entropy accumulation in the organism. (Negative geotaxis curves from Miquel *et al.*, 1976.)

of integrated neuromuscular function such as the negative geotaxis of fruit flies (Figs. 10, 11) and the neuromuscular coordination and vigor of mice (Fig. 12) show marked age-related decrements. Similarly, human aging results in a slowing of reflex responses and poorer muscular coordination. This special vulnerability of neuromuscular functions to aging is probably related to the fact that, in contrast to relatively nonaging systems such as the digestive apparatus and the bone marrow (which contain actively dividing cells), the nervous and muscular tissues contain great amounts of differentiated cells which have lost the ability to regenerate through the process of growth followed by division.

Previous research on the two models of metazoan aging used in our lab-

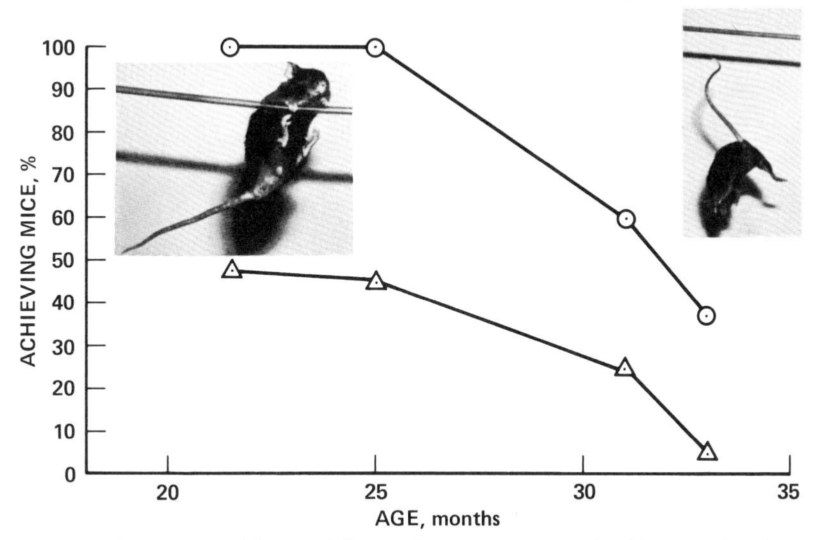

Figure 12. When mice are left on a tightrope they try to move to the sides, grasping the rope with the four legs and the tail. The percent of "achieving" mice (able to reach a side pole in less than one minute) decreases with age, since senescent mice are unable to hold on the rope for longer than a few seconds. The lower curve shows the age-related decline of this parameter of neuromuscular coordination in control mice. The upper curve obtained for mice of the same age fed thioproline shows that it is possible to delay physiological decline by administration of this antioxidant (Miquel and Blasco, 1978; Miquel and Economos, 1979; Economos *et al.*, 1981, 1982).

oratory, i.e., the fruit fly and the mouse, has shown that, while *Drosophila* sustains a striking cellular degeneration in all its organs (Miquel, 1971; Miquel *et al.*, 1979a,b, 1981), the organs of old mice often contain side by side aging and nonaging cells (Fig. 4). The adult fruit fly does not have any dividing cells in its tissues. Therefore, the intensity of the age-related disorganization of its cells (that, as noted above, is accompanied by fast physiological decline and short life span) is in agreement with Minot's (1907) concept that the main source of aging and death in metazoan animals is cell differentiation with concomitant loss of the ability to regenerate the tissues through the process of growth and cell division. It is also central to Minot's view on the causes of cellular aging that a change in the nucleo–cytoplasmic ratio is an important index of senescence. He notes that factors which produce cell rejuvenation cause an increase in the relative size of the nucleus, while aging is the result of an increase in the cytoplasmic mass relative to the nucleus. Our comparative fine structural studies on cell aging in insects and mammals (Miquel *et al.*, 1981) have provided data in agreement with Minot's views, as illustrated by the electron micrographs in Figs. 6, 13, and 14.

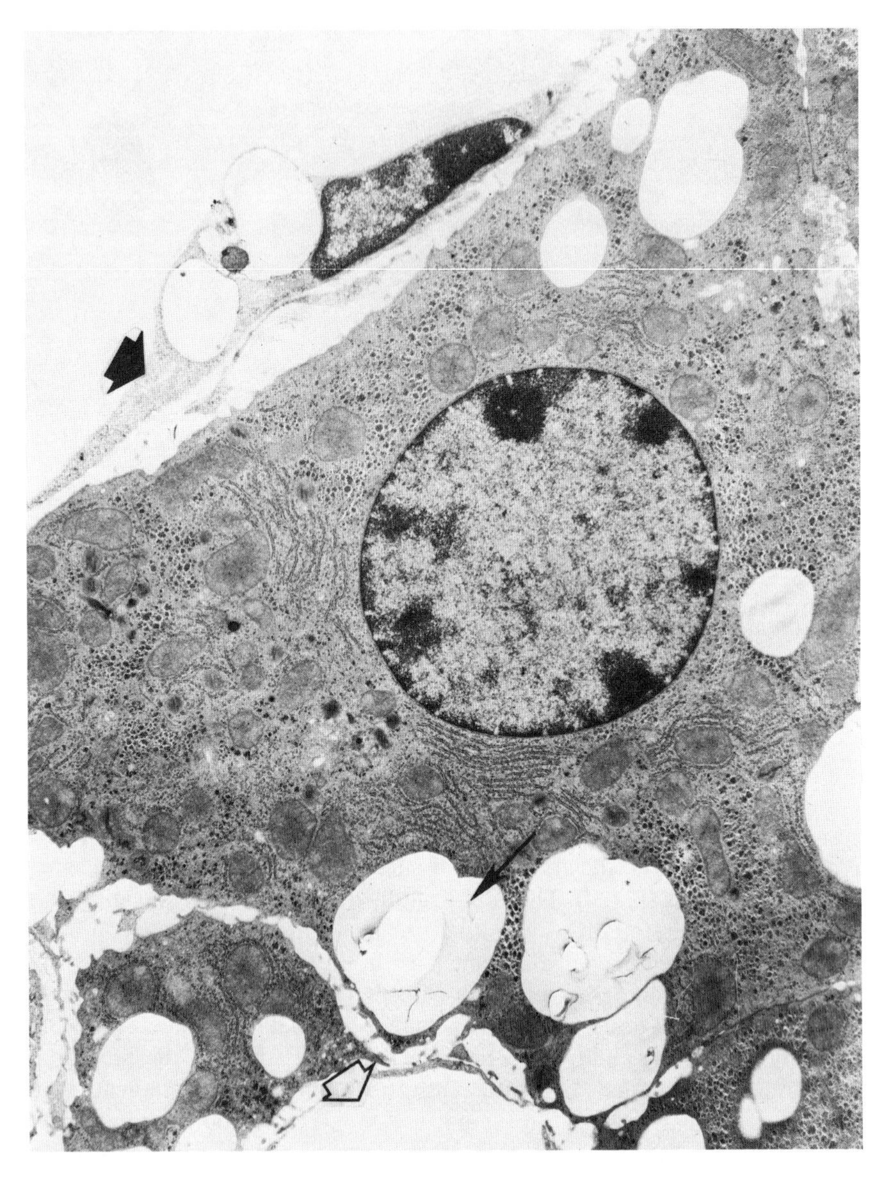

Figure 13. This electron micrograph illustrates the characteristic large nucleocytoplasmic ratio of mammalian cells which have not lost the ability to replicate. The larger cell is a mouse hepatocyte showing pinocytotic activity (arrow) which is in agreement with the "open" thermodynamic character of dividing cells. The thick arrow points out another type of replicating cell (Kupfer cell), which is part of the sinusoidal endothelium. ×7500.

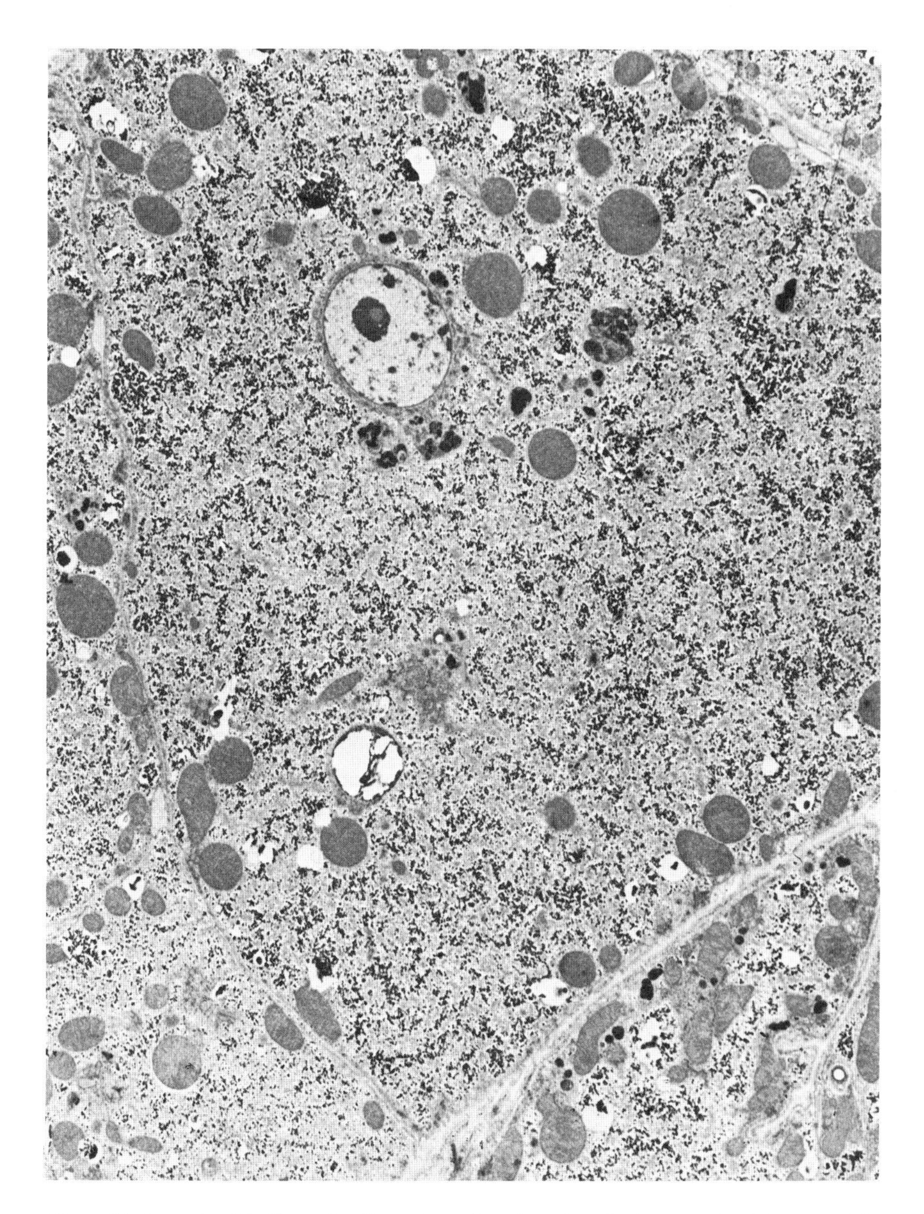

Figure 14. In contrast to the replicating nature of the hepatic cells of mammals, this *Drosophila* cell from the liverlike fat body of *Drosophila* is irreversibly differentiated and shows the nucleo-cytoplasmic ratio typical of fixed postmitotic cells. Our studies reviewed elsewhere (Miquel *et al.*, 1981) show that, while mammalian-liver cells experience minimal senescent injury, the fat body cells of old *Drosophila* contain degenerating mitochondria, in addition to aging changes. ×7000.

In apparent contrast with the above concepts on the key role of terminally differentiated cells as the pacemakers of physiological aging, the lung and blood vessels show striking age-related functional decrements, despite the fact that they do not contain fixed postmitotic cells. As discussed in detail by Kohn (1977), the fast tempo of the senescent disorganization of the lung and blood vessels is the result of the increased rigidity of its elastic components. The mechanical properties of lungs and blood vessels are dependent on the rubber-like protein elastin. Biochemical studies have shown that this extracellular macromolecule is subjected to progressive cross-linkage and fragmentation with age.

The changes in elastin and the progressive cross-linking of other extracellular biopolymers such as collagen are one of the few universal manifestations of mammalian senescence, which has led to the suggestion that the primary aging change does not occur in the terminally differentiated cells, as postulated by Minot (1907), but in the extracellular compartment. The concept of a senescent increase in tissue rigidity, proposed by Leonardo da Vinci on the basis of his dissection studies, is in agreement with a number of experimental observations (see review by Kohn, 1977). As maintained by the proponents of the collagen theory of aging, the age-related increase in rigidity of the extracellular materials would hinder the function of the lungs and blood vessels. This, in turn, would result in poor perfusion of organs and inadequate transfer of nutrients and waste between blood capillaries and cells.

In our opinion, although it is undeniable that extracellular aging contributes a great deal to the senescent involution of many metazoan organs, including those of the cardiovascular system, it is not a common denominator of metazoan aging, since many free living protozoa show aging that is totally unrelated to any extracellular material. Moreover, despite the fact that the mammalian brain is almost totally lacking collagen and is very well perfused even in old age, there is considerable evidence that its nonreplaceable neurons experience age-related changes such as ribosomal loss (Miquel and Johnson, 1978) and accumulation of lipofuscin (Miquel *et al.,* 1977).

The observation that human fibroblasts maintained in culture have a limited replication potential was interpreted as experimental proof that cell aging may be related to an exhaustion of the mitotic potential of replicating cell pools (Hayflick, 1965). However, Strehler (1979) concluded from his extensive review of *in vitro* cell aging that "There is no compelling reason to believe that clonal senescence (loss of division potential) is an important cause of organismic aging and no reason to believe that the phenomenon occurs in the important self-replenishing cell lines, such as epidermis, gut lining, and blood forming tissues." This was recently admitted by Hayflick (1979), who stated that "the likelihood that animals age because one or more important cell populations lose their proliferative capacity is unlikely."

Recent light-microscopic studies suggest that the loss of proliferative capacity of fibroblasts *in vitro* may be a manifestation of differentiation instead of *direct* aging of dividing cells. In the words of Martin *et al.* (1977): "such cultures undergo a sort of terminal differentiation *in vitro* analogous to the kind of terminal differentiation one observes with many types of stem cells *in vivo* such as haematopoietic cells or myoblasts." This concept is supported by the electron microscopic finding that cells *in vitro* often show the fine structural characteristics of differentiating cells (Johnson, 1979) and by biochemical studies demonstrating that cytochrome oxidase (which is a marker enzyme for the inner mitochondrial membrane) increases its specific activity by 300% in WI-38 fibroblasts aging *in vitro* (Sun *et al.,* 1975). Since it is known that *in vivo* cell differentiation is accompanied by an increase in respiratory enzymes, this observation by Sun *et al.* (1975) supports the hypothesis that cells senesce *in vitro* as a result of differentiation. Thus, paradoxically, *in vivo* cell aging in nematodes and insects (which contain only terminally differentiated cells in their tissues) may be a simpler model for elucidation of the fundamental mechanisms of cell aging than the popular cell culture model (Figs. 3, 15, 16). (For review of recent experiments on *in vitro* aging studies, see Kelley and Vogel as well as Pool and Metter; this volume.)

In summary, all available information on cell aging both *in vitro* and *in vivo* seems to be in agreement with Minot's view (1907) that aging is "the price" paid for cell differentiation or, in a thermodynamic framework, the result of an inability of differentiated nonreplaceable cells to dissipate entropy with a 100% effectiveness. Since all metazoan organs contain terminally differentiated cells, primary senescent change in this cell type could underly all manifestations of aging, including the decrease in number and slower turnover of dividing cells *in vivo* which, as discussed elsewhere (Miquel *et al.,* 1979) are the probable result of humoral changes linked with extracellular and differentiated-cell aging.

5. DISORGANIZING MOLECULAR PROCESSES IN THE SUBCELLULAR ORGANELLES

In agreement with the aforementioned concepts, a considerable number of studies reviewed elsewhere (Miquel *et al.,* 1977, 1978, 1979b, 1981) suggest that senescence may result in fine structural and biochemical changes in the nuclei, ribosomes, and mitochondria of fixed postmitotic cells. Moreover, it seems that aging is associated with changes in the cytoplasmic membrane (Zs.-Nagy, 1979). This universality of organelle senescent change makes imperative the use of a systems analysis approach in order to discriminate between primary and secondary aging processes at the organelle and molecular levels.

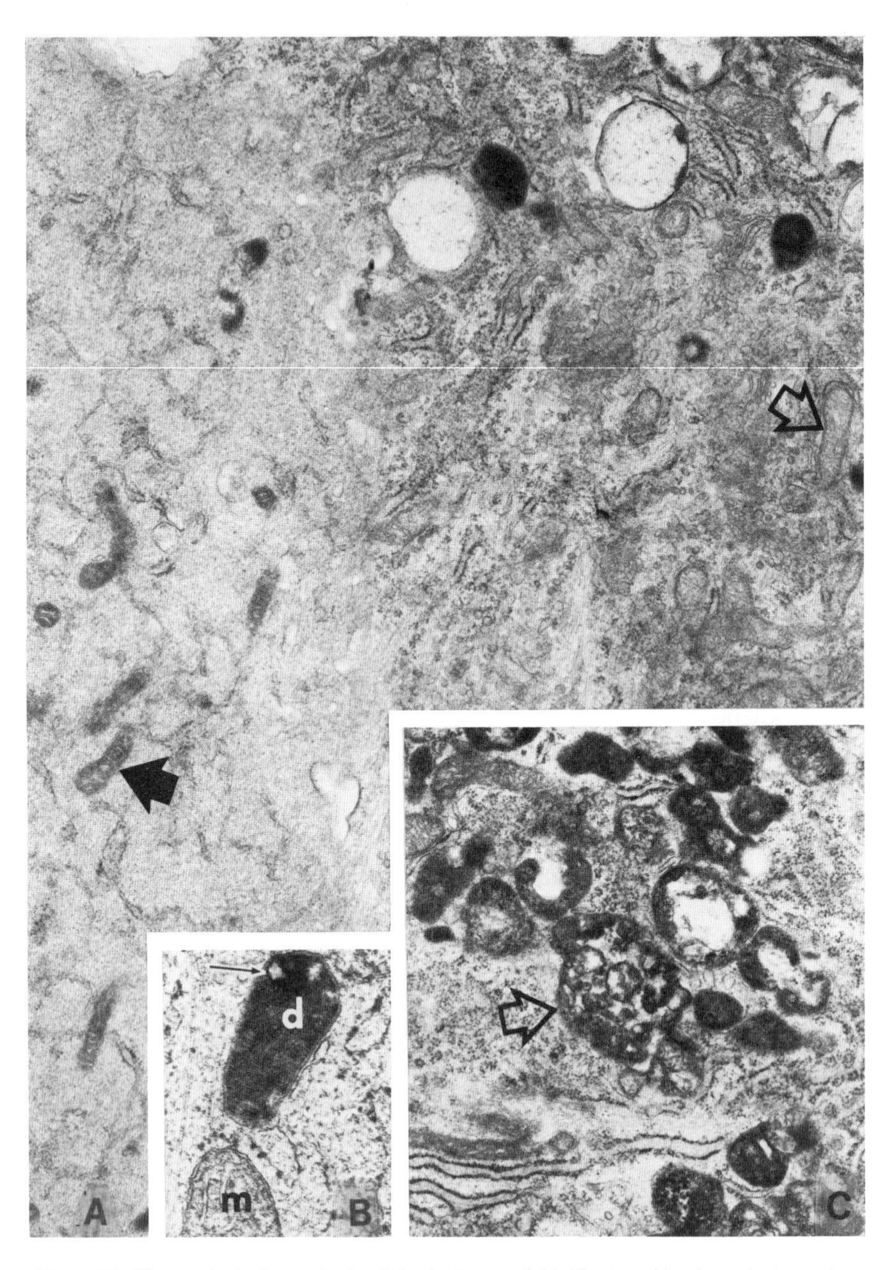

Figure 15. The cytological complexity of the *in vitro* model is illustrated by these electron micrographs from a population of IMR-90 human fetal lung fibroblasts at the 49th population doubling level (PDL). In (A), two cells (boundaries indistinct) are seen, each with a different type of mitochondria. One type is osmiophilic and thin (solid arrow) while the other is wider and less dense (hollow arrow). At later PDL, more cells are found to contain the dense mitochondria. In (B), a mitochondrion (m) lies next to a dense body (d) whose mitochondrial origin is suggested by the presence of a double membrane (arrow). In (C) a cell contains numerous mature dense bodies (arrow). (×19,200; modified from Johnson, 1979).

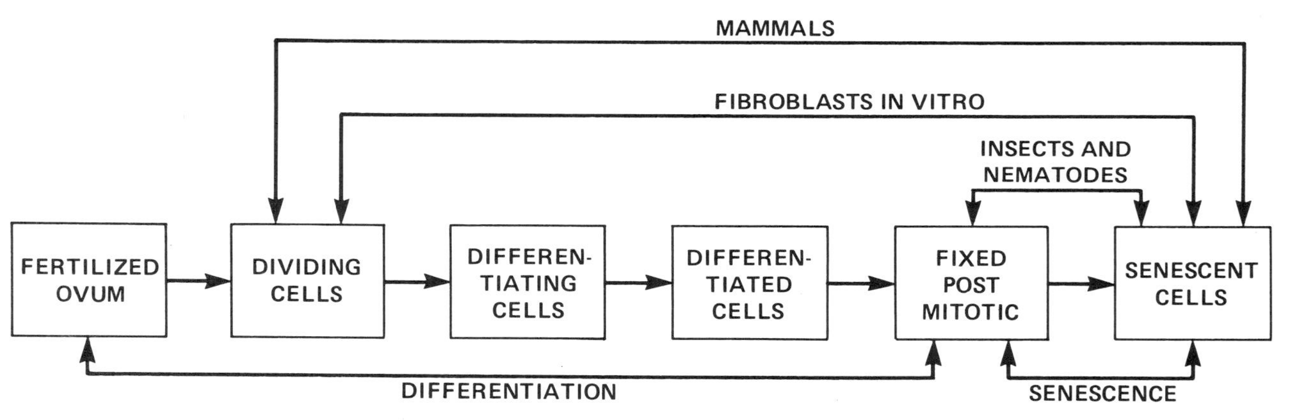

Figure 16. This diagram points out that popular models for study of animal aging can be ranged, in increasing order of cytological complexity, as follows: (1) lower animals such as insects and nematodes, which only contain in their somatic tissues irreversibly differentiated (fixed postmitotic) cells; (2) dividing and differentiating fibroblasts maintained *in vitro* under standard culture conditions; (3) mammals. Insects, such as fruit flies and house flies, show a normal aging process that is quite similar at the fine structural and biochemical levels to that of the mammalian differentiated cells (neurons, Sertoli cells, etc.). Therefore a parsimonious explanation of metazoan cell aging is that, as postulated by Minot (1907), "senescence is the price paid for differentiation." In agreement with this concept, age-related changes seen in the dividing cell pools of mammals are probably the result of more fundamental changes occurring in the differentiated cells and in the extracellular tissues (see Miquel *et al.*, 1979).

In view of the key role of the nuclear genome in controlling structure and function, senescent changes in cytoplasmic organelles could very well be the end result of a more fundamental lesion in the informational macromolecules of the nucleus. Since intensive research has failed to demonstrate the presence of freshly synthesized abnormal proteins in old organisms or late passage cell cultures (Rothstein, 1981), the postulated genome injury would be quantitative (exhaustion) rather than qualitative (mutation). Thus, the fundamental senescent change might be a decrease in redundant or metabolic DNA (Medvedev, 1972; Cutler, 1978). Since metabolic DNA consists of extra copies of those genes which are active in order to support the function of a specialized cell, the apparent decrease in hybridizable ribosomal DNA of old fixed postmitotic cells (Strehler and Chang, 1979; Strehler *et al.*, 1979) could have serious implications for the "rejuvenation" of the cytoplasmic organelles through the process of organelle replication and turnover of their macromolecules. According to Cutler (1978), the specific biochemical mechanism responsible for this postulated inactivation of the nuclear DNA with age would be cross-linking reactions involving free radicals. In his opinion, although free radicals are known to be associated with mitochondrial reactions and other energy-generating reactions present in the cells, these reactions are fairly well contained and, therefore, " . . . the free radicals that are likely to be of the most damaging type are those free to diffuse throughout the cell. Many of these radicals are probably initiated by oxygen." Random reactions by free radicals or by molecular oxygen have also been proposed by Tas (1976, 1978) as the mechanism responsible for the age-related increase in disulfide bonds that results in a more compact state of the nuclear chromatin.

In our view (Miquel *et al.*, 1980), the concept that random free radical injury to the informational macromolecules of the nucleus plays the primary role in the senescence of fixed postmitotic cells is hard to reconcile with the resistance of these cells to very high doses of ionizing radiation (Zeman *et al.*, 1959; Miquel and Haymaker, 1965; Atlan *et al.*, 1970). This type of radiation is known to release great amounts of free radicals from water radiolysis (Bileski and Gebicki, 1977). Furthermore, random free radicals from extramitochondrial sources are less likely to produce injury than intramitochondrially produced free radicals, in view of the effectiveness of the peroxide-detoxifying mechanisms of the extramitochondrial microsomes and peroxisomes (Chance *et al.*, 1979). By contrast, the detoxifying mechanisms of mitochondria may be relatively inefficient, as suggested by the observation that potentially injurious oxygen species, including superoxide and H_2O_2, exist in tightly coupled mitochondria from rat, pigeon, and beef heart at concentrations that depend on the metabolic state of the organelles (Flohe *et al.*, 1977; Chance *et al.*, 1979). The intramitochondrial presence of superoxide is of great significance to mechanisms of aging because, as discussed by Fridovich (1975), this oxygen radical is very reactive and able to injure the mitochondrial membranes. Moreover, the superoxide radical has the capacity to generate other reactive oxygen species such as the hydroxyl radical and singlet molecular oxygen which, in turn, initiate chain reactions leading to extensive lipid peroxidation (Harman, 1956; Tappel, 1965, 1973; Pryor, 1976; see also monograph by Puig-Musset, 1976).

The fact that, according to Nohl and Hegner (1978), submitochondrial particles from the heart of 24-month-old rats produce superoxide at a higher rate than those of the heart of 3-month-old rats, suggests that disorganizing free radical chain reactions may exceed the homeostatic protection of the mitochondria in aging animals.

In view of the above, it is suggested that the mitochondria rather than the nuclei may be the primary target in the chain of subcellular aging changes. Further, in contrast to previous concepts that emphasize the role of membrane injury in mitochondrial senescence, we maintain that free radical injury to the mitochondrial genome plays the primary role (Fig. 17). This hypothesis of respiration-linked peroxidative injury to mtDNA (Economos *et al.*, 1980; Miquel *et al.*, 1980; Fleming *et al.*, 1982) is supported by recent observations on the relative inefficiency of the mtDNA repair mechanisms, as compared to the mechanisms protecting the nuclear genome (Fukanaga and Yielding, 1979). Moreover, it seems that DNA synthesis occurs on the inner mitochondrial membrane and that inner membrane–DNA complexes can synthesize DNA *in vivo* (Shearman and Kalf, 1975; Nass and Nass, 1964). It has also been found that all the enzymes necessary for the replication of mtDNA are attached to

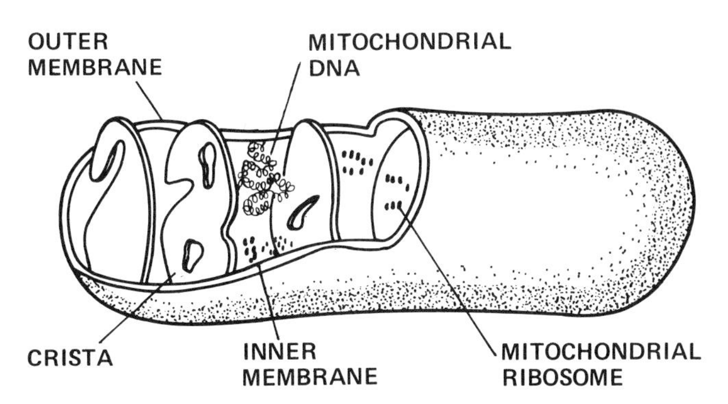

Figure 17. Diagram of a "generalized" mitochondrion showing the respective locations of the mitochondrial DNA and the free radical-generating inner membrane. Since the mitochondria of fixed postmitotic cells generate high amounts of disorganizing oxygen radicals in order to produce energy through respiration, we have proposed that free radical injury to the mitochondrial genome (with resulting loss of the ability to regenerate the mitochondrial population by normal organelle replication) may be the primary cause of cell aging (Economos *et al.*, 1980; Miquel *et al.*, 1980; Fleming *et al.*, 1982).

the inner mitochondrial membrane at or near the origin of replication (Albring *et al.*, 1977).

In view of the close spacial relationship of the mtDNA with the inner membrane, which is the main cellular site of release of disorganizing free radicals, it is likely that mtDNA is exposed to continuous peroxidative attack leading to cross-linking reactions. The result might be deletion errors in transcription of the mitochondrial genome because of polymerization of the mtDNA (Fleming *et al.*, 1982). Mitochondrial division may thus be inhibited with concomitant loss of the ability to rejuvenate the mitochondrial population and eventual decrease in the production of energy. This concept is supported by numerous observations on age-related decline in the number of mitochondria of old differentiated cells (Miquel *et al.*, 1980), senescent decrease in the spe-

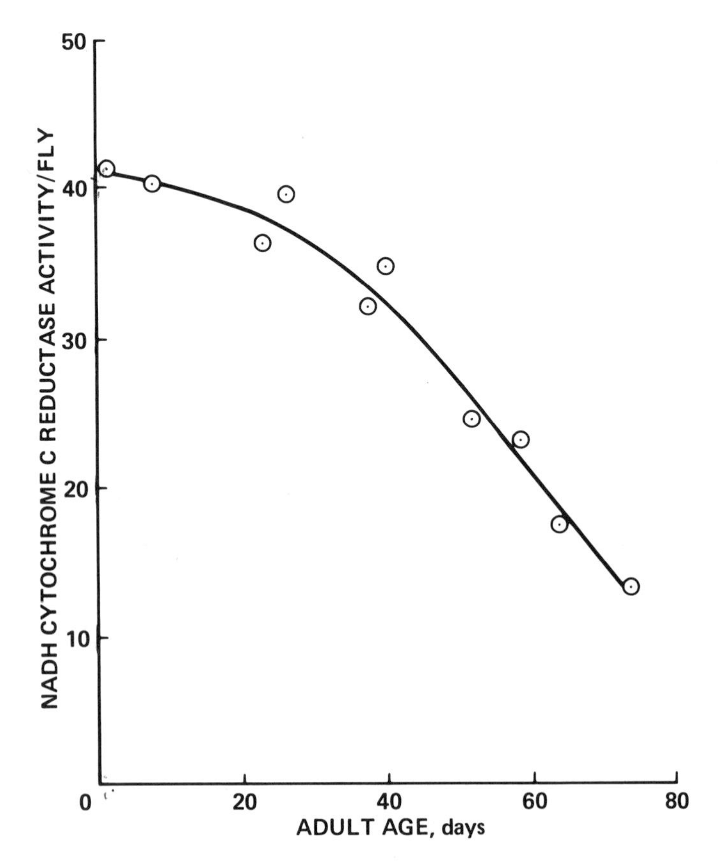

Figure 18. Our concept on the key role of mitochondrial senescence in cell and organismic aging is supported by the observation of a striking age-related loss in activity of the respiratory enzyme cytochrome-c-reductase. (Courtesy of Dr. Stephen F. Cottrell.)

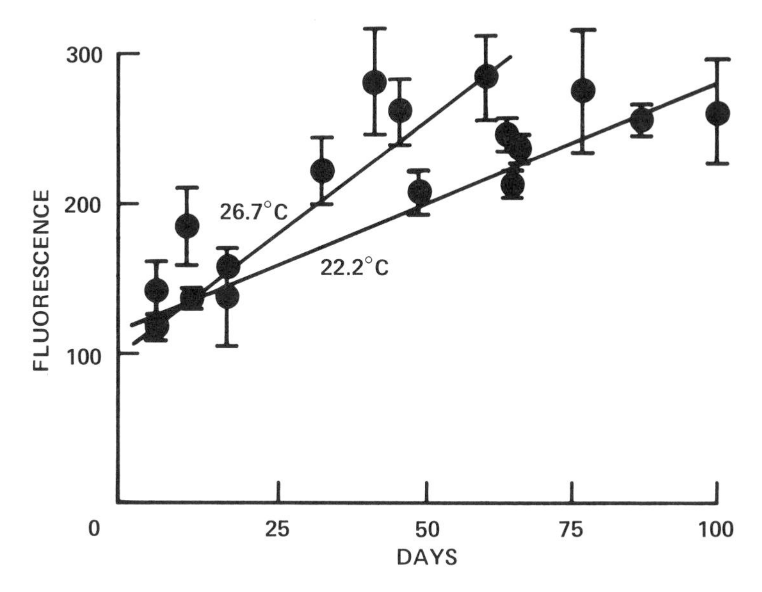

Figure 19. In parallel with the age-related loss of organization (negentropy) suggested by structural disorganization and by decreased enzyme activity, aging flies show an increase in inert material (entropy), such as the age pigment lipofuscin. As is the case for function (negative geotaxis) loss, the rate of entropy gain may be influenced by environmental parameters. In this particular study performed by Sheldahl and Tappel (1974) on flies from our laboratory, the accumulation of age pigment was faster at near 27°C than at near 22°C.

cific activity of some respiratory enzymes of the mitochondrial inner membrane (Fig. 18), accumulation of peroxidized debris from mitochondrial membranes (Fig. 19), and loss of mtDNA in old cells (Massie *et al.*, 1975, 1981).

As pointed out in Fig. 20, the postulated injury to mtDNA could result, through its effects on ATP synthesis, in reduction of protein synthesis and decline in physiological performance. Thus, our concept of primary free radical injury to mitochondrial DNA offers a possible mechanism for the senescent changes found at the higher levels of biological organization discussed in the preceding section of this chapter.

6. CONCLUSIONS

Our preliminary systems analysis of the aging process suggests that it is possible to develop a parsimonious explanation of metazoan aging by linking together, in a cause–effect, rate-limiting senescent changes at several levels of biological organization.

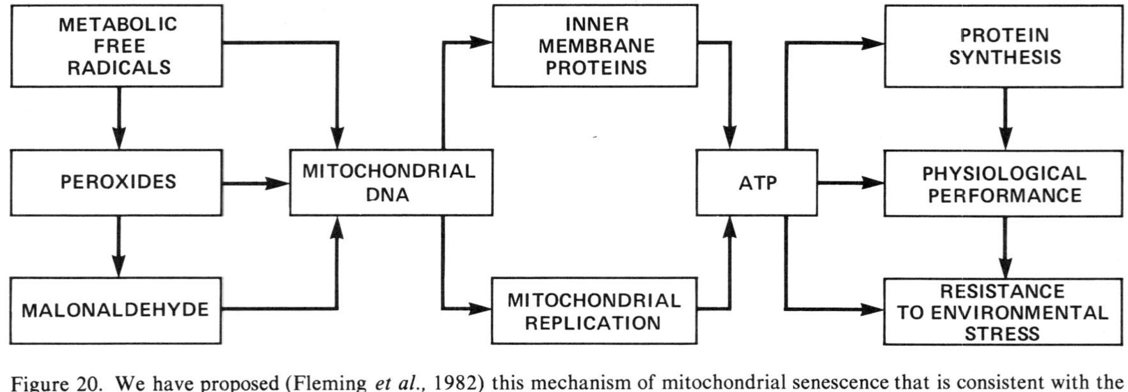

Figure 20. We have proposed (Fleming *et al.*, 1982) this mechanism of mitochondrial senescence that is consistent with the finding that free radicals, peroxides, and malonaldehyde (which are capable of mutating or inactivating DNA) are generated in the mitochondrial inner membrane. An accumulation of unrepaired injury to the mitochondrial genome is likely to lead to inadequate synthesis of inner membrane (hydrophobic) proteins and interference with mitochondrial division. The resulting decline in ATP synthesis would set up a vicious circle of mitochondrial loss and concomitant declines in energy production, protein synthesis, and physiological performance.

Our systems view, which includes the often forgotten interaction of the organism with its environment, calls attention to the entropic changes occurring in aging organisms. As pointed out by Schrödinger (1947), entropy is synonymous with disorder, and, as summarized above, there is ample evidence of increasing disorder with age at each level of organization. This makes advisable an analysis of aging in the framework of the principles enunciated by Prigogine (1947) for the study of irreversible phenomena in open systems. As discussed above, living organisms can be considered open systems, that is systems which exchange both matter and energy with the environment (Fig. 6), and, therefore, are typical examples of Prigogine's "dissipative" structures. By suitable exchanges with the environment, a system can reach a state of low entropy, that is, a state of greater order, than that from which it started. (This is apparently what happens in metazoan organisms during the developmental stages.) However, according to Glensdorff and Prigogine (1971), the open system can sustain an entropy increase because of irreversible processes within the system. In a gerontological context this *entropic change* is synonymous with the senescent changes listed in Table I.

Table I. Main Senescent Processes at Several Levels of Organization of Multicellular Animals

Level	Senescent (entropy-gaining) process
1. Population	Mortality kinetics reflecting initial differences in vitality among the individual animals Age-related increase in the interindividual differences of physiological and structural parameters
2. Organism	Decreased physiological performance and resistance to stress
3. Physiological system and organ	Functional loss in all systems, with the neuromuscular, the neuroendocrine, and the connective tissues leading the way
4. Cell	Primary involutional changes in fixed postmitotic cells. Other cell types may show secondary and less striking senescent degeneration
5. Subcellular organelle	Age-related changes in all subcellular organelles as the result of primary molecular injury to the mitochondrial genome, leading to lack of mitochondrial turnover and replenishment
6. Informational macromolecule	The fundamental molecular process underlying aging is the loss or mutation (because of respiration-linked free radical injury) of the mitochondrial DNA

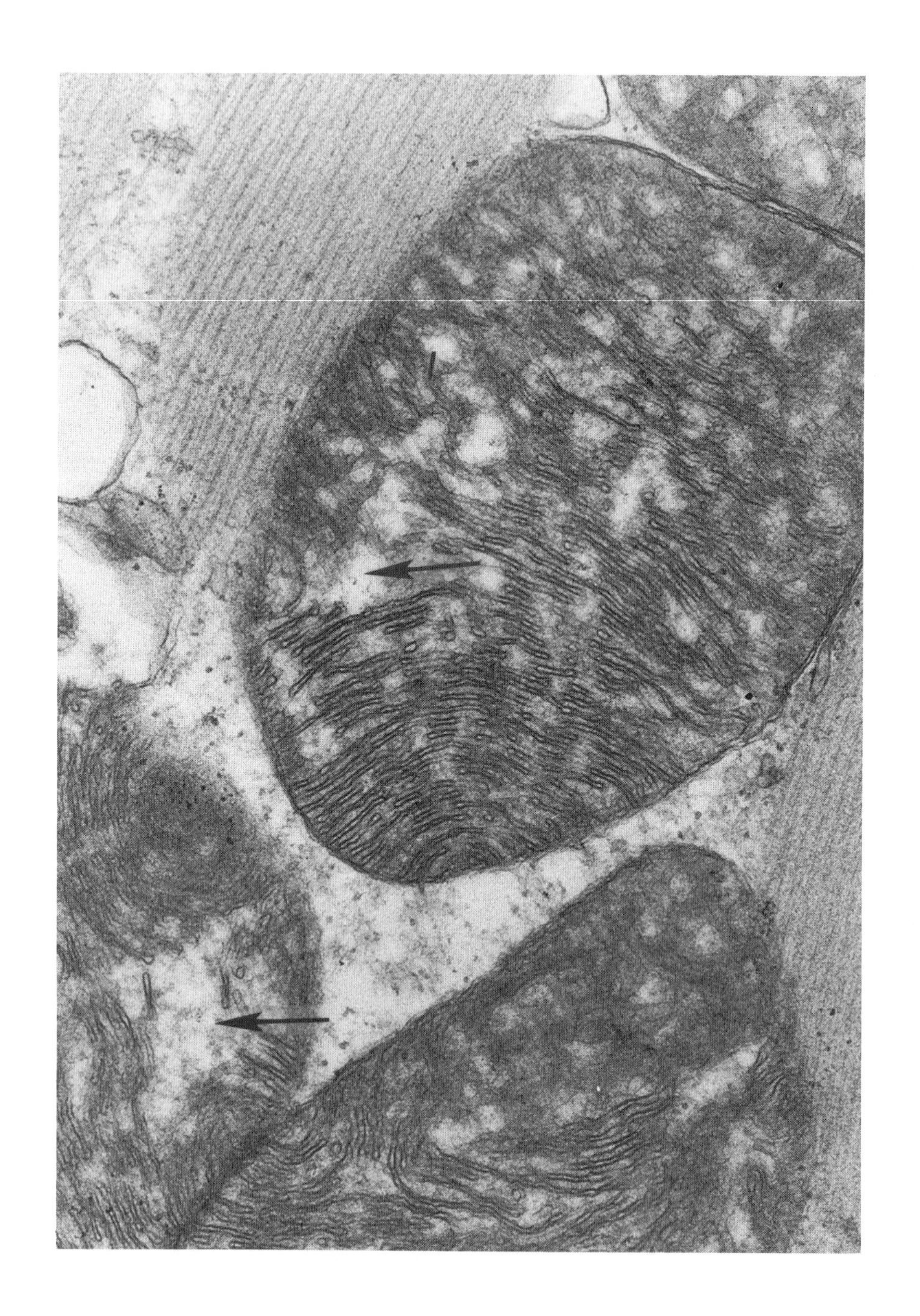

Table II. Main Molecular Mechanisms of Aging

1. Random reactions of oxygen radicals and peroxides (which are a by-product of mitochondrial respiration) with macromolecules of the mitochondria including the mtDNA
2. Hydrolytic decomposition of biomolecules
3. Heat denaturation of biomolecules
4. Cross linkage

In agreement with the above, it is clear that the maximum life span of the members of a species (a question of perennial interest to gerontologists and nongerontologists alike) is a function of deleterious side-effects of aerobic metabolism (Table II and Fig. 21). This fundamental entropic gain, which in agreement with the principles of hierarchical systems, results in disruption at all levels, is counteracted more or less efficiently in each species and individual by a host of counterentropic (life-maintaining, antisenescent) mechanisms (Table III), which delay but do not prevent the ultimate senescence of the organism.

The Achilles' heel or design flaw which makes the metazoan vulnerable to senescent disorganization is probably found in the mitochondria of fixed postmitotic cells (Fig. 21). Although it may seem surprising, high rates of mitochondrial respiration are not required to support cell growth (Sun *et al.,* 1981). Thus, actively growing and replicating cells, such as those found in the stem pools of the intestine and bone marrow, are protected against the injury resulting from excessive production of the free radicals associated with the respiratory process. On the other hand, the high quotas of oxygen utilization that have evolved in order to support the functional activity of the differentiated somatic cells may make total entropy dissipation impossible in this cell type and, therefore, the soma sustains irreversible aging (Fig. 22).

←

Figure 21. High magnification of the flight muscle of an 84-day-old fruit fly, showing a slight disruption in the packing of the cristae (arrows) that is a probable effect of aging. The high number of cristae in this type of mitochondria, with resulting high production of oxygen radicals, may be responsible for the age-related structural disorganization of mitochondria (Miquel *et al.,* 1980) and for the senescent decrease in the activity of respiratory enzymes (Fig. 18). In our view, a similar close packing of cristae prevents complete entropy dissipation in differentiated mammalian cells such as those of the muscle and the nervous tissues. In contrast, the low levels of respiratory activity associated with the "empty" appearance of the mitochondria of the germ cells lines (see spermatogonium in Fig. 4), in addition to other mechanisms such as very effective repair of nuclear DNA, makes these reproductive cells (as well as other replicating cells found in the stem cell pools) relatively immune to aging.

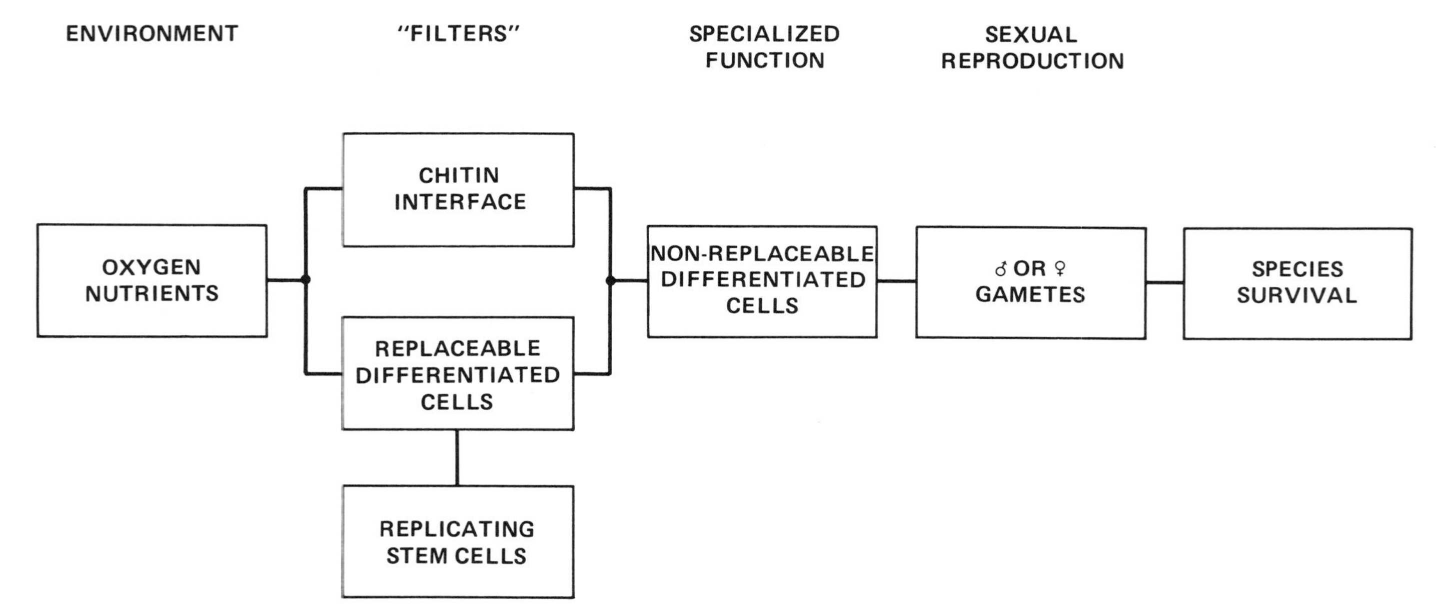

Figure 22. Biological evolution has provided metazoans with a soma of "shields" and "filters" made of inert materials and differentiated cells, which in mammals (but not in flies or nematodes) are replenished from pools of replicating amoebalike cells. In a bioengineering context, the somatic structures can be considered as "disposable devices" evolved for nourishment, transport, delivery, and general protection of the germ cells (see Kirkwood, 1971). In contrast to the well-protected germ cells, the repair mechanisms of the soma are not adequate. The differentiated cells that are the main components of the somatic tissues show high quotas of energy production (in order to support specialized function), which is associated with entropic release of oxygen radicals and with reduced (entropy-dissipating) macromolecular turnover in the subcellular organelles. This results in progressive accumulation of entropic injury in the disposable soma, i.e., aging, and eventual death.

Table III. Examples of Organization-Maintaining Mechanisms at the Species, Organism, and Cell Levels

Species
  Sexual reproduction
Organism and organ
  Body and organ growth
  Stem cell pools which provide "disposable" differentiated cells
  "Shields" of inert material such as chitin (insects) or cornified epithelia (mammals) (Figs. 2 and 3)
  "Interfaces" or "filters" of ancillary cells such as the glia of the nervous system and the Kupfer cells of the liver (Fig. 13)
  Depression of metabolic rate (underfeeding, hybernation, sleep, etc.)
Cell
  Nuclear-DNA repair
  Cell growth and division
  Meiosis (in the germ cell lines)
  Turnover of organelles and macromolecules
  Microsomal mixed-functional oxidative systems
  Peroxysomes
  Free radical scavengers and antioxidants (vitamins C and E, selenium, glutathione, superoxide dismutases, glutathione peroxidase, etc.).

REFERENCES

Albring, M., Griffith, J., and Attardi, G., 1977, Association of a protein structure of probable membrane derivation with HeLa cell mitochondria DNA near its origin of replication, *Proc. Natl. Acad. Sci. USA* **74:**1348.

Alpatov, W. W., and Pearl, R., 1929, Experimental studies on the duration of life. XII. Influence of temperature during the larval period and adult life on the duration of life of the imago of *Drosophila melanogaster, Amer. Naturalist* **63:**37.

Atlan, H., Miquel, J., and Welch, G., 1970, Effects of low and high LET cyclotron accelerated particles on longevity of *Drosophila melanogaster. Int. J. Radiat. Biol.* **18:**423.

Atlan, H., Miquel, J., Helmle, L. C., and Dolkas, C. B., 1976, Thermodynamics of aging in *Drosophila melanogaster, Mech. Ageing Dev.* **5:**371.

Beard, R. E., 1959, Note on some mathematical models, *CIBA Found. Colloq. Ageing* **5:**302.

Bielski, B. H. J., and Gebicki, J. M., 1977, Application of radiation chemistry to biology, in: *Free Radicals in Biology* (W. A. Pryor, Ed.), Vol. 3, pp. 1–51, Academic Press, New York.

Chance, B., Sies, H., and Boveris, A., 1979, Hydroperoxide metabolism in mammalian organs, *Physiol. Rev.* **59:**527.

Cutler, R. G., 1978, Cross-linkage hypothesis of aging: DNA adducts in chromatin as a primary aging process, in: *Proteins and Other Adducts to DNA: Their Significance to Aging, Carcinogenesis and Radiation Biology* (K. C. Smith, ed.), p. 443, Plenum Press, New York.

Dobzhanski, T., 1964, Introduction *Proc. Natl. Acad. Sci. USA* **51:**907.

Economos. A. C., 1979, A non-Gompertzian paradigma for mortality kinetics of metazoan animals and failure kinetics of manufactured products, *Age* **2:**74.

Economos, A. C., 1980, Kinetics of metazoan mortality, *J. Soc. Biol. Struct.* **3:**317.

Economos, A. C., 1982, Rate of aging, rate of dying and the mechanisms of mortality, *Arch. Geront. Geriatr.* **1**:3.

Economos, A. C., and Miquel, J., 1977, Systems analysis of the effect of temperature, oxygen and radiation on mortality kinetics of fruit flies: rate of living and rate of dying, *Proc. 30th ACEMB*, p. 448.

Economos, A. C., and Miquel, J., 1978, Non-Gompertzian mortality kinetics, *Age* **1**:76 (abstract), 1978.

Economos, A. C., and Miquel, J., 1979, Analysis of population mortality kinetics with application to the longevity follow-up of the Navy's "1000 aviators," *Aviat. Space Environ. Med.* **50**:397.

Economos, A. C., and Miquel, J., 1980, Usefulness of stochastic analysis of body weight of mice as a tool in experimental aging research, *Exp. Ageing Res.* **6**:417.

Economos, A. C., Miquel, J., Binnard, R., and Kessler, S. 1979, Quantitative analysis of mating behavior in aging male *Drosophila melanogaster, Mech. Ageing Dev.* **10**:233.

Economos, A. C., Miquel, J., Fleming, J., and Johnson, J. E., Jr., 1980, Is there intrinsic mitochondrial aging? *Age* **3**:117.

Economos, A. C., Miquel, J., Ballard, R. C., and Johnson, J. E., Jr., 1981, Variation, principles and application in the study of cell structure and aging, *Aging Cell Struct* **1**:187.

Economos, A. C., Ballard, R. C., Miquel, J., Binnard, R. and Philpott, D. E., 1982, Accelerated aging of fasted *Drosophila:* Preservation of physiological function and cellular fine structure by thiazolidine carboxylic acid (TCA). *Exp. Gerontol.* **75**:105.

Fleming, J. E., Miquel, J., Cottrell, S. F., Yengoyan, L. S., and Economos, A. C., 1982, Is cell aging caused by respiration-dependent injury to the mitochondrial genome? *Gerontology* **28**:44.

Flohe, L., Loschen, G., Azzi, A., and Richter, Ch, 1977, Superoxide radicals in mitochondria in: *Superoxide and Superoxide Dismutases* (A. M. Michelson, J. M. McCord, and I. Fridovich, eds.), pp. 323–334, Academic Press, New York.

Fukanaga, M., and Yielding, K. L., 1979, Fate during cell growth of yeast mitochondrial and nuclear DNA after photolytic attachment of the monoazide analog of ethidium bromide, *Biochem. Biophys. Res. Commun.* **90**:582.

Gerschman, R. 1964, Biological effects of oxygen, in: *Oxygen in the Animal Organism* (F. Dickens and E. Neil, eds.), pp. 478–479, The MacMillan Co., New York.

Glansdorff, P., and Prigogine, I., 1971, *Structure, Stability and Fluctuations,* Wiley, New York.

Gompertz, B., 1825, On the nature of the function expressive of the law of human mortality, and on a new mode of determining the value of life contingencies, *Philos. Trans. R. Soc. London,* 252.

Haran, D., 1956, Aging, a theory based on free radical and radiation chemistry, *J. Gerontol.* **11**:298.

Hayflick, L., 1965, The limited *in vitro* life time of human diploid cell strains, *Exp. Cell Res.* **37**:614.

Hayflick, L., 1979, Progress in cytogerontology, *Mech. Ageing Dev.* **9**:393.

Johnson, J. E., Jr., 1979, Fine structure of IMR-90 cells in culture as examined by scanning and transmission electron microscopy, *Mech. Ageing Dev.* **10**:405.

Kirkwood, T., 1979, Senescence and the selfish gene, *New Scientist* **81**:1040.

Kohn, R. R., 1977, *Principles of Mammalian Aging,* Prentice Hall, Inc., Englewood Cliffs, New Jersey.

Loeb, J., and Northrop, J. H., 1917, On the influence of food and temperature upon the duration of life, *J. Biol. Chem.* **32**:103.

Martin, G. M., Sprague, C. A., Norwood, T. H., Pendergrass, W. R., Bernstein, P., Hoehn, H., and Arend, W. P., 1973, Do hyperplastoid cells differentiate themselves to death?, in: *Cell*

Impairment in Ageing and Development (V. J. Cristofalo and E. Holeckova, eds.), pp. 67–90, Plenum Press, New York.

Masoro, E. J., 1981, *CRC Handbook of Physiology in Aging,* CRC Press, Inc., Boca Raton, Florida.

Massie, H. R., Baird, M. B., and McMahon, M. M., 1975, Loss of mitochondrial DNA with aging, *Gerontology* **21**:231.

Massie, H. R., Colacicco, J. R., and Williams, T. R., 1981, Loss of mitochondrial DNA with aging in the Swedish C strain of *Drosophila melanogaster, Age* **4**:42.

Medvedev, Z., 1972, Repetition of molecular-genetic information as a possible factor in evolutionary changes in life span, *Exp. Gerontol.* **7**:227.

Miller, A., 1965, The internal anatomy of the imago of *Drosophila melanogaster,* in: *Biology of Drosophila* (M. Demerec, ed.), pp. 420–534, Hafner, New York.

Miller, J. G., 1978, *Living Systems,* McGraw-Hill Book Co., New York.

Minot, C. S., 1907, The problem of age, growth and death, *Popular Sci. Monthly,* **71**:496.

Miquel, J., 1971, Aging of male *Drosophila melanogaster:* Histological histochemical and ultrastructural observations, *Adv. Gerontol. Res.* **3**:39.

Miquel, J., and Blasco, M., 1978, A simple technique for evaluation of vitality loss in aging mice, by testing their muscular coordination and vigor, *Exp. Gerontol.* **13**:389.

Miquel, J., and Economos, A. C., 1979, Favorable effects of the antioxidants sodium and magnesium thiazolidine carboxylate on the vitality and life span of *Drosophila* and mice, *Exp. Gerontol.* **14**:279.

Miquel, J., and Haymaker, W., 1965, Astroglial reactions to ionizing radiation (with emphasis on glycogen accumulation). *Progress Brain Res.* **15**:89.

Miquel, J., and Johnson, J. E., Jr., 1978, Senescent changes in the ribosomes of animal cells *in vivo* and *in vitro, Mech Ageing Dev.* **9**:247.

Miquel, J., Calvo, W., and Rubinstein, L. J., 1968, A simple and rapid stain for the biopsy diagnosis of brain tumors, *J. Neuropath. Exp. Neurol.* **27**:517.

Miquel, J., Lundgren, P. R., Bensch, K. G., and Atlan, H., 1976. Effects of temperature on the life span, vitality and fine structure of *Drosophila melanogaster, Mech. Ageing Dev.* **5**:347.

Miquel, J., Oro, J., Bensch, K. G., and Johnson, J. E., Jr., 1977. Lipofuscin: Fine structural and biochemical studies, in: *Free Radicals in Biology,* Vol. 3 (Pryor, W., ed.), pp. 133–182, Academic Press, New York.

Miquel, J., Economos, A. C., Bensch, K. G., Atlan, H., and Johnson, J. E., Jr., 1979, Review of cell aging in *Drosophila* and mouse, *Age* **2**:78.

Miquel, J., Economos, A. C., Fleming, J., and Johnson, J. E., Jr., 1980, Mitochondrial role in cell aging, *Exp. Gerontol.* **15**:575.

Miquel, J., Economos, A. C., and Bensch, K. G., 1981, Insect vs. mammalian aging, *Aging Cell Struct.* **1**:347.

Miquel, J., Fleming, J., and Economos, A. C., 1982, Antioxidants metabolic rate and aging in *Drosophila, Arch. Gerontol. Geriatr.* **1**:159.

Nass, S., and Nass, M. M. K., 1964, Intramitochondrial fibers with deoxyribonucleic acid characteristics: observations of Ehrlich ascites tumor cells, *J. Natl. Cancer Inst.* **33**:777.

Nohl, H., and Hegner, D., 1978, Do mitochondria produce oxygen radicals *in vitro? Eur. J. Biochem.* **82**:563.

Pearl, R., 1928, *The Rate of Living,* Knopf, New York.

Prigogine, I., 1947, *Étude Thermodynamique des Phénomènes Irreversibles,* Desoer, Liege, 1947.

Pryor, W. A., 1976, The role of free radical reactions in biological systems, in: *Free Radicals in Biology,* Vol. 1 (W. A. Pryor, ed.), pp. 1–50, Academic Press, New York.

Puig-Muset, P., 1976, *Oxygeno(s),* Oikos-tau, S. A., Ediciones, Barcelona.

Rosenberg, B., Kemeny, G., Smith, L. G., Skurnik, I. D., and Bandurski, M. J., 1973, The kinetics and thermodynamics of death in multicellular organisms, *Mech. Ageing Dev.* **2**:275.

Rothstein, M., 1981, Postranslational alteration of proteins, in: *CRC Handbook of Biochemistry of Aging* (Florini, J. R., ed.,), pp. 103–111, CRC Press, Inc., Boca Raton, Florida.

Rubner, M., 1908, *Das Problem der Lebensdauer und seine Beziehungen zum Wachstum und Ernährung,* Oldenburg, Munich.

Sacher, G. A., 1967, The complementarity of entropy terms for the temperature-dependence of development and aging, *Ann. N. Y. Acad. Sci.* **138**:680.

Sacher, G. A. 1977, Life table modification and life prolongation, in: *Handbook of the Biology of Aging* (C. E. Finch and L. Hayflick, eds.), pp. 582–639, Van Nostrand Reinhold, Co., New York.

Schrödinger, E., 1945, *What is Life?* Cambridge U. P., Cambridge.

Shearman, C. W., and Kalf, G. F., (1977), DNA replication by a membrane-DNA complex from rat liver mitochondria. *Arch. Biochem. Biophys.* **182**:573.

Sheldahl, J., and Tappel, A., 1974, Fluorescent products from aging *Drosophila melanogaster:* An indicator of free radical lipid peroxidation damage. *Exp. Gerontol.* **9**:33.

Shock, N. W., 1970, Physiological aspects of aging, *J. Am. Diet. Associat.* **56**:491.

Sohal, R. S., 1981, Metabolic rate, aging, and lipofuscin accumulation, in: *Age Pigments* (R. S. Sohal, ed.), pp. 303–316, Elsevier–North Holland, New York.

Sohal, R. S., 1982, Oxygen consumption and life span in the adult male house fly, *Musca domestica, Age* **5**:21.

Strehler, B. L., 1967, Environmental factors in aging and mortality, *Env. Res.* **1**:46.

Strehler, B. L., 1977, *Time, Cells and Aging,* pp. 103–124, Academic Press, New York.

Strehler, B. L., and Chang, M-P, 1979, Loss of hybridizable ribosomal DNA from human postmitotic tissues during aging: II. Age-dependent loss in human cerebral cortex—hippocampal and somatosensory cortex comparison. *Mech. Ageing Dev.* **11**:379.

Strehler, B. L., Chang, M-P., and Johnson, L. K., 1979, Loss of hybridizable ribosomal DNA from human postmitotic tissues during aging. I. Age-dependent loss in human myocardium, *Mech. Ageing Dev.* **11**:371.

Sun, A. S., Aggarwal, B. B., and Packer, L., 1975, Enzyme levels of normal human cells: Aging in culture, *Arch, Biochem. Biophys.* **170**:1.

Tappel, A. L., 1965, Free radical lipid peroxidation damage and its inhibition by vitamin E and selenium, *Fed. Proc.* **24**:73.

Tappel. A. L., 1973, Lipid peroxidation damage to cell components, *Fed. Proc.* **32**:1870.

Tas, S., 1976, Disulfide bonding in chromatin proteins with age and a suggested mechanism for aging and neoplasia, *Exp. Gerontol.* **11**:17.

Tas, S., 1978, Involvement of disulfide bonds in the condensed structure of facultative heterochromatin and implications for cellular differentiation and aging, *Gerontology* **24**:358.

Zeman, W., Curtis, H. J., Gebhard, E. L., and Haymaker, W., 1959, Tolerance of mouse brain tissue to high-energy deuterons, *Science* **130**:1760.

Zs.-Nagy, I., 1979, The role of membrane structure and function in cellular aging: A review, *Mech. Ageing Dev.* **9**:237.

Index